AF581725

Pharmaceutical Particulates and Membranes for Delivery of Drugs and Bioactive Molecules

Pharmaceutical Particulates and Membranes for Delivery of Drugs and Bioactive Molecules

Special Issue Editors

Diganta B. Das

Mostafa Mabrouk

Hanan Beherei

Arthanareeswaran Gangasalam

MDPI • Basel • Beijing • Wuhan • Barcelona • Belgrade

Special Issue Editors
Diganta B. Das
Loughborough University
UK

Mostafa Mabrouk
National Research Centre
Egypt

Hanan Beherei
National Research Centre
Egypt

Arthanareeswaran Gangasalam
National Institute of Technology
Tiruchirappalli, India

Editorial Office
MDPI
St. Alban-Anlage 66
4052 Basel, Switzerland

This is a reprint of articles from the Special Issue published online in the open access journal *Pharmaceutics* (ISSN 1999-4923) from 2018 to 2020 (available at: https://www.mdpi.com/journal/pharmaceutics/special_issues/particulates_and_membranes).

For citation purposes, cite each article independently as indicated on the article page online and as indicated below:

LastName, A.A.; LastName, B.B.; LastName, C.C. Article Title. *Journal Name* **Year**, *Article Number*, Page Range.

ISBN 978-3-03936-392-6 (Hbk)
ISBN 978-3-03936-393-3 (PDF)

Cover image courtesy of Diganta B. Das.

Contents

About the Special Issue Editors

Diganta B. Das is an Associate Professor in the Department of Chemical Engineering at Loughborough University, UK. His research involves the application of principles of fluid flow and mass transport behaviour in porous media (both biological and non-biological) to solve bioengineering and water engineering problems. He currently serves as the editor of *Water Science and Technology* and *Water Supply* (IWA Publishers), and is an Editorial Board member of *Biotechnology Letters* (Springer-Verlag), *Pharmaceutics* (MDPI) and *Clean Technologies* (MDPI), besides being associated with a number of other journals. Prior to his current appointment, Dr. Das was a Lecturer (Assistant Professor) at the University of Oxford, UK, a postdoctoral research fellow at Delft University of Technology in the Netherlands, and a visiting fellow at Princeton University. He has also held visiting appointments at Eindhoven University, Netherlands, Texas Southern University, USA, and Selcuk University, Turkey. Dr. Das has published more than 100 peer-reviewed journal articles, more than 40 proceedings of international conferences and edited books, co-edited 2 scientific books and co-authored 3 books.

Mostafa Mabrouk is an Associate Professor in the Department of Refractories, Ceramics and Building Materials at the National Research Centre (NRC), Egypt. His research involves the synthesis and characterisation of various forms of bioactive materials such as bioactive glass, ceramics, and their polymer composites. The final delivered form of the biomaterials was one of the most important discoveries of Dr. Mabrouk as he was able to deliver different types of biomaterials through carriers including nanomaterials, scaffolds, microspheres and membranes. Dr. Mabrouk has been an Academic Visitor at Loughborough University, UK. He also spent two years as a postdoctoral fellow in the WADDP research group at Witwatersrand University, South Africa. He got his Ph.D. in Chemistry from the Biomaterials Group at Rennes 1 University, France. His current research is devoted to the preparation of different porous inorganic carriers and 3D scaffold for the tissue engineering field.

Hanan Beherei is currently a Professor of the Biomedical Materials and Regenerative Medicine group in the Ceramics Department at the National Research Centre, Egypt. Prof. Beherei's research focuses on the design and development of new types of nanomaterials using a number of advanced technologies and evaluating their use in biotechnological and biomedical applications, including the development of novel systems for effective drug delivery, tissue engineering, and renewable medicine in vitro and in vivo. She prepared and characterised new biomaterials based on bioactive glasses with electrospinning assisted methods developed for the fabrication of biomedical nanostructures. She also designed specific nanomaterials being considered in current research projects on bone, nerve, skin and cardiac tissue through local projects in Egypt. She has worked in close collaboration with a number of researchers and academics with expertise in cell biology, biochemistry, biotechnology, and medical sciences in Japan and in the UK. She has designed, prepared and characterised many novel biocompatible ceramic composite-based hydroxyapatites and others calcium phosphates with metal oxides like TiO_2, ZrO_2, Al_2O_3, MgO, and ZnO, as well as calcium aluminate and calcium silicate for use in bone grafts. She designed, prepared and characterised many glass systems by melting and sol gel methods for use as bone grafts. She was a postdoctoral fellow in the Tissue Engineering Department of GBF, Braunschweig Germany. She was also a Lecturer (Associate Professor) at Taif

University, KSA from 2008 to 2009 and a Lecturer (Professor) at Taif University, KSA from 2010 to 2015.

Arthanareeswaran Gangasalam is a Professor in the Department of Chemical Engineering at the National Institute of Technology, Tiruchirappalli, India. His current research involves improving the performance of polymer membranes by mixing with nanomaterials for ultrafiltration, nanofiltration, gas separation, and membrane distillation applications. The prime novelty of this work is the selection of organic and inorganic membrane materials for membrane fabrication, as well as to achieve high permeability and anti-fouling properties with rejection of molecules. He is the lead investigator of various funded projects from DST-India, DBT-India. CNPq-Brazil, MST-Korea, RAE-UK, AISTDF-India, and NRDI-Hungary. Arthanareeswaran received a Research Exchange Award from RAE-UK, an Endeavour Executive Award from the Australian Government and a Brainpool Award from the South Korean Government. Hiyoshi Corporation, Japan awarded him the Hiyoshi Environmental Award for outstanding research in the field of environmental conservation. Dr. Arthanareeswaran was an academic visitor of the University of Sao Paulo, Brazil, Monash University, Australia, Loughborough University, UK, Universiti Teknologi Malaysia, Malaysia, Konkuk University, South Korea, University of Szeged, Hungary, and Prince of Songkla University, Thailand. He served as a Guest Editor for the journals *Desalination*, *Environmental and Ecological Safety*, and *Desalination and Water Treatment*. Arthanareeswaran has authored over 110 original articles in peer-reviewed international journals, more than 45 papers in the proceedings of international conferences, two scientific books, and seven book chapters.

Editorial

Pharmaceutical Particulates and Membranes for the Delivery of Drugs and Bioactive Molecules

Diganta B. Das [1,*], Mostafa Mabrouk [2], Hanan H. Beherei [2] and G. Arthanareeswaran [3]

[1] Department of Chemical Engineering, Loughborough University, Loughborough LE113TU, Leicestershire, UK

[2] Refractories, Ceramics and Building Materials Department, National Research Centre, 33 El Bohouth St (Former EL Tahrir St), Dokki, Giza P.O.12622, Egypt; mostafamabrouk.nrc@gmail.com (M.M.); hananh.beherei@gmail.com (H.H.B.)

[3] Department of Chemical Engineering, National Institute of Technology, Tiruchirappalli 620015, Tamil Nadu, India; arthanareeg@nitt.edu

* Correspondence: D.B.Das@lboro.ac.uk

Received: 22 April 2020; Accepted: 29 April 2020; Published: 1 May 2020

The delivery of drugs and bioactive molecules using pharmaceutical particulates and membranes are of great significance for various applications such as the treatment of secondary infections, cancer treatment, skin regeneration, orthopedic applications and others. Several techniques can be utilized for the preparation of the particulates and membranes, which include, but are not limited to, lyophilization, microemulsification, nano-spray-drying, nano-electro-spinning, slip casting and 3D printing. The particulates and membranes possess great prospects to improve the existing strategies and develop new ones for sustained and controlled drug delivery. Therefore, the development of this subject as an area of research has been rapid in the last two decades, and their applications are now very broad. However, these depend on the properties of the designed particulates and membranes.

In addressing these points, this Special Issue (SI) of the journal *Pharmaceutics* seeks to highlight the recent trends and innovative developments in the pharmaceutical particulates and membranes for the delivery of drug and bioactive molecules.

We received in total twenty submissions for the SI, all of which went through a rigorous peer review process. Eight papers were declined at the peer review stage, and, the remaining twelve papers have now been published, all as open access papers as per the policy of the journal. The published papers are also being compiled as an edited e-book, to be published by MDPI. We introduce the published twelve papers briefly in this guest editorial.

To begin with, we collaborated with other experts in the field of pharmaceutical particulates and membranes to review the current state of inorganic nanoparticulates and nanomembranes based on their design, and the key factors for adjusting their morphology and size for their possible medical applications, especially as drug carrier materials (Mabrouk et al. [1]). A very good example of these points can be seen in the paper by Kumar et al. [2] who have developed a prolonged release device for site-specific delivery of a neuroprotective agent (nicotine). The device has been formulated as a novel reinforced crosslinked composite polymeric system with the potential for intrastriatal implantation for Parkinson's disease interventions. These have been developed in the form of membranes with minimal rates of matrix degradation and retarding nicotine release. This has led to the zero-order release for 50 days following exposure to simulated cerebrospinal fluid (CSF).

Mora-Espíet et al. [3] have investigated the effects of specific targeting of microparticles on their internalization by cells under fluidic conditions. For this purpose, two isogenic breast epithelial cell lines, one overexpressing the human epidermal growth factor receptor 2 (HER2) oncogene (D492HER2) and highly tumorigenic, and the other expressing HER2 at much lower levels and nontumorigenic (D492) were cultured in the presence of polystyrene microparticles of 1 μm in diameter, biofunctionalized with either a specific anti-HER2 antibody or a nonspecific secondary antibody. The authors have come to conclude that the biofunctionalization of microparticles with a specific targeting molecule remarkably increases their internalization by cells in fluidic culture conditions (simulating the blood stream).

Huang et al. [4] have reported a modified coaxial electrospraying technique, which explored how to create ibuprofen-loaded hydroxypropyl methylcellulose nanoparticles for accelerating the drug dissolution rate. During this process, it was shown that a key parameter, i.e., the spreading angle of atomization could provide a linkage among the working process, the property of generated nanoparticles and their functional performance. They confirmed that the nanoparticle diameter (prepared based on a modified technique) has a profound influence on the drug release performance. It is envisaged that the clear process–property–performance relationship should be useful for optimizing the electrospraying process, and, in turn, for achieving the medicated nanoparticles with desired functional performances. Shah et al. [5] designed and optimized a nano-emulsion-based system to improve therapeutic efficacy of moxifloxacin in ophthalmic delivery. Their findings suggest that optimized nanoemulsion can enhance the therapeutic effect of moxifloxacin and, therefore, it can be used as a safe and effective delivery vehicle for ophthalmic therapy. In addition, Wan et al. [6] developed a novel sustained release pellet of loxoprofen sodium (LXP) by coating a dissolution-rate controlling sublayer containing hydroxypropyl methyl cellulose (HPMC) and citric acid, and a second diffusion-rate controlling layer containing aqueous dispersion of ethyl cellulose (ADEC) on the surface of a LXP conventional pellet in order to compare its performance in vivo with an immediate release tablet (Loxinon®). Their results identified both the citric acid (CA) and ADEC as the dissolution- and diffusion-rate controlling materials significantly decreasing the drug release rate. The optimal formulation for a pH-independent drug release in media has been suggested as at a pH above 4.5 and at slightly slow release in acid medium. The pharmacokinetic studies have revealed that a more stable and prolonged plasma drug concentration profile of the optimal pellets has been achieved, with a relative bioavaibility of 87.16% compared with the conventional tablets.

Iglesias et al. [7] have reported the synthesis and characterization of magnetic nanoparticles of two distinct origins, one inorganic (MNPs) and the other biomimetic (BMNPs). The authors have declared that the BMNPs are better suited to be loaded with drug molecules positively charged at neutral pH (notably, doxorubicin for instance) and released at the acidic tumor environment. In turn, MNPs may provide their transport capabilities under a magnetic field. However, in this study the authors have used a mixture of both kinds of particles at two different concentrations, trying to derive the best from each of them. Also, they have studied which mixture performs better from different points of view, considering factors such as stability and magnetic hyperthermia response, while keeping suitable drug transport capabilities. The authors have recommended this as a close to ideal drug vehicle with enhanced hyperthermia response. Savin et al. [8] have discussed the antitumoral potential of three gel formulations loaded with carbon dots prepared from N-hydroxyphthalimide (CD-NHF) on two types of skin melanoma cell lines as well as two types of breast cancer cell lines in 2D (cultured cells in normal plastic plates) and 3D (Matrigel) models. Antitumoral gels based on sodium alginate (AS), carboxymethyl cellulose (CMC), and the carbomer Ultrez 10 (CARB) loaded with CD-NHF. The in vitro results for the tested CD-NHF-loaded gel formulations have revealed that the new composites can affect the number, size, and cellular organization of spheroids and impact individual tumor cell ability to proliferate and aggregate in spheroids.

Guadarrama-Acevedo et al. [9] have prepared a novel biodegradable wound dressing by means of alginate membrane and polycaprolactone nanoparticles loaded with curcumin for potential use in wound healing. The membrane has exhibited a diverse range of functional characteristics required to perform as a substitute for synthetic skin, such as a high capacity for swelling and adherence to the skin, evidence of pores to regulate the loss of transepidermal water, transparency for monitoring the wound, and drug-controlled release by the incorporation of nanoparticles. The incorporation of the nanocarriers aids the drug in permeating into different skin layers, solving the solubility problems of curcumin. The paper by Rancan et al. [10] is related to the production of PVP nanofibrous mats and foils loaded with a poorly soluble antibiotic, ciprofloxacin, for the treatment of topical wound infections. The research has revealed that nanofiber mats reach the highest amount of delivered drug concentration after 6 h, whereas foils maintain a maximum drug concentration over a 24 h period. The treatment has had no effect on the overall skin metabolic activity, but influenced the wound healing process. Importantly, a complete eradication of wound infections with *P. aeruginosa* (10^8 CFU) could be achieved.

Lian et al. [11] introduced red blood cell membrane-camouflaged ATO-loaded sodium alginate nanoparticles (RBCM-SA-ATO-NPs, RSANs) to relieve the toxicity of ATO while maintaining its efficacy. The average particle size of RSANs has been found to be 163.2 nm with a complete shell-core bilayer structure, and the average encapsulation efficiency is 14.3%. Compared with SANs, RAW 264.7 macrophages reduced the phagocytosis of RSANs by 51%, and the in vitro cumulative release rate of RSANs is 95% at 84 h, which have revealed a prominent sustained release. Furthermore, it has been demonstrated that RSANs have lower cytotoxicity when compared to normal 293 cells and exhibited antitumor effects on both NB4 cells and 7721 cells. In vivo studies have further showed that ATO can cause mild lesions of main organs while RSANs can reduce the toxicity and improve the antitumor effects. Thus, the developed RSANs system has provided a promising alternative for ATO treatment safely and effectively.

Finally, this SI has included the paper by Adeleke et al. [12] that has formulated and evaluated a reconstitutable dry suspension (RDS) containing isoniazid, a first-line antitubercular agent used in the treatment and prevention of TB infection in both children and adults. These formulations have been prepared by direct dispersion emulsification of an aqueous-lipid particulate interphase coupled with lyophilization and dry milling. The dug release behavior has been characterized with an initial burst up to 5 min followed by a cumulative release of 67.88% ± 1.88% (pH 1.2), 60.18% ± 3.33% (pH 6.8), and 49.36% ± 2.83% (pH 7.4) over 2 h. An extended release at pH 7.4 and 100% drug liberation have been achieved within 300 min. RDS has been dispersible and stable in the dried and reconstituted states over 4 months and 11 days respectively, under common storage conditions.

As evident, the SI and the forthcoming e-book demonstrate a range of articles with different research concerns. We hope that both the authors of the papers and ourselves as guest editors have been able to motivate future research in the field of pharmaceutical particulates and membranes for delivering drug and bioactive molecules.

Finally, we would like to acknowledge the contributions made by the authors of each paper irrespective of whether their submissions have been accepted for publication or not, as these have determined the success of this SI and the forthcoming e-book. We also acknowledge the Editorial Office of the Journal *Pharmaceutics* for their continued interest and support in bringing out the SI and the edited e-book.

References

1. Mabrouk, M.; Rajendran, R.; Soliman, I.E.; Ashour, M.M.; Beherei, H.H.; Tohamy, K.M.; Thomas, S.; Kalarikkal, N.; Arthanareeswaran, G.; Das, D.B. Nanoparticle-and Nanoporous-Membrane-Mediated Delivery of Therapeutics. *Pharmaceutics* **2019**, *11*, 294. [CrossRef] [PubMed]
2. Kumar, P.; Choonara, Y.E.; Du Toit, L.C.; Singh, N.; Pillay, V. In Vitro and In Silico Analyses of Nicotine Release from a Gelisphere-Loaded Compressed Polymeric Matrix for Potential Parkinson's Disease Interventions. *Pharmaceutics* **2018**, *10*, 233. [CrossRef]
3. Mora-Espí, I.; Ibáñez, E.; Soriano, J.; Nogués, C.; Gudjonsson, T.; Barrios, L. Cell Internalization in Fluidic Culture Conditions Is Improved When Microparticles Are Specifically Targeted to the Human Epidermal Growth Factor Receptor 2 (HER2). *Pharmaceutics* **2019**, *11*, 177. [CrossRef] [PubMed]
4. Huang, W.; Hou, Y.; Lu, X.; Gong, Z.; Yang, Y.; Lu, X.-J.; Liu, X.-L.; Yu, D.-G. The Process–Property–Performance Relationship of Medicated Nanoparticles Prepared by Modified Coaxial Electrospraying. *Pharmaceutics* **2019**, *11*, 226. [CrossRef] [PubMed]

5. Shah, J.; Nair, A.B.; Jacob, S.; Patel, R.K.; Shah, H.; Shehata, T.M.; Morsy, M.A. Nanoemulsion Based Vehicle for Effective Ocular Delivery of Moxifloxacin Using Experimental Design and Pharmacokinetic Study in Rabbits. *Pharmaceutics* **2019**, *11*, 230. [CrossRef] [PubMed]
6. Wan, D.; Zhao, M.; Zhang, J.; Luan, L. Development and In Vitro-In Vivo Evaluation of a Novel Sustained-Release Loxoprofen Pellet with Double Coating Layer. *Pharmaceutics* **2019**, *11*, 260. [CrossRef] [PubMed]
7. Iglesias, G.R.; Jabalera, Y.; Peigneux, A.; Checa Fernández, B.L.; Delgado, Á.V.; Jimenez-Lopez, C. Enhancement of Magnetic Hyperthermia by Mixing Synthetic Inorganic and Biomimetic Magnetic Nanoparticles. *Pharmaceutics* **2019**, *11*, 273. [CrossRef] [PubMed]
8. Savin, C.-L.; Tiron, C.; Carasevici, E.; Stan, C.S.; Ibanescu, S.A.; Simionescu, B.C.; Peptu, C.A. Entrapment of N-Hydroxyphthalimide Carbon Dots in Different Topical Gel Formulations: New Composites with Anticancer Activity. *Pharmaceutics* **2019**, *11*, 303. [CrossRef] [PubMed]
9. Guadarrama-Acevedo, M.C.; Mendoza-Flores, R.A.; Del Prado-Audelo, M.L.; Urbán-Morlán, Z.; Giraldo-Gomez, D.M.; Magaña, J.J.; González-Torres, M.; Reyes-Hernández, O.D.; Figueroa-González, G.; Caballero-Florán, I.H.; et al. Development and Evaluation of Alginate Membranes with Curcumin-Loaded Nanoparticles for Potential Wound-Healing Applications. *Pharmaceutics* **2019**, *11*, 389. [CrossRef] [PubMed]
10. Rancan, F.; Contardi, M.; Jurisch, J.; Blume-Peytavi, U.; Vogt, A.; Bayer, I.S.; Schaudinn, C. Evaluation of Drug Delivery and Efficacy of ciprofloxacin-Loaded Povidone Foils and Nanofiber Mats in a Wound-Infection Model Based on Ex Vivo Human Skin. *Pharmaceutics* **2019**, *11*, 527. [CrossRef] [PubMed]
11. Lian, Y.; Wang, X.; Guo, P.; Li, Y.; Raza, F.; Su, J.; Qiu, M. Erythrocyte Membrane-Coated Arsenic Trioxide-Loaded Sodium Alginate Nanoparticles for Tumor Therapy. *Pharmaceutics* **2020**, *12*, 21. [CrossRef] [PubMed]
12. Adeleke, O.A.; Hayeshi, R.K.; Davids, H. Development and Evaluation of a Reconstitutable Dry Suspension Containing Isoniazid for Flexible Pediatric Dosing. *Pharmaceutics* **2020**, *12*, 286. [CrossRef] [PubMed]

Review

Nanoparticle- and Nanoporous-Membrane-Mediated Delivery of Therapeutics

Mostafa Mabrouk [1,*,†], Rajakumari Rajendran [2,†], Islam E. Soliman [3], Mohamed M. Ashour [4], Hanan H. Beherei [1], Khairy M. Tohamy [3], Sabu Thomas [2], Nandakumar Kalarikkal [2], Gangasalam Arthanareeswaran [5] and Diganta B. Das [6,*]

1. Refractories, Ceramics and Building Materials Department, National Research Centre, 33 El Bohouth St (former EL Tahrirst)-Dokki, Giza 12622, Egypt; hananh.beherei@gmail.com
2. International and Inter-University Centre for Nanoscience and Nanotechnology, Mahatma Gandhi University, Kottayam, Kerala 686560, India; rajimpharm@gmail.com (R.R.); sabuthomas@mgu.ac.in (S.T.); nkkalarikkal@mgu.ac.in (N.K.)
3. Biophysics Branch, Faculty of Science, Al-Azhar University, Cairo 11884, Egypt; islam_biophysics@yahoo.com (I.E.S.); already_a555@yahoo.com (K.M.T.)
4. Faculty of Engineering, Badr University, Cairo 11829, Egypt; mohamedashour739@gmail.com
5. Department of Chemical Engineering, National Institute of Technology, Tiruchirappalli 620015, India; arthanareeg@nitt.edu
6. Department of Chemical Engineering, Loughborough University, Loughborough LE113TU, UK

* Correspondence: mostafamabrouk.nrc@gmail.com (M.M.); D.B.Das@lboro.ac.uk (D.B.D.); Tel.: +201097302384 (M.M.); +44-1509-222509 (D.B.D.)

† These authors contributed equally to this work.

Received: 1 May 2019; Accepted: 14 June 2019; Published: 21 June 2019

Abstract: Pharmaceutical particulates and membranes possess promising prospects for delivering drugs and bioactive molecules with the potential to improve drug delivery strategies like sustained and controlled release. For example, inorganic-based nanoparticles such as silica-, titanium-, zirconia-, calcium-, and carbon-based nanomaterials with dimensions smaller than 100 nm have been extensively developed for biomedical applications. Furthermore, inorganic nanoparticles possess magnetic, optical, and electrical properties, which make them suitable for various therapeutic applications including targeting, diagnosis, and drug delivery. Their properties may also be tuned by controlling different parameters, e.g., particle size, shape, surface functionalization, and interactions among them. In a similar fashion, membranes have several functions which are useful in sensing, sorting, imaging, separating, and releasing bioactive or drug molecules. Engineered membranes have been developed for their usage in controlled drug delivery devices. The latest advancement in the technology is therefore made possible to regulate the physico-chemical properties of the membrane pores, which enables the control of drug delivery. The current review aims to highlight the role of both pharmaceutical particulates and membranes over the last fifteen years based on their preparation method, size, shape, surface functionalization, and drug delivery potential.

Keywords: pharmaceutical particulates; membranes; drug delivery systems; bio-imaging; bioactive molecules

1. Introduction

Nanotechnology has emerged as one of the most versatile and powerful technologies in the development of drug delivery techniques. Nanomaterials include particulates which have at least one dimension of 100 nm or less. Scale differences and modification to the material surfaces result in different physico-chemical properties of the materials which make them suitable for biomedical applications such as drug delivery, disease diagnosis, and therapy. Currently, there are many applications of nanotechnology in general and nanoparticle (NP)-based drug delivery specifically. Similarly, membrane technologies that focus on both nanoporous membrane preparation and applications have developed significantly. In this context, a membrane involves a porous system which is made up of either inorganic, organic, or a combination of both inorganic and organic materials. The utilization of membranes to convey medications/bioactives is opening up new treatment preferences of interest, e.g., enhancing the solubility of bioactives shielding an active ingredient from corruption, enhancing the bioavailability of medication, lowering lethal impacts, offering suitable structures for all courses of administration, permitting advancement and offering fitting structures for courses of drug administration, and permitting fast formulation improvement. These points are discussed further below.

1.1. Why Nanoparticles?

There are many different types of NPs that have promising biomedical applications, e.g., polymeric NPs, polyethylene glycol (PEG)-ylation modified particles, micelles, liposomes, dendrimers, and nanosized inorganic materials [1]. Organic NPs vary in their activity in different biological systems including penetration depths in tissues and their toxicity and targeting efficiency [2]. Inorganic NPs regularly display novel physical properties as their size is close to nanometer scale measurements. For instance, the extraordinary physical and chemical properties of these NPs may prompt future applications in drug delivery and biomedical imaging. Plenty of sophisticated and different applications, including diagnosis and therapy and flow investigation, are related to configuration of the high surface-to-volume proportions of NPs as a potential system for these strategies. Structures such as core/shell NPs can show improved properties and increased usefulness due to their changed chemical distinction and nanostructured parts. This article therefore features an assortment of structures and properties that can be acknowledged in materials dependent on inorganic NPs and focuses on discussing major inquiries brought up in controlling these properties.

Typical characteristics of inorganic nanomaterials, such as ease of fabrication, modification and functionalization, simple preparation methods, resistance to microbial attacks, high stability and suitable size for cells (plasma membrane below 100 nm), a low toxicity profile, biocompatibility, and having a hydrophilic nature, make them suitable as drug carriers [3]. Moreover, these NPs can be biodegradable, non-toxic, non-immunogenic responsive, have a high loading capacity, and have the capacity for controlled drug release. These characteristics are desired for biomedical applications in addition to properties such as magnetism. However, in practice, it is not easy to fabricate ideal NPs which have all the desired properties [4,5] and some NPs are indeed not suitable for biomedical applications. To improve their biocompatibility, protective coatings which can be non-toxic, such as natural polymers (carbohydratesand peptides) or synthetic polymers (e.g., PEG, polyvinyl alcohol (PVA), and polyglycolic acid (PGA), etc.) can be used. Figure 1 shows a cut-out model of an inorganic NP functionalized with biomolecules for biomedical applications [6].

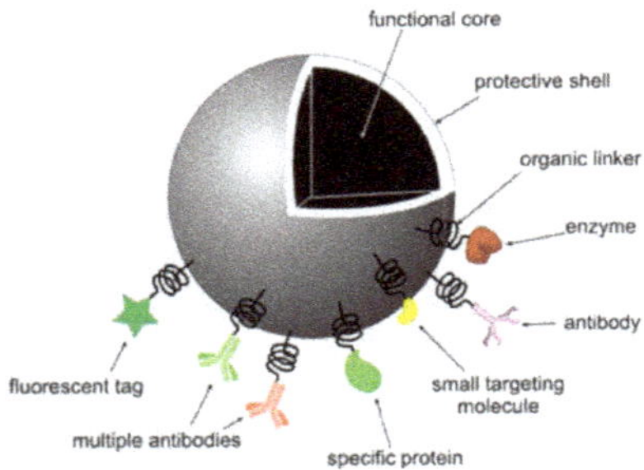

Figure 1. A model design of an inorganic nanoparticle (NP) functionalized with biomolecules for biomedical applications [6]. Reproduced with copyright permission from Springer Nature, 2010.

1.2. *Why a Nanoporous Membrane?*

Membranes have several biological functions which are useful in sensing, sorting, imaging, separating, and releasing bioactive/drug molecules. Engineered nanoporous and microporous membranes have been developed for their usage in drug delivery. The latest advancement in technology is therefore possible for use in regulating the physico-chemical properties of membrane pores, which make them attractive for controlling drug delivery rates. In addition, different types of materials are used for the fabrication of membranes and their properties and surface modification in order to improve the functions of the membranes, providing different invitro and in vivo applications of therapeutic delivery. In spite of the extensive work carried out for preparation, characterization, and biological evaluation of membranes, there are still a number of challenges which need to be overcome to develop biological membranes.

Membranes are used to control the rate of delivery of drugs to the body as well as drug permeation from the reservoir to attain the required rate of drug delivery. Therefore, drug delivery is controlled by both passive diffusion and biodegradation mechanisms. Membranes can carry one or more bioactive agents and have been developed into different classes of carriers. These different carriers can be carbon-based nanomaterials, polymeric membranes, and inorganic membranes, where the bulk properties of the membrane are governed by its building blocks, i.e., the NPs. Keeping this in mind, the current review aims to highlight the role of both pharmaceutical NPs and membranes during the last fifteen years based on their preparation method, size, shape, surface functionalization, and drug delivery potential. The following classification gives an overview structure of the article (Figure 2).

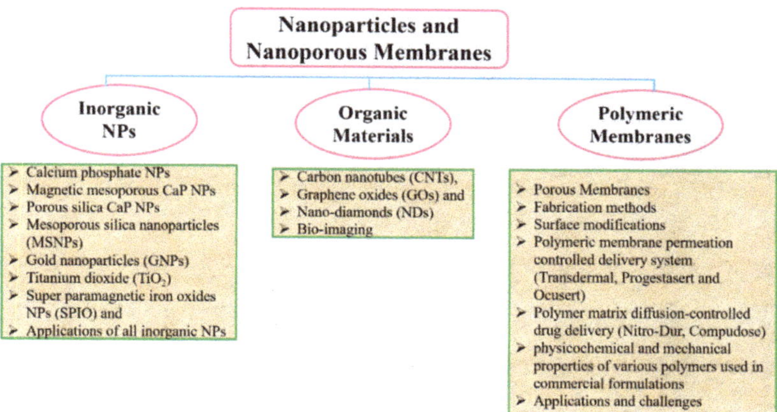

Figure 2. Structural overview of the article.

2. Drug Delivery System

Since the emergence of controlled drug delivery systems (DDS) in early the 1970s, these systems have attracted increasing attention. DDS are aimed at delivering drugs using pre-defined doses and drug delivery rates. Moreover, the area of drug delivery is an expanding domain focused on targeting genes or drug formulations to a group of diseased cells or tissues. The objective of this technique is to carry an appropriate amount of drug to the target sites (such as diseased tissues and tumors, etc.) while limiting undesirable reactions of the drugs within healthy tissues [7]. The propensity of the carrier material to cure cancer has been influenced by various parameters that relate to the carrier material, e.g., the immune response to the carrier material and uncontrollable drug behavior [8]. The morphology of NPs also performs a significant role in drug loading and release and achieving the maximum cell viability and minimum cell morbidity [9]. In order to illustrate how NPs have been used as DDS, we provide a number of key examples below.

2.1. Calcium Phosphate NPs

Calcium phosphate (CaP) is a common NP used in biological systems and medical applications, especially within those related to diagnosis and treatment. In particular, CaP NPs are extensively used in imaging, bone/tooth repair, and DNA delivery in cell biology [10,11]. The widespread uses of CaP NPs are due to their presence as a natural component in the body, as CaP is well enhanced and easily absorbed in the circulatory system [12]. Hence, previous studies have shown that CaP NPs can be used as a drug carrier. Kester et al. [13] have assessed the ability of CaP NPs to be encapsulating hydrophobic antineoplastic chemotherapeutics. Furthermore, these NPs have shown their ability to encapsulate both fluorophores and chemotherapeutics. These NPs have a diameter of 20–30 nm and pH sensitivity, as well as little degree of disparity. In addition, these NPs have been observed to be steady in physiological solution for a time duration at a surrounding temperature of 37°C. Bastakoti et al. [14] have claimed that colloidal NPs smaller than 100 nm filled with fluorescent pigments and anticancer drugs have resulted in the successful enhancement of robust biocompatible nanocomposites carriers for simultaneous release of drug molecules and imaging agents.

Shinto et al. [15] have utilized hydroxyapatite porous ceramic blocks (a CaP-based material) NPs in order to formulate nanocarriers for sustained delivery of antibiotics. Firstly, the cylindrical cavities in the hydroxyapatite blocks were loaded with the antibiotic and they were further implanted in the bone defect sites. The results revealed that a higher concentration of the antibiotic was released after seven days of implant. Then, there was a gradual decrease in the concentration after 12 weeks. Overall, the release of antibiotics with an efficiency of 70% was obtained.

Zhao et al. [16] have investigated the drug delivery ability of anticancer drug docetaxel-loaded lipid-calcium phosphate hybrid NPs, where the NPs showed a high drug loading capacity and biocompatibility. The NPs used in this study had a diameter of 72 nm. Mukesh et al. [17] have also proposed the use of CaP NPs as a carrier for the anticancer drug methotrexate; the size of the NPs in their case was 262 nm, with an encapsulation efficiency of 58%.They showed a low release rate of methotrexate at physiological pH and the authors observed that over 90% release was obtained in 3 to 4 h at endosomal pH. Liang et al. [18] have examined the in vitro delivery of the anticancer drug doxorubicin (DOX) hydrochloride from CaP hybrid NPs with particle sizes smaller than 50nm. Heparin/$CaCO_3$/CaP NPs were loaded with this anticancer drug. It was observed that the unloaded hybrid NPs showed high biocompatibility while the anticancer loaded NPs exhibited a strong cell inhibitory effect. These examples suggest a clear potential for CaP NP use in biomedical applications. CaP NPs have demonstrated successful delivery of drugs and bioactive molecules alone or in combination with polymers, owing to their biocompatibility. However, more research efforts should be focused in the future on investigating CaP NP degradation products in vivo and their effects on the vital organs.

2.1.1. Magnetic Mesoporous CaP NPs

Magnetic mesoporous CaP NPs with high water solubility and a diameter of 41 nm have been fabricated by Rout et al. [19]. These CaP NPs were composed of platinum pharmacophorecis-diaquadiamine platinum, folic acid, and rhodamine isothiocyanate, in order to use them against human cervical carcinoma cells. It was observed that the targeting of these cancer cells and delivery of cisplatin could be achieved by utilizing magnetic CaP NPs, as confirmed by cell apoptosis that was followed by cell death. However, no information was reported about their clearance from the body on both bases of animal models and clinical trials.

2.1.2. Porous Silica CaP NPs

El-Ghannam et al. [8] have used porous silica CaP NP as a carrier for 5-fluorouracil, which is considered very cytotoxic for 4T1 mammary tumor cells. The invitro study demonstrated that the NPs loaded with the anticancer drug possessed characteristics of burst release (the maximum release rate) in the first 24 h, and, following this, a sustained release, which was observed for the next 32 days. Meihuaet al. [20] have reported the preparation of mesoporous silica NPs coated with a CaP-hyaluronic acid hybrid. The core shell was further coated with another hyaluronic acid layer in order to target CD44 over-expressed cancer cells. The authors proved that the anticancer release in an acidic subcellular environment could be controlled using their NPs. These studies exhibited the superiority of core shell nano-systems over normal nanomaterials in the delivering of anticancer in a sustained manner.

2.2. Carbon-Based Nanomaterials

Other types of nanomaterials that may be utilized in therapeutic delivery, and especially for anticancer therapy, are carbon-based nanomaterials, including carbon nanotubes (CNTs), graphene oxides (GOs), and nano-diamonds (NDs), as discussed below.

2.2.1. Carbon Nanotubes

When carbon atom nanostructures are arranged in a tube-like hollow cylindrical shape, they may be called carbon nanotubes (CNTs) [21]. These are divided into three categories: single-walled carbon nanotubes (SWNTs), double-walled carbon nanotubes (DWNTs), and multi-walled carbon nanotubes (MWNTs). Their diameters may range between several angstroms and tens of nanometers while their lengths may reach half a meter [22,23].

One of the biggest benefits of carbon nanomaterials in biomedical applications is their ability to be a drug carrier. Pastorin et al. [24] have reported that, as an anticancer drug, methotrexate linked via a covalent bond to carbon nanotubes with fluorescein isothiocyante (FITC) was more effectively internalized through folate receptors into cells in comparison to the free drug. CNTs have also been designed to be tumor targetable through functionalization to be used as drug carriers. Moreover, SWNTs have also been used as drug carriers in their water-soluble PEG-ylated form. They have demonstrated a loading capacity for the anticancer drug DOX [25]. MWNTs have been examined as a drug carrier, and in particular oxidized MWNTs have been PEG-ylated for anticancer drug specific delivery to the brain for the treatment of glioma. Other polymers have also been utilized as functionalizing agents for CNTs for drug delivery applications, as has been reported previously [26]. MWNTs have been functionalized with a polyethylenimine (PEI) composite, which has shown impressive biocompatibility. Furthermore, these composites were designed to be accompanied by FITC and prostate stem cell antigen (PSCA) monoclonal antibody (mAb). Finally, a CNT-PEI (FITC)-mAb composite has been obtained which showed impressive biocompatibility with a cancer-cell-targeted delivery system [27].

Cellular uptake of CNTs has been explained by Bhatt et al. [28]. They have suggested mechanisms for CNT cell-internalization via two pathways, namely, dependent and independent endocytosis pathways, where the endocytosis-independent pathway is divided into two types: receptor-mediated endocytosis and non-receptor-mediatedendocytosis (see Figure 3). CNTs with their available types either functionalized or not have demonstrated great success in the delivery of bioactive molecules, especially anticancer agents, and the targeting of tumor cells.This has encouraged scientists to explore their applicability within most common diseases other than cancer. Nevertheless, CNTs have not been explored widely, owing to their difficult and expensive preparation methods.

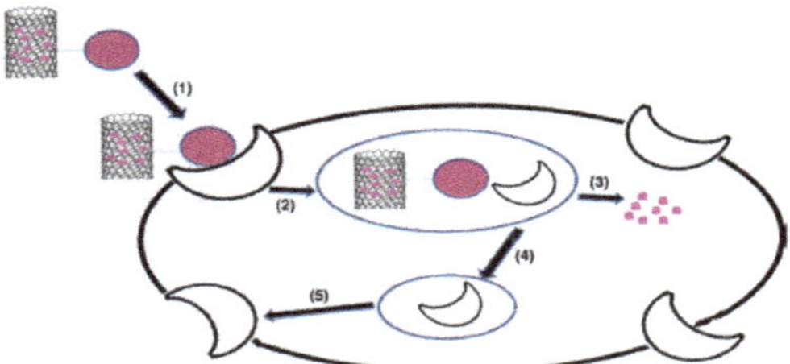

Figure 3. Receptor-mediated endocytosis of CNTs. (**1**) Association of ligand conjugated drug-loaded CNTs with receptor; (**2**) endosomal internalization of conjugates, (**3**) drug release, and (**4**,**5**) receptor regeneration [28]. Reproduced with copyright permission from Elsevier, 2016.

2.2.2. Graphene Oxide

Another carbon-based nanomaterial for use in biomedical applications is graphene oxide. Nano-graphene oxide complexes, especially metal nanocomposites (nanogold and nanosilver), have been found to be a good choice of material to treat cancer. Due to their greater drug entrapment capability and photothermal and synergizing effects, it is worth highlighting them for their targeted treatment of cancer based on their chemo-thermal behavior.

Chauhan et al. [29] selected gold NPs (AuNPs) as the composite metal and folic acid (FA) as the graphene oxide surface functionalization moiety for active tumor targeting of the model anticancer drug DOX (see Figure 4). Yang et al. [30] have examined the uptake of graphene oxide which has been functionalized with PEG, noting that it has good solubility and biocompatibility. In addition to this, PEG-ylated graphene oxide, which accumulates in tumors and has low retention in the reticuloendothelial system, has also been reported [30].

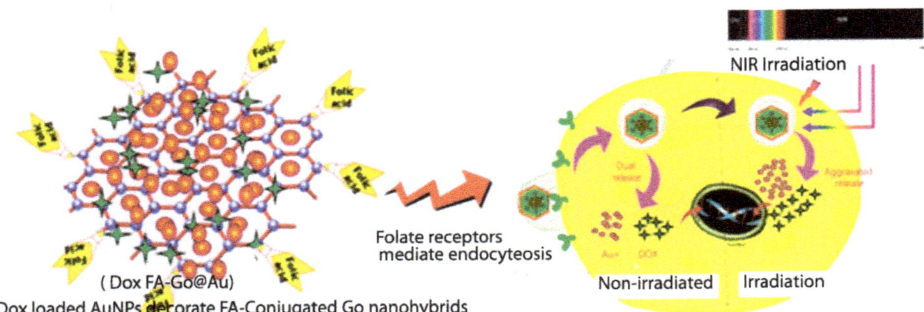

Figure 4. Tumor-localized DOX delivery with simultaneous photothermal ablation [29]. Reproduced with copyright permission from Elsevier, 2017. Legend: FA, folic acid; GO, graphic oxide; AuNPs, gold NPs.

Moreover, Huang et al. [31] have reported that the magnetic functionalized graphene oxide composite may be used as a nanocarrier for the delivery of anticancer drugs. Magnetic graphene oxide was produced by the authors by chemical co-precipitation of Fe_3O_4 magnetic NPs on graphene oxide nano-platelets. Furthermore, magnetic graphene oxide was modified by chitosan and mPEG-NHS through covalent bonds to synthesize mGOC-PEG. Irinotecan (CPT-11) or DOX was loaded to mGOC-PEG. Another experiment has been reported by Barahuie et al. [32] in which graphene oxide was synthesized in order to explore its potential use as a nanocarrier for an active anticancer agent, chlorogenic acid (CA).

DOX-loaded PEG-ylated nanographene oxide (NGO) preparation has been reported by Zhang et al. [33] where this combination caused destruction of a tumor without recurrence. Qin et al. [34] have used FA-conjugated NGO-PVP as a carrier for DOX, where folate receptors of cancer cells possessed great affinity to bind with folic acid. The FA-NGO-PVP was delivered within the cytoplasm of folate receptor positive human cervical cancer cells (HeLa). The minimal cellular uptake for the nanocarrier was an indication of the adhering of the FA-NGO-PVP composites to the folate receptors on the surface of the A549 cells. Thus, the nanocarrier was shown to have photo-thermal sensitivity and could be used as an anticancer carrier for chemo-photothermal therapy. Furthermore, camptothecin (CPT) and DOX were involved in the sulfonated NGO sheets, where folic acid was attached to carboxyl groups of the sulfonate NGO sheets to deliver DOX. The cytotoxicity of FA-NGO/CPT/DOX to MCF7 human breast cancer cells has been highlighted elsewhere [35]. Zhao et al. [36] have also prepared pH-responsive chitosan/ graphene oxide nanohybrids to be used for the delivery of DOX anticancer.

Liu et al. [37] have functionalized NGOs with PEG, using these to deliver the SN38 anticancer drug for colon cancer. Masoudipoura et al. [38] have reported a nanographene oxide conjugated with dopamine (DA-nGO) and laden with the anticancer drug methotrexate (MTX) to be supplied to positive human breast adenocarcinoma though adhering to DA receptors. GOs have demonstrated better characteristics when compared to CNTs, owing to their greater drug encapsulation ability. However, their use in biomedical applications is still limited because of their complicated synthesis and lower production yield, which makes them uneconomical.

2.2.3. Nano-Diamond

Nano-diamond is considered to be another type of carbon-based nanomaterial and is rarely used in cancer treatment. ND is considered an allotrope of carbon with an average particle size between 4 to 6 nm and which in aggregation could be recorded as having particle sizes of 100 to 200 nm. The surface of each ND possesses functional groups which allow for a different series of active molecules to be conjugated to it, including anticancer drugs [39,40]. The appearance of functional groups on ND surfaces works in the form of a covalent bond between the ND and other material, such as a polymer, which improves the solubility of the ND powder in various solvents [39]. NDs have an excellent biocompatibility, meaning they can be used in chemotherapy. For example, the biocompatibility of a fluorescent ND powder with a particle size of 100 nm has been confirmed for kidney cells by Yu et al. [41].

Some trials which have reported the utilization of ND in biomedical applications are summarized below. Carboxylated NDs have been utilized to improve the solubility in aqueous media of some anticancer drugs. Purvalanol A and 4-hydroxytamoxifan, known to be effective drugs for breast and liver cancers, have been conjugated with NDs [42]. Moreover, efficient drug loading capacity has been confirmed for NDs with negative charge which are accompanied by both NaCl and cationic DOX ions to achieve better treatment for HT-29 colorectal cancer via a more efficient method, in comparison to other carbon-based materials [43].

Upal et al. [44] have studied, using *invitro* testing, modified and surface-unmodified (–NH_2 and –COOH) ND for its ability to carry the anti-HIV-1 drug efavirenz and its cytotoxicity. It was found that there was a highly significant drug loading capacity for the unmodified ND conjugated drug formulation in comparison to the surface-modified ND, which had a very low toxic effect. NDs have shown few successful trials on an in vitro basis but animal studies need to be carried out in the near future to figure out their effect on the vital organs and their clearance mechanism. However, NDs suffer from the same restrictions as carbon-based materials, as mentioned above.

2.3. *Mesoporous Silica Nanoparticles (MSNPs)*

One common types of NP which is used as a drug delivery carrier is the solid mesoporous silica nanoparticle. Mesoporous NPs are particles in which the core of the particle is solid in nature while the shell or the particle's surface is porous in nature. Alternatively, the particle itself is porous in nature.

MSNPs are inorganic particles comprising of a honeycomb-like porous structure with empty channels (pores) of nanosized dimensions [45]. There are two types of mesoporous silica according to their particle size: mesoporous silica microspheres and mesoporous silica nanospheres. Although the use of mesoporous silica microspheres is important in numerous non-biological applications, they are not appropriate for use in many biomedical applications. However, the average particle diameter of a typical MSNP is ~100 nm with pore volumes ~0.9 cm^3/g, pore volume surface areas ~900 m^2/g, and pore sizes ~2 nm [46]. A hexagonal crystal structure has been recorded for MSNP porous channels (Figure 5).

Slowing et al. [47] have studied the implications of surface functionalization of MCM-41-Type MSNPs on the mechanism and efficiency of endocytosis different charge profiles on HeLa. The study concluded that the ability of MSNPs to escape endosomal entrapment was influenced by surface functionalities as this is considered to be a key factor in the designing of compounds that are effective for intracellular delivery.

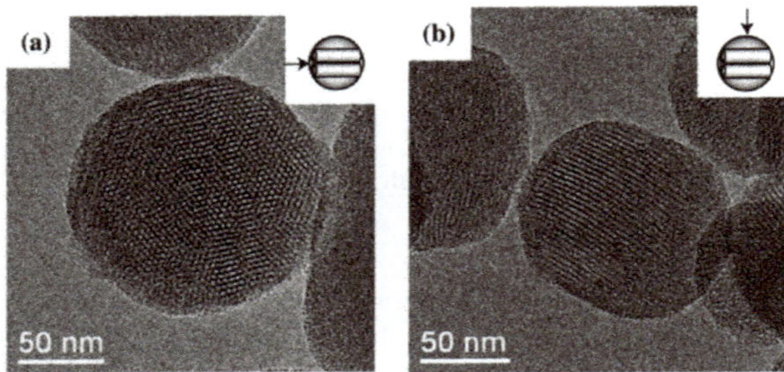

Figure 5. TEM images of mesoporous silica nanoparticle (MSNP) material recorded from the direction (**a**) parallel or (**b**) perpendicular to the long axis of the meso-channels [45]. Reproduced with copyright permission from Elsevier, 2008.

2.4. Gold Nanoparticles (GNPs)

Gold nanoparticles are considered excellent carriers for anticancer drug delivery because of their many unique features, such as their tunability in surface characteristics and particle size, their inertness, and their biocompatibility. Sabu et al. [48] have demonstrated that gold NPs can be loaded with anticancer agents. Paclitaxel (an anticancer drug) has been conjugated covalently by Gibson et al. [49] to gold NPs of size 2 nm, and the system has been found suitable to be used as a self-therapeutic. Dhar et al. [50] have conjugated gold NPs with DNA (producing DNA-AuNPs) to be employed as carriers of platinum compounds for the treatment of cancer. Cellular internalization of the tethered platinum was more effective than free cisplatin in killing cancer cells.

Brown et al. [51] have reported that oxaliplatin in its active form conjugated with gold NPs has possessed an effective killing ability with regard to cancer cells compared to pure gold NPs. It has also been proven that BSA-capped gold NPs can effectively be used as a nanocarrier for anticancer drug MTX and has been considered to be an anticancer for MCF-7 breast cancer cells when compared to non-capped NPs, as earlier declared by Yu-Hung et al. [52]. Furthermore, core/shell gold NPs have been fabricated and evaluated in order to be used as carriers for dual treatment where a radiosensitizer was presented by GNPs and the anticancer drug (DOX) [53].

Manivasagan et al. [54] have demonstrated a new method for the production of gold NPs to be used as a drug carrier or to be useful in photo-acoustic imaging (PAI). In particular, they have synthesized multifunctional doxorubicin-loaded fucoidan-capped gold NPs (DOX-Fu AuNPs). Dhamecha et al. [55] have also reported that GNPs work as an effective drug carrier for targeting DOX to fibrosarcoma cancer cells. They have been observed to subsequently reduce DOX-associated cardiac toxicity and myelo-suppression. DOX was seen to be absorbed onto GNPs with a high drug loading capacity, and industrial scalability of the process was demonstrated in their research [55]. Scientists working with GNPs face very challenging issues related to their removal from the human body along with their accumulation in the liver and kidney.

2.5. Other NPs Used as Drug Delivery Carriers

Some other inorganic materials (Table 1) have been used as drug delivery systems, and include functional multilayer fluorescent nonporous silica SNPs with an external shell, e.g., those containing primary amino groups [56]. AL-Ajmi et al. [57] have also prepared crystalline ZnO NPs using organic precursor techniques. These ZnO NPs were used as a drug carrier to deliver 5-Fluorouracil (5 Fu).

Table 1. Inorganic nanocarriers for drug delivery.

Inorganic Carrier	Drug Loaded	Purpose	References
CaP	Docetaxel	Breast, lung, and ovarian cancer.	[8]
CaP, dopamine nanographene oxide (NGO)	Methotrexate	Anti-rheumatic drug, breast adenocarcinoma.	[38]
Heparin/CaCo/CaP	Doxorubicin hydrochloride	Breast, lung, bladder, stomach, and ovarian cancer, and leukaemia.	[18]
Porous silica CaP	5-fluorouracil	Mammary tumors	[8,20]
NGO	SN38	Colon cancer	[31–36]
Carboxylate NDs	Purvalanol and 4-hydroxytamoxitan	Liver and breast cancer	[40–42]
Single-walled carbon nanotubes (SWNTs)-polyethylene glycol (PEG), NGO-PEG and MSNPs-GO-chitosan (CHI)	Doxorubicin	Leukaemia, breast cancer, gastric cancer, head and neck cancer, Hodgkin's lymphoma, liver cancer, kidney cancer, ovarian cancer, small cell lung cancer, soft tissue sarcoma, thyroid cancer, bladder cancer, uterine sarcoma.	[53–55]
NDs-NaCl		HT-29 colorectal cancer cells.	[43]
Gold NPs	Paclitaxel	Breast, lung, and pancreatic cancer.	[49]

3. Inorganic NPs for Hard Tissue Regeneration

The ideal nanostructured composite presented in the human is bone, which is considered the typical model of a hierarchical microstructure [58]. Nonstoichiometric nanocrystalline hydroxyapatite (HAP), ($Ca_{10}(PO_4)_6(OH)_2$), which is 20–40 nm in length, is the major inorganic component of bone, and Type-I collagen, which is ~300 nm in length, is its major organic counterpart. It has been revealed that HAP nanocrystals cover bone collagen molecules to form nanocomposites fibers. HAP C-axes are generally aligned along the collagen molecules in these nanocomposites [59].

Recently, the synthesis of nanomaterials for bone repair has been explored based on the simulation of various bone properties. One significant characteristic of materials used for bone regeneration is their mechanical property, and this property is possessed by different nanomaterials [60,61]. The other approach is to carry out surface modification at the nano-level, as this can provide an improved matrix for osteoblasts to grow and function (Figure 6) [62]. Several types of materials are suitable as biocompatible materials for bone healing with applications possible for metallic oxides (aluminium oxide, zirconium dioxide, and titanium dioxide), the CaPs family (hydroxyapatite, tricalcium phosphate (TCP), and calcium tetraphosphate) and glass ceramics (bioglass and ceravital) [63].

In addition, nano-scaffolds have superior properties in comparison to micro-scaffolds, owing to their higher porous microstructures that can mimic a real extracellular matrix (ECM), their porosity, and their other physical properties. This in turn improves their osteoblast adherence and viability [60,64,65]. Moreover, NPs enable the delivery of drugs and growth factors to promote healing and functional recovery. Modifying the surface of these biocompatible materials at the nano-scale for bone healing purposes has also been studied [66].

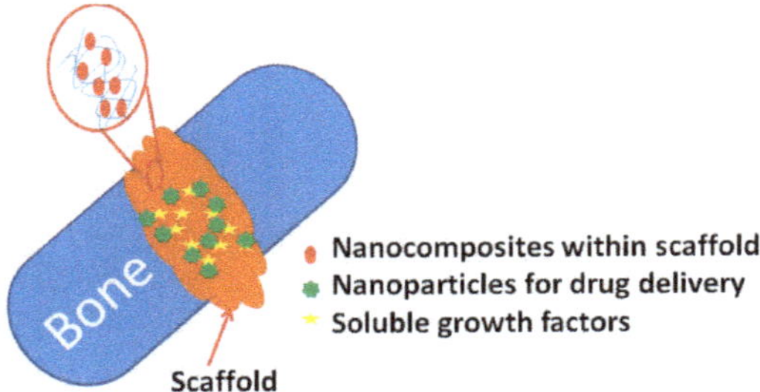

Figure 6. NPs used to repair a bone fracture: in cases of bone fracture, nanomaterials have been implanted into the target area (adapted from [62]).

3.1. Carbon Nanotubes

Owing to the impressive mechanical characteristics of carbon nanotubes, the idea of reinforcing brittle ceramic such as HAP using CNTs for the treatment of bone disease and fractures has been considered in the last three decades. Zhang and Kwok [67] have investigated the bone healing ability of HAP-TiO_2-CNT NPs. They noted the formation of an apatite layer on the surfaces of monolithic HAP and HAP-CNT and HAP-TiO_2-CNT NP coatings after 4 weeks of soaking in Hanks' solution. In addition, no remarkable effect was recorded on the formation of the apatite layer on the surfaces of TiO_2 and CNT when added to HAP. Furthermore, it has been reported that the addition of CNTs to HAP has improved the mechanical properties of the final composite as recorded by nano-indentation technique. These results revealed that the higher the addition of CNTs the greater the mechanical properties which were achieved compared to pure HAP [68].

A different application for the CNTs that has been reported aside from their being used as reinforcing material during bone regeneration is as a biosensor for bone regeneration [69]. Using NPs as a delivery system for bio-active molecules from bone healing, for example, both genes and proteins can be delivered to enhance propagation of osteoblasts, angiogenesis, and a reservoir for the necessary calcium salts [69]. Furthermore, one study has been conducted using multi-walled carbon nanotubes (MWCNTs) impregnated into HAP and the dip coating of a nanocomposite onto a titanium alloy (Ti-6Al-4V) plate to enhance the surface roughness of the implant. The authors noted that the mechanical properties of the HAP coating were enhanced and the surface roughness was reduced upon the addition of low concentrations of MWCNTs in comparison with pure HAP. Moreover, normal cell attachment and the growth process were confirmed by cell studies [70].

Other studies have discussed the cytotoxicity of titanium alloy implants coated with plasma-sprayed CNT-reinforced HAP embedded in rodents' bones, and it has been observed that there is no adverse effect or cytotoxicity associated with the addition of CNTs to bone tissues [71]. Balani and Agarwal have also confirmed the non-toxicity of a HAP-CNT coating on Ti-6Al-4V implants and have observed the ultimate growth of human osteoblast cells near the CNT regions [72].

CNT-based 3D networks have been found to be suitable for cell growth as a biomatrix scaffold [73]. In [73] a 3D network was produced by exerting chemically-induced capillary forces in order to manipulate the vertically aligned CNT array and the authors found that those networks seemed to be suitable for cell growth as a scaffold (Figure 7).

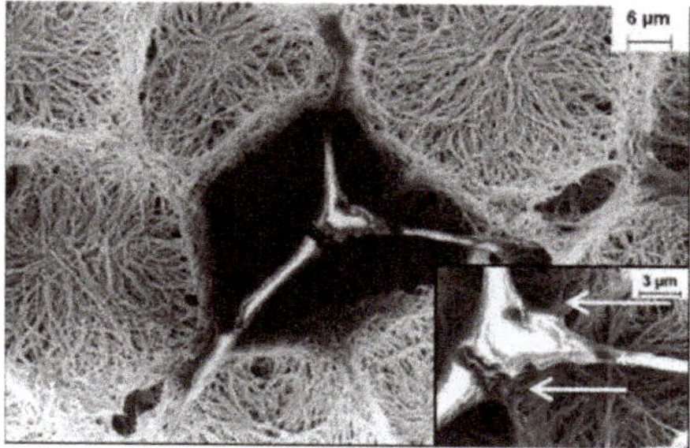

Figure 7. Scanning electron micrographs of L929 mouse fibroblasts growing on a multi-walled carbon nanotube (MWCNT)-based network [73]. Reproduced with copyright permission from the American Chemical Society, 2004.

Another scaffold based on MWCNTs has been reported by Abarrategi et al. [74]. Here, the MWCNTs were functionalized by chitosan (MWNT/CHI) and these nanocomposites were affirmed to be a scaffold for tissue engineering. In addition, this study showed that the MWNT/CHI scaffolds implanted subcutaneously accompanied by recombinant human bone morphogenetic protein-2 (rhBMP-2) for 3 weeks had increased the growth of C_2C1_2 cell lines. Afterwards, the MWNT/CHI scaffold composite was replaced by cells and bone regeneration was observed. In this study the authors confirmed the suitability of the MWNT/CHI scaffolds for utilization in bone reconstruction. Apart from hard (bone) tissue regeneration, CNTs have also been utilized for soft tissue regeneration (nerve) in order to assess cells behavior [75]. For this purpose, Keefer et al. [76] designed a neural network made of tungsten and stainless-steel wires and the electrode's wires were coated with CNTs by electrochemical methods.

3.2. Titanium Dioxide (TiO_2)

Titanium dioxide or titania (TiO_2) has been used as a bioactive coating and there has been consideration given to using TiO_2 within the coating as reinforcement as a technique for improving the mechanical reliability of HAP. Also, it has been proven that TiO_2 has the ability to induce osteoblast cell adhesion and growth [77,78]. It has been determined that the smaller the titanium grain size the greater the osteoblast adhesion. This may be because of the higher surface area attributed to the TiO_2 nanoparticles. Various techniques have been used to improve the TiO_2-HAP nanocomposite coatings for biomedical applications in order to enhance the strengths of coating and adhesion, reducing a loss of implants because of potential failure at the metal-coating interfaces and achieving an increase in biological response [79,80].

Kuwabara et al. [80] tested the bone formation ability on TiO_2-HAP nanocomposite coatings with a thickness of 100 nm. They cultured the osteoblast cells, which were derived from rat bone marrow, on the surface of coatings with different electrical charges and found that the HAP- and TiO_2-coated surfaces demonstrated a higher influence with regard to adhering to and propagation of cellsin comparison to pure HAP or pure TiO_2-coated surfaces. Ahmed et al. [81] revealed that the proliferation of mesenchymal stem cells (MSCs) was enhanced after using a Se/Ti nanocomposite. Furthermore, another study has unveiled a new composite which was generated using HAP/chitosan bioactive nanocomposites [82].

4. Bio-Imaging

Due to the magnificent properties of metal inorganic NPs, they have been directed to plenty of biomedical applications, especially in the bio-imaging field. The size of NPs plays an important role in the bio-imaging process, as NPs are preferred to other small molecules owing to their fast penetration in biological tissues and their ability to pass through the circulatory system. In addition, NPs exhibit limited renal excretion and prolonged blood circulation time, which allows repeated passing of NPs through tumors' vessels. Unlike organic NPs, inorganic NPs exhibit inefficient extra vasation inside tumors, which involves inorganic NPs tending to remain in the vasculatures of tumors without being in the interstitial spaces. This reduces the marking of nonspecific tumor cells within the imaging spaces [83–86].

An important approach for improving the imaging process which has been conducted by researchers has been to load one NP with different contrast agents related to several imaging techniques for multimodal imaging which attempts to overcome the limitations related to single imaging techniques [87,88]. Moreover, imaging and treatment can be achieved by the same NP [89,90]. NPs such as semiconductor quantum dots (QDs) have been considered for photoluminescence and have been widely used in bio-imaging [88–90]. In addition, luminescent UCNPs and SERS nanoprobes based on gold and silver NPs can be used for bio-imaging [91–93]. Magnetic NPs with sizes 1–100 nm can display superpara-magnetism meaning they can be widely used as a contrast agent in magnetic resonance imaging (MRI); iron oxide magnetic NPs coated with dextran can also be used in MRI [94,95]. Because of the optical properties of single-walled CNTs, such as high optical absorption and photoluminescence in the near IR region and strong resonance Raman-scattering, single-walled CNTs are widely utilized for bio-imaging [96].

4.1. Quantum Dots

QDs are NPs composed of semiconductor materials or atoms from groups II-IV or III-V, includingcds, cdse, cdte, zns, znse, zno, GaAs, InAs, and InP, owing to their unique optical properties. Each semiconductor material is covered by another semiconductor material, and these have a large spectral bond-gap which allows for an increase in the photo stability and quantum yield for the emission process; QD NPs show stability against an aggregation with capping agents [97]. In addition, these materials have a high molar extinction coefficient and high absorption from UV to near IR [88], where the changing diameter of NPs can modulate the excitation and emission peaks of QDs, with QDs showing sharp emission peaks that are considered ideal for multi-color imaging [97–99] (Figure 8).

Cai et al. [98] have reported the in vivo integrin $\alpha v\beta 3$ imaging of RGD peptide-conjugated QDs. Their results revealed that RGD peptides when conjugated with PEG-ylated QDs demonstrate maximum emission at 705 nm when injected intravenously into mice bearing U87MG tumors. In addition, the results showed a tumor contrast 20 min after injection and reached a maximum 6 h from injection. Another in vivo study conducted by Chen et al. [87] attempted to use optical and polyethylene terephthalate (PET) imaging of VEGFR in vasculature tumors by using QDs, where, they reported that, the amine-functionalized QDs conjugated with VEGF protein and then were exposed to radiation to be radio-labelled for VEGFR-targeted NIR fluorescence and PET imaging of tumor vasculatures. Furthermore, Ostendorpet al. [93] have utilized cyclic Asn-Gly-Arg (cNGR), which was seen to conjugate with paramagnetic QDs (pQDs) as a tumor nanoprobe, where cNGR targeted (the aminopeptidase N) CD13 on the endothelium of a tumor and was used for fluorescence/MRI dual evaluation of tumor activity.

Yong et al. [89] prepared InP-ZnS QDs that were used for in vivo detection of pancreatic cancer because of the low cytotoxicity exhibited by InP-ZnS. These were considered to be biocompatible nanoprobes for diagnosis of pancreatic cancer cells. The results showed that in primary and metastatic pancreatic cells the antigen receptors for anti-claudian 4 were over-expressed and QDs conjugated to anti-claudian 4; in addition, a polyclonal antibody was conjugated to QDs and no damage to the cells was observed. However, Cai et al. [98] have explored PET imaging of radio-labelled QD-RGD, which showed predominant uptake of QDs in the spleen and liver. Another problem related to QDs could be their presence within the body, where long retention time for accumulation of QDs has been observed. There is in vivo toxicity related to the presence of II-IV groups inside the body, but there are strategies to overcome this problem where there is a new generation of QDs that has been developed to reduce toxicity, including Cd-free QDs. Repeated demands of the product development of QDs and their related fields should be considered for their side effects and in the long run, substantiated by prototype modules and pilot-line fabricating, especially in light of arrangement producers and subsidizing organizations frequently setting their needs in relation to being dependent on these demands.

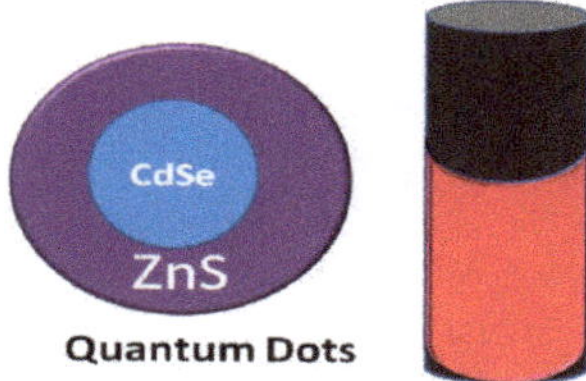

Figure 8. A semiconductor quantum and a quantum dot (QD) aqueous solution under UV light showing bright pink fluorescence. QDs are widely used in fluorescence imaging (adapted from [99]).

4.2. Carbon Nanotubes

As mentioned above, there are two types of CNTs, namely, SWNTs and MWNTs. SWNTs have strong optical absorption in the visible and NIR regions due to the optical properties of CNTs, such as resonance. Raman-scattering and photoluminescence in the near IR region make CNTs useful for biomedical imaging. SWNTs exhibit different optical absorption peaks from the UV region to the near IR region, which made these NPs suitable for utilization as photo-thermal agents and photo-acoustic imaging agents [99] (Figure 9).

Liu et al. [100] have reported how in vivo integrin $\alpha v\beta 3$ imaging was able to utilize SWNTs and SWNTs functionalized by PEG-ylated phospholipids to increase water solubility which was labelled by a 64-Cu radioisotope for micro PET imaging. RGD peptide conjugated 64 Cu radiated SWNTs to PEG coating (SWNT-PEG 5400-RGD) was injected intravenously into the glioblastoma U87MG tumor in mice and then monitored by a micro PET. The results demonstrated higher uptake of SWNT-PRG5400-RGD by the tumor (~13% of dose ID/g) when compared to the tumor uptake of SWNTs without combination with RGD (4~5%) of injected dose per gram tissue. Moreover, Smith et al. [97] have used RGD conjugated with PEG-ylated SWNTs as a Raman nanoprobe for in vivo tumor imaging in mice, where SWNT-RGD was injected intravenously into the mice's tumors with high integrin $\alpha v\beta 3$ expression. Raman microscopy showed strong signals for tumor cells injected with SWNT RGD while weak signals were recorded for the tumors injected with non-targeted SWNTs. They also utilized SWNT-RGD as a contrast agent for tumor photoacoustic imaging. PET imaging and Raman imaging showed strong photoacoustic signals in U87MG tumors. CNTs with their impressive biocompatibility have demonstrated great success in the imaging field, thus encouraging scientists to explore their applicability in the diagnosis of most common diseases other than cancer. Nevertheless, CNTs have not been explored widely owing due to their difficult and expensive preparation methods.

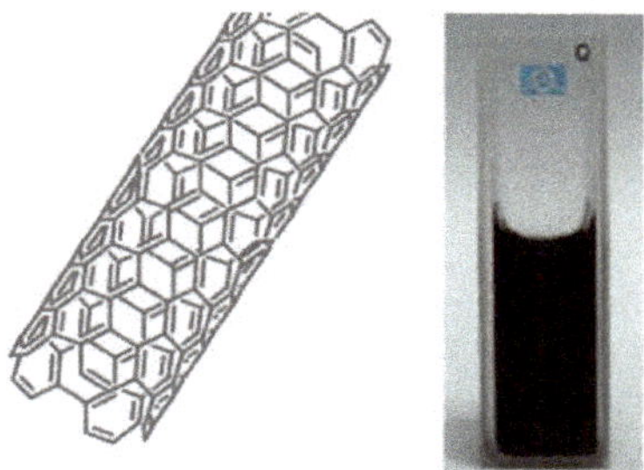

Figure 9. A single-walled carbon nanotube and an aqueous solution of SWNTs functionalized by PEG-SWNTs with highly optical properties, which are considered excellent platforms for biomedical imaging [99]. Reproduced with copyright permission from Springer Nature, 2010.

4.3. *Super Paramagnetic Iron Oxides NPs*

Superparamagnetic iron oxide (SPIO) NPs have been widely utilized as a contrast agent in MRI imaging; FeCo alloys have the best magnetic properties but because of oxidation and toxicity their appearance in biomedical applications has been limited [101]. Iron-based oxide NPs have been investigated by Won et al. [95], in which coated FeCo nanocrystals with a single layer of graphite carbon (GC) was used as a contrast agent for MRI. These NPs were PEG-ylated to increase the solubility of the FeCo/GC NPs. The water-soluble FeCo/GC NPs exhibited higher relaxivities on a per metal atom basis compared to other materials which utilized MRI as a contrast agent. Afsaneh et al. [102] have reported on recently developed MRI and PET imaging probes using SPIO NPs as a contrast agent where iron oxide NPs were coated with various coats, such as poly aspartic acid, PEG, dopamine, and dextran. The effects of these coats were discussed with regards to the particle size, targeted organ, final uptake, and time retention in the studied organ.

One of the methods by which magnetic NPs are delivered to the diseased organ is by intravenous infusion or by means of a blood circulatory system. Another method relies on utilizing magnetic nanoparticles suspensions for infusion. A steady uniform solution is required to avoid the aggregation of the NPs. Particle size and surface functionalization are two parameters required for the stability of the magnetic colloidal suspension. Determination of appropriate magnetic nanoparticles is the principal significant advancement for bioapplication. For applications in drug delivery, the magnetic nanoparticles are required to be steady in water at neutral pH, which relies on their size, charge, and surface functionalization. However, super magnetic materials, for example, cobalt and nickel, are not utilized in biomedical applications because of their lethal properties and oxidation susceptibility.

4.4. *Gold NPs*

Gold NPs have been highlighted among contrast agent materials for their bioimaging application. Monodispersity, stability, and higher attenuation coefficients for X-rays are the impressive properties which make these NPs suitable for this process. Peng et al. [91] have found that a way to synthesize dendrimer-stabilized gold NPs is by use of amine-terminated fifth-generation poly(amidoamine) (PAMAM) dendrimers. These were modified by diatrizoic acid as stabilizers for enhanced CT imaging. Li et al. [103] prepared Au-coated iron oxide (Fe_3O_4-Au) nanoroses to be used as a probe for multi-function as well as aptamer-based targeting, MRI, optical imaging, photo thermal therapy, and chemotherapy. Zhang et al. [104] prepared PEG with PEI-stabilized gold NPs for blood pool, lymph node, and tumor CT imaging. Moreover, Yigit et al. [105] synthesized gold NPs conjugated to 3,3-diethylthiatricarbocyanine iodide (AuNP-DTTC) which is used as a contrast agent for in vivo MRI and Raman spectroscopy. Here the probe consisted of MRI-active super paramagnetic iron oxide NPs coupled with AuNP-DTTC. Tailoring the properties of gold NPs and utilizing chemical techniques has grown extensively in the last two decades, especially in the field of bio-imaging. Developed tests and probing techniques rely on advances in gold nanoparticles conjugates, new optically-controlled functional materials, new highly specific color-coded probes of cellular function, and new optically-based therapeutic methods.

5. Porous Membranes

The second objective of this review is to present an overview of the therapeutic applications of porous membranes and their key challenges. Porous membranes have been used in numerous engineering applications such as molecular separation, catalysis, and filtration, etc. [106,107]. Literature shows that porous membranes are frequently regarded as nanoporous structures because the pore size of such membranes lies between 1 and 100 nm, although several terminologies have been used to explain these porous membranes and the terms nanoporous and microporous are used based on pore size [108,109]. In this section, different types of membranes, types of materials, etched membranes, and the fabrication of micro- and nanoporous membranes are briefly explained (Figure 10).

In the following sections, the properties, surface modification techniques, biocompatibility, and drug delivery applications are also discussed. Lastly, we describe the key challenges and future prospects of these membranes.

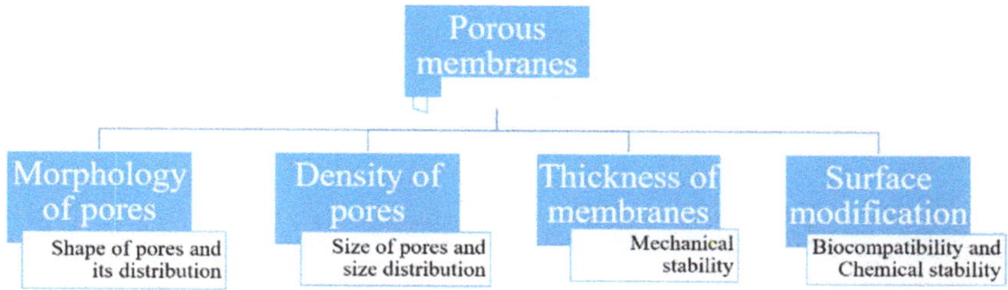

Figure 10. Key characteristics of porous membranes.

5.1. Membrane Classifications

Membranes can be classified based on the following characteristics.

5.1.1. Material

Membranes comprise of inorganic, organic, polymeric, and composite materials. Mostly, the organic/polymeric membranes are made of materials such as polycarbonate (PCL), polyethylene terephthalate (PET), and polysulfones, etc. (Table 2). These composite materials contain two different materials and combine apolymer and an inorganic material (e.g., a polymer with a ceramic). They are developed in particular to improve the stability, permeability, and selectivity [110].

5.1.2. Size, Shape, and Order of Pores

Membranes are classified according to their size, distribution of size, order, and shape. Hence, membranes pores<2 nm are called microporous, those with pores 2–50 nm mesoporous, and those with pores>50 nm macroporous. The shape of the pores can be cylindrical, slit-like, conical, or irregular in shape. The pores should be well arranged in a vertical order instead of in a tortuous network [111]. The porous network for drug delivery depends upon the following properties:

- Optimum pore size with narrow distribution;
- Less flow resistance to allow high flux;
- Sufficient mechanical strength with adequate chemical and thermal stability, and;
- Biocompatibility.

5.1.3. Fabrication Method

There are various methods available for fabricating membranes and they include ion-track etching, lithography, selective electrochemical leaching, focused ion beam etching, the sol-gel process, and the phase separation technique [111,112].

Ion-Track Etching

The polymeric membrane plays an essential role in drug delivery. One of the most common methods applied for formulating polymeric membranes with ordered pores is ion-track technology. This process is achieved by exposing a thin film of polymer with heavy ions which form ion tracks. The ion tracks are allowed to form pores via a chemical etching process which selectively attacks the ion tracks. The pores formed by this process are cylindrical or conical in shape with a diameter from 10 nm to within the μm range. The limitation of this process is that manufacturing of pores of a diameter less than within the nanometer range is not possible [113].

Micromolding

Another important method used to prepare membranes is micromolding. This technique is similar tothephase separation method and involves a solution of the polymer being poured into a mold, which is solidified by the phase separation process. Langley et al. and Ultricht et al. have discusseda porous structure developed by a phase separation micromolding technique which considers the structural, physical, and chemical properties of the polymers used [109,110].

Lithography

Micro/nanofabrication technologies have been recently developed as the most interesting methodology used to produce an ordered array of micro/nanopores on the surfaces of silicon. Conventional methods have drawbacks such as broad size distribution of pores, poor mechanical strength, and poor chemical stability. Micro/nanofabrication techniques are widely used to overcome these problems. One example is the work of Desai et al. in which biocapsules with controlled pores of about 7 nm were developed. The authors established that these nanoporous membranes allow exchange of nutrients, waste products, and therapeutic proteins, and that they are biocompatible, providing immune support to cells. In addition, they used their fabricated membrane for an implantable pancreas and oral drug delivery formulation [114,115].

Focused Ion Beam Etching

Polymeric membranes with a well-ordered array of pore structures and a size range of about 100 nm with a thickness of 1–5 μm have been able to be prepared. The membranes formed with this technique are called nanosieves and are also effectively achieved by photolithography. An excellent example of this method has been prepared by Tong et al. which has included an ordered structure of a cylindrical membrane of diameter 25 nm [116].

Electrochemical Etching

The most extensively studied membrane-producing electrochemical etching method is the anodic membrane technique. The membranes fabricated with this technique produce a honeycomb-like and ordered structure. The pore morphology and size can be controlled through the process of anodization [117]. The organization of the porous structure depends on the voltage and chemicals used in the fabrication. These ordered channel assemblies of membranes are obtained by an anodization process. The most effective technique for preparing well-arranged porous membranes with large dimensions involves this process. During the anodization procedure, thin ordered structures are engraved within a larger dimension membrane [118]. Within the small ordered structures biomolecules are encapsulated, and these membranes are used for diffusion-controlled delivery systems. The drug molecules are encapsulated inside the micelles as biocapsules and it has been shown that membranes containing biocapsulesare able to release the drugs in a controlled fashion [119].

5.1.4. Surface Modifications

When membranes come into contact with physiological fluids, three main processes occur which create problems, namely, biofouling, degradation of the membrane, and immune reactions caused by the membrane. There are various approaches to addressing these problems, and these are discussed as follows. If cells, proteins, and other materials accumulate on a membrane surface when the membrane is in contact with a biological environment, a process is occurring which is called biofouling [106,120]. The second issue is that this also causes tissue encapsulation, which leads to fibroblast proliferation, collagen synthesis, and proliferation of blood vessels. These processes in turn lead to vascular tissue capsule formation, which delay the transport of glucose molecules in biological environments due to steric hindrance [121]. Wound healing occurs via the processes of hemostasis, inflammation, and formation of scars and repair. During the implantation of the membrane, cells of epithelial connective tissue and the basement membrane may be damaged. Firstly, the bruised area is filled with coagulated blood, and this clotting allows neutrophils. The day after wound injury, the inflammatory cells disturb the function of the membrane by taking nutrients and releasing proteolytic enzymes and free radicals. On the third day, there is occurrence of macrophages and granulation tissue in the wounded area, followed by neovascularization on fifth day. Finally, at the end of the first month, the formation of a mature scar in the epithelial layer of the tissues is observed. Hence, the biocompatibility depends on the surface of the implanted membrane, which influences each stage of the process. A thick vascular fibrous scar formation leads to diffusion of analytes and uptake of nutrients, resulting in reduced response of the membrane [115,122].

In order to reduce biofouling, several reports have demonstrated the use of coatings and other methods of surface treatment. For this, the membrane surface is treated with cross-linked polymers of poly (hydroxyethyl methacrylate) or poly (ethylene glycol). Polymer coatings on the membrane surface result in electrically neutral, polar, flexible, and swellable membranes in water, and create an interface between the membrane surface and the physiological environment. Coatings of poly (hydroxyethyl methacrylate) or poly (ethylene glycol) are considered the most attractive for membranes because water-soluble drugs diffuse through the water-swollen polymer layer [123–125].Biocompatibility of the implantable membrane is characterized by inflammatory responses. Surfactants are also called surface-active agents and have hydrocarbon tails attached to polar head groups. Many membranes used for drug delivery contain these surfactant molecules as a plasticizing agent. Unexpectedly, these surfactants diffuse out of the membranes and to the surface, until they get depleted. Accordingly, these agents need calcium chelating plasticizers which limit the enzyme in the coagulation cascade [126,127]. Keeping in mind the characteristics of the membrane and the physiological environment, the biocompatibility of the material should also be addressed. In addition, it is very essential to prove that there is very little or no leaching, degradation, biofouling, and inflammatory responses. Therefore, selection of the perfect surface coating is important to figure out problems in a biological environment [128].

5.2. Drug Delivery by Membranes

The use of membranes to deliver drugs/bioactives is opening up new therapeutic advantages like increasing the solubility of bioactives, protecting bioactive molecules from degradation, providing a sustained release of the actives, improving drug bioavailability, providing targeted delivery, decreasing lethal effects, offering an appropriate form for all routes of administration, and allowing for rapid formulation development. Moreover, they can carry one or more bioactive agents and have been developed into different classes of carriers. The different carriers can be carbon-based nanomaterials, polymeric membranes, and inorganic membranes [129].

5.2.1. Polymeric Membranes

Here, the drug formulation is encapsulated in the compartment of a drug reservoir, in which the surface of the drug-releasing layer is covered by a rate-controlled polymeric membrane. The drug reservoir could be solid, a solid dispersion, or a drug solution in liquid form (Figure 11). The encapsulation process for preparing the drug formulation inside the reservoir compartment includes fabrication by microencapsulation, coating, and molding techniques. The polymeric membrane is manufactured from a nonporous, microporousor semipermeable membrane (Figure 12) [130].

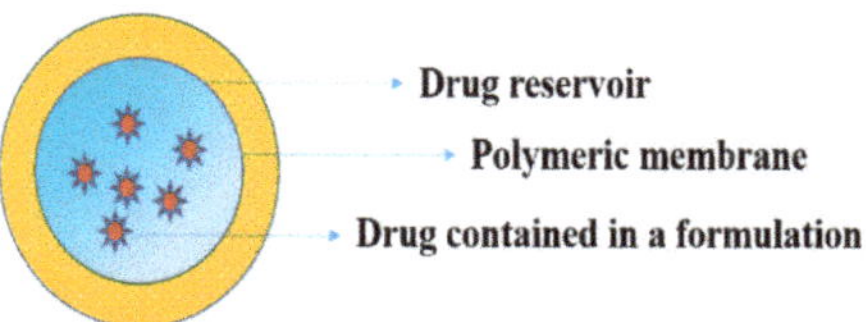

Figure 11. Polymeric membrane used for drug delivery system.

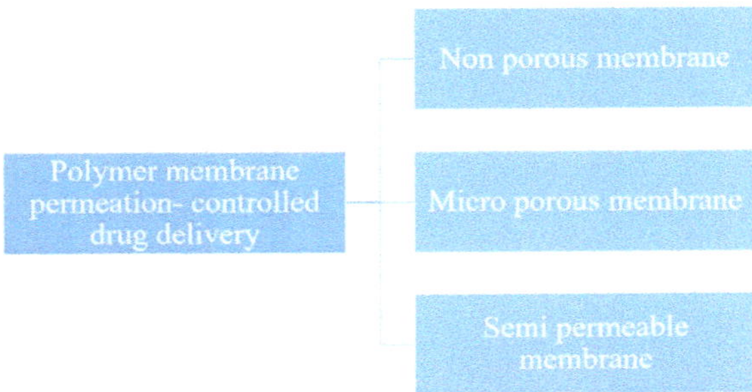

Figure 12. Classification of polymeric membranes for drug delivery.

Drug release from the polymeric membrane must be at a constant rate (Q/t) which is well-defined by the Equation (1),

$$\frac{Q}{t} = \frac{K_{m/r}K_{a/m}D_dD_m}{K_{m/r}D_mh_d + K_{a/m}D_dh_m} X\, C_R, \quad (1)$$

where, $K_{m/r}$ and $K_{a/m}$ are the partition coefficient of the drug molecule from the reservoir to the rate-controlling membrane and from the membrane to the aqueous layer, respectively, D_d and D_m are the diffusion coefficient of the rate-controlling membrane and the aqueous diffusion layer, respectively, h_m and h_d are the thickness of the rate-controlling membrane and the aqueous diffusion layer, respectively, and CR is the drug concentration in the reservoir compartment. The drug release from this system is controlled at a preprogramed rate by controlling the partition coefficient, diffusivity of the drug molecule, the rate-controlling membrane, and the thickness of the membrane. There are several controlled release polymeric membrane drug delivery systems which have been successfully marketed and some of the examples of these are outlined below [131].

5.2.2. Transdermal Systems

In this type of system, nitroglycerin and lactose are triturated in silicone fluid which is then encapsulated in a thin membrane of ellipsoid shape. The reservoir of the drug layer is introduced in between the metallic plastic laminate (backing sheet) which is impermeable and ethylene-vinyl acetate copolymer porous membrane which is the rate controlling membrane. This is formulated by the injection molding process and a thin layer of adhesive (silicone) is again coated on the permeable drug membrane for the immediate contact of the drug-releasing membrane with the skin's surface to be maintained and achieved (Figure 13). Nitroglycerin (0.5 mg/cm^2)/day is delivered transdermally for treating angina [132]. Other examples are Estraderm-controlled delivery of estradiol for 3–4 days to relieve postmenopausal syndrome and osteoporosis [133], Duragesic-controlled delivery of fentanyl for 72 h to relieve chronic pain [134] and Androderm-controlled delivery of testosterone(24 h) used as an additional therapy for patients with testosterone deficiency [135].

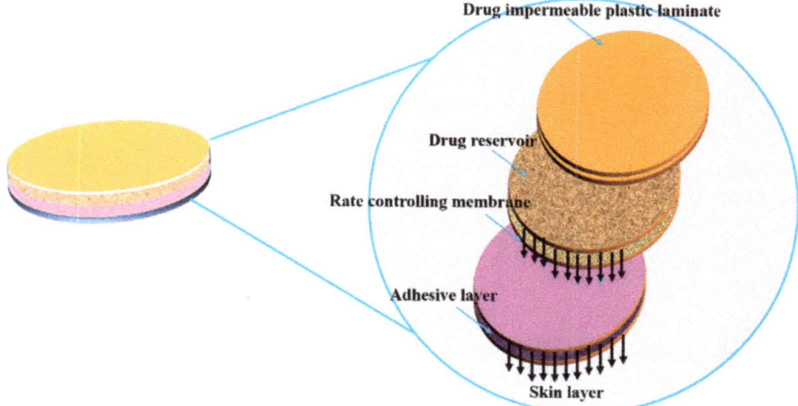

Figure 13. Diagrammatic representation of membrane permeation-controlled system in which the drug reservoir is sandwiched between the membrane layers and the adhesive layers facing the skin's surface (adapted from [134]).

5.2.3. Ocusert

In this type of system, in between two transparent microporous membranes a drug reservoir layer containing a pilocarpine-alginate complex is fabricated (Figure 14). The porous membrane permits the tears to permeate into the layer containing the drug. When the microporous membrane comes into contact with the tear fluid, it will dissolve and constantly deliver the pilocarpine at a rate of 20 to 40 mcg/h for treating glaucoma disease [136].

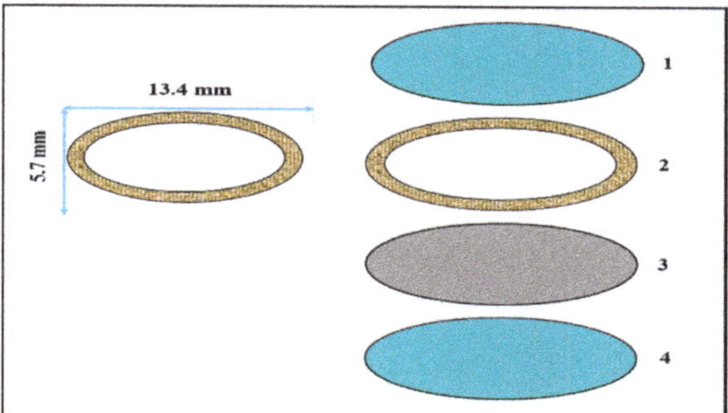

Figure 14. Ocusert: (1) and (4) show transparent polymer membranes, (2) shows a titanium dioxide white ring, and (3) shows a pilocarpine core reservoir (adapted from [136]).

5.2.4. Progestasert

In this type of system, a drug reservoir layer is used which consists of $BaSO_4$ and progesterone dissolved in silicone fluid and encapsulated into a T-shaped device using an ethylene-vinyl acetate copolymer (nonporous membrane) (Figure 15). This system is particularly designed to continuously deliver progesterone to the uterus with a dose of 65 μg/day to attain contraception for a year [137]. Another example of this system is the Mirenaplastic T-shaped device(steroid reservoir), which contains 52 mg of levonorgesterol for achieving contraception for 5 years [138].

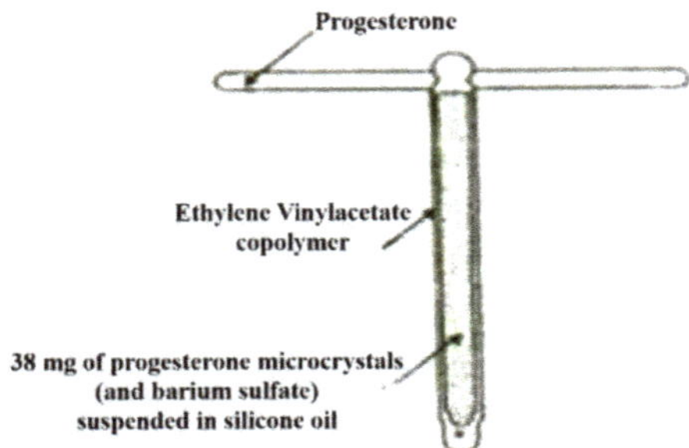

Figure 15. Progestasert IUD with structural components shown (adapted from [137]).

5.2.5. Polymer Matrix Diffusion-Controlled Drug Delivery

In this type of pre-programmed delivery system, the drug reservoir is prepared by homogenous dispersion of the drug in a rate-controlling polymer matrix fabricated by a hydrophilic or hydrophobic polymer. Secondly, the drug dispersion in the matrix is fabricated by blending the drug particles with a polymer or highly viscous base polymer followed by the cross-linking of polymer chains. Next, the drug solids are mixed in a rubbery polymer at an elevated temperature. The resulting drug-polymer dispersion is then extruded or molded to form a drug delivery device of varying shapes/sizes (Figure 16). It can also be prepared by dissolving the drug and the polymer in a solvent, followed by solvent evaporation, or under vacuum conditions [139].

Release of drug molecules from this type of controlled release drug delivery system is controlled at a pre-programmed rate by controlling the:

- Loading dose;
- Polymer solubility of the drug, and;
- Diffusivity in the polymer matrix.

An example of this polymer matrix diffusion-controlled drug delivery system is given in the subsection on the Nitro-Dur system.

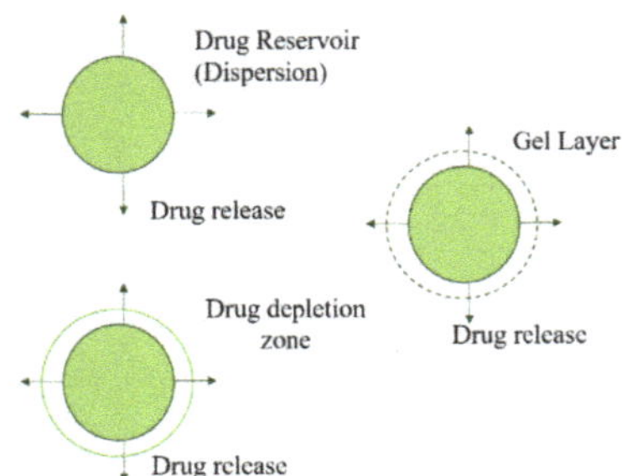

Figure 16. A polymer matrix diffusion-controlled system.

The drug release rate from the polymer matrix is not constant and can be expressed by the equation

$$\frac{Q}{t^{1/2} = \left(2AC_{R}D_{p}\right)^{1/2}}, \tag{2}$$

where, $Q/t^{1/2}$ is the rate of release of the drug, A is the initial drug loading dose in the polymer matrix, C_R is the drug solubility in the polymer and D_p is the diffusivity of the drug in the polymer matrix.

5.2.6. Nitro-Dur System

This type of system is fabricated by heating a hydrophilic polymer, glycerol, and polyvinyl alcohol and then lowering their temperature to develop a polymer gel. Nitrogylcerin and lactose are mixed and dispersed in the polymer, which is solidified at room temperature to form a medicated disc by molding technique. This is assembled into a metallic plastic laminate where the transdermal patch is developed with an adhesive rim. This patch is designed for application to the skin for continuous release of 0.5 mg/cm^2 for angina treatment. Poly (isobutylene) or poly (acrylate) adhesive are the polymers used in this method and this polymeric adhesive is spread by solvent casting technique over the area of the drug impermeable membrane (Figure 17) [140]. The marketed formulations are Frandol tape by Toaeiyo/Yamanouchi in Japan—isosorbide dinitrate [141]; Nitro-Dur II by Key in the United States—nitroglycerin, which replaces Nitro-Dur, a first generation patch from the market;Habitrol and Nicotrol—nicotine (smoking cessation) [142]; Minitran—nitroglycerin [143], Testoderm—testosterone (testosterone replacement therapy) [144], and Climara—estradiol (vasomotor system treatment associated with menopause) [145].

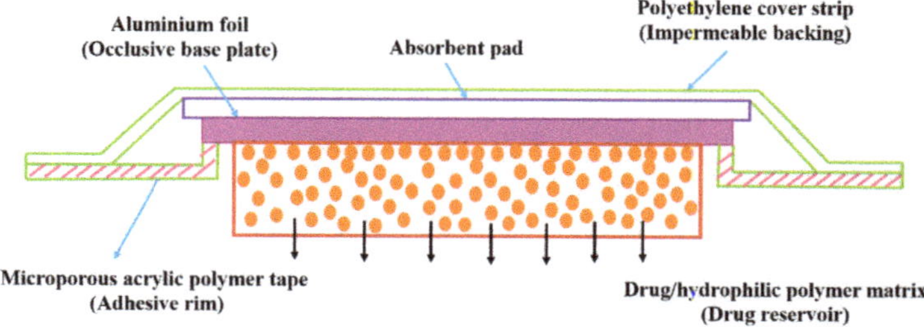

Figure 17. Nitro-Dur system (adapted from [140]).

5.2.7. Compudose

This is a subdermal implant formulated on the basis of dispersing estradiol crystals in a viscous silicone elastomer and coating the dispersion over a silicone rod by extrusion technique to form a cylindrically-shape implant (Figure 18). Compudose is subcutaneously implanted for a duration of 200 to 400 days for the delivery of estradiol [146,147].

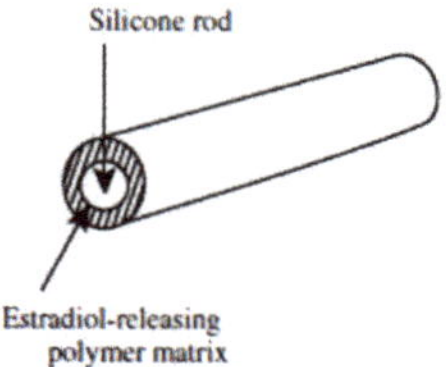

Figure 18. Compudose implant.

5.2.8. Polymeric Membrane-Based Drug Delivery Systems as Commercialized Formulations

Polymers form the basis of a significant number of membrane-based drug delivery systems. In fact, advances in the field of polymer sciences have paved the way for a large number of membrane-based drug delivery system designs that have considerable flexibility and functions [148–150]. Table 2 summarizes various formulations of drug delivery systems emphasizing the physico-chemical and mechanical properties of various polymers being used in commercially available membrane drug delivery systems.

Table 2. Various formulations of drug delivery systems emphasizing the physico-chemical and mechanical properties of various polymers being used in commercially available membrane drug delivery systems.

Polymers	Structure	Fabrication Method	Commercial Products/Literatures	Comments	Reference
Polycarbonate (PC)		Ion-track etching	Estrogen	Excellent stability against oxidation and biodegradation and improves antifouling properties	[151]
Polyethylene (PE)		Ion-track etching	Catapress (Clonidine), Boehringer IngelheimClimara (Estradiol), Berlex	Physico-chemical stability andordered pore formation with superior membrane performance	[152]
Polyethylene terephthalate (PET)		Lithography	Ketoprofen	Biostable, antifouling, has better performance of membranes, in useful in preparing surgical meshes and ligaments	[153]
Polystyrene (PS)		Lithography	d-limonene, ibuprofen	Chemical resistance, easy processing, lower cost, exhibits enhancements in strength, stiffness, toughness, and ductility	[154]
PC, PE	-	Ion-track etching	Estraderm (Nitroglycerin), Rotta Research	Cost-effective and biocompatibility is fairly good	[155]
PC, PE, PET, PS	-	Phase separation	Deponit (Nitroglycerin), Pharma SchwarzHabitrol (Nicotine), Novartis	Cost-effective and biocompatibility is fairly good	[154,156]
Polyurethane (PU)		Sol-gel/solvent casting	Vivelle (Estradiol), Novartis	Good elasticity, biodegradable, suitable for hydrophilic drugs, biocompatibility is fairly good	[154]

Table 2. *Cont.*

Polymers	Structure	Fabrication Method	Commercial Products/Literatures	Comments	Reference
Polysiloxane (silicone)		Sol-gel/solvent casting	Prostep (Nicotine), Lederle, Transderm Nitro (Nitroglycerin), AlzaSyncro-Mate-C (Norgestomet)	Better insulation, excellent biocompatibility, and fabricated easily for hydrophilic drugs	[157]
Polyisobutylene (PIB)		Solvent casting	Aminopyrene, Mitsubishi Petrochem Co., Japan	Good adhesive drug impermeable layer and high degree of tack or self-adhesion	[158]
Polymethyl methacrylate (PMMA), poly (2-hydroxy ethyl methacrylate)		Layer by layer deposition	Androderm (Testosterone), SmithKline Beecham	Physical strength and transparency	[159]
Polyvinyl alcohol (PVA), Poly (ethylene-co-vinyl acetate)		Solvent casting	Nitro-Dur I (Nitroglycerin), Key PharmaTestoderm TTS (Testosterone), Alza	Rate-controlling membranes, high membrane permeability, hydrophilicity and strength, suitable for lipophilic drugs	[160]
Polyacrylic acid, polyacrylate, polyacrylamide	Polyacrylate	Layer by layer deposition	Epinitril (Nitroglycerin), Rotta ResearchMonsanto (Fentanyl), Dow Corning	Good adhesivity and spreadability and contains a drug impermeable layer	[160]

Table 2. *Cont.*

Polymers	Structure	Fabrication Method	Commercial Products/Literatures	Comments	Reference
Polylactides (PLA), polylactic-co-glycolic acid (PLGA), polyglycolides (PGA)	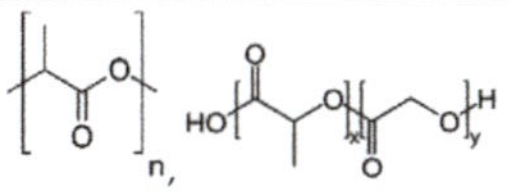	Sol-gel/solvent casting	Propranalol, Exxon Chemical Co.	Good biocompatibility; lactic and glycolic acids are the degradation products and they are easily eliminated from the body	[161,162]
Polyvinyl pyrrolidone (PVP), poly (N-vinyl pyrrolidone)	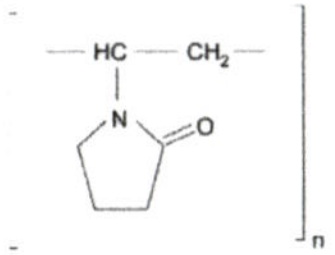	Sol-gel/solvent casting	Cytarabine, ara-ADA, Polyscience	Superior biocompatibility, has suspension capabilities, antinucleating agent, and enhances release rate	[163]
Polyethylene glycol (PEG)		Sol-gel/solvent casting	Miconozale, Rohm, Germany	Chemically inert and free of leachable impurities	[164]
Oxide plus polymer	-	Sol-gel/solvent casting		Superior biocompatibility and has narrow pore size	
Polymer coating on support membrane	-	Layer by layer deposition			[165]

5.3. Future Challenges

The next generation of inorganic particulates and membrane materials could be intended for flow regulation, screening of size, and dynamic pore sizing by means of external controls. Another possibility for fabricating these inorganic particulates and membranes is by the use ofa lab on chip microfluidic model to be used in medical diagnostics. Surface modification of inorganic particulates and membranes with organic and inorganic materials is now being explored. Moreover, when the surface of the inorganic particulates and membranes is modified with grafted polymers, in vitro testing shows promising results. Surface modification using polymers which undergoes conformational changes with a response to stimuli such as temperature, pH, and concentration of ions has shown acceptable results as well. The in vivo tests are necessary to confirm effective delivery and biocompatibility. Many challenges are there in manufacturing biocompatible inorganic particulates and membranes for in-vivo drug delivery applications. The key challenge is to fabricate inorganic particulates and membranes which respond to various stimuli. These inorganic particulates and membrane systems could therefore be used in micro/nanoscale chips for rate-controlled programmable delivery of drugs.

6. Conclusions

Inorganic nanomaterials have emerged as promising materials in different fields, including drug delivery, hard tissue regeneration and bio-imaging. Their exceptional properties, such as tunable stability, functionality, high surface area, and low inherent toxicity, make them the materials of choice for efficient use in targeting. The development of NPs for the enhancement of suitable biomedical applications and decreasing the permeability of relapse is still to be achieved. Membrane materials are essential for biomedical applications such as drug delivery, targeted drug delivery, and other medical applications. The main properties of these membranes include having an average pore size of about a few μm to nm or less, with a narrow size distribution and even smaller thickness. In addition, the membranes also possess high flux, adequate mechanical strength, and chemical stability. It is also obvious that an interdisciplinary approach is important for developing membranes for diverse biomedical applications. As discussed in this article, advanced fabrication techniques are necessary to develop well-ordered, monodisperse, and ultrathin membranes. The market for inorganic NPs and membrane systems is expected to grow steadily in the future.

Funding: One of the authors (R.R.) wishes to express her sincere thanks for being in receipt of the DST-Inspire Fellowship, Department of Science and Technology, New Delhi, India.

Conflicts of Interest: The authors declare no conflict of interest.

References

1. Kim, K.; Fisher, J.P. Nanoparticle technology in bone tissue engineering. *J. Drug Target.* **2007**, *15*, 241–252. [CrossRef] [PubMed]
2. Goldberg, M.; Langer, R.; Jia, X. Nanostructured materials for applications in drug delivery and tissue engineering. *J. Biomater. Sci. Polym. Ed.* **2007**, *18*, 241–268. [CrossRef] [PubMed]
3. Zhang, S.; Uludag, H. Nanoparticulate Systems for Growth Factor Delivery. *Pharm. Res.* **2009**, *26*, 1561–1580. [CrossRef] [PubMed]
4. Sharma, C.P. *Biointegration of Medical Implant. Materials; Science and Design, Inorganic Nanoparticles for Targeted Drug Delivery*; Woodhead Publishing: Cambridge, UK, 2010.
5. Zhang, R.; Olin, H. Carbon nanomaterials as drug carriers: Real time drug release investigation. *Mater. Sci. Eng. C* **2012**, *32*, 1247–1252. [CrossRef]
6. Öchsner, A.; Shokuhfar, A. Novel Principles and Techniques, (Inorganic–Organic Hybrid Nanoparticles for Medical Applications. In *New Frontiers of Nanoparticles and Nanocomposite Materials*; Olariu, C.I., Yiu, H.P., Bouffier, L., Eds.; Springer: Berlin, Germany, 2010.
7. Javad, S.; Zohre, Z. Advanced drug delivery systems: Nanotechnology of health design A review. *J. Saudi Chem. Soc.* **2014**, *18*, 85–99.

8. El Ghannam, A.; Ricci, K.; Malkawi, A.; Jahed, K.; Vedantham, K.; Wyan, H.; Allen, L.D.; Dréau, D. A ceramic-based anticancer drug delivery system to treat breast cancer. *J. Mater. Sci. Mater. Med.* **2010**, *21*, 2701–2710. [CrossRef]
9. Uskoković, V.; Batarni, S.S.; Schweicher, J.; King, A.; Desai, T.A. The Effect of Calcium Phosphate Particle Shape and Size on their Antibacterial and Osteogenic Activity in the Delivery of Antibiotics in vitro. *ACS Appl. Mater. Interfaces* **2013**, *5*, 2422–2431. [CrossRef]
10. Sharma, S.; Verma, A.; Teja, B.V.; Pandey, G.; Mittapelly, N.; Trivedi, R.; Mishra, P. An insight into functionalized calcium based inorganic nanomaterials in biomedicine: Trends and transitions. *Colloids Surf. B Biointerfaces* **2015**, *133*, 120–139. [CrossRef]
11. Epple, M.; Ganesan, K.; Heumann, R.; Klesing, J.; Kovtun, A.; Neumann, S.; Sokolova, V. Application of calcium phosphate nanoparticles in biomedicine. *J. Mater. Chem.* **2010**, *20*, 18–23. [CrossRef]
12. Chen, F.; Zhu, Y.J.; Zhang, K.H.; Wu, J.; Wang, K.W.; Tang, Q.L.; Mo, X.M. Europium-doped amorphous calcium phosphate porous nanospheres: Preparation and application as luminescent drug carriers. *Nanoscale Res. Lett.* **2011**, *6*, 67–76. [CrossRef]
13. Kester, M.; Heakal, Y.; Fox, T.; Sharma, A.; Robertson, G.P.; Morgan, T.T.; Altinoğlu, E.I.; Tabaković, A.; Parette, M.R.; Rouse, S.M.; et al. Calcium Phosphate Nanocomposite Particles for In Vitro Imaging and Encapsulated Chemotherapeutic Drug Delivery to Cancer Cells. *Nano Lett.* **2008**, *8*, 4116–4121. [CrossRef] [PubMed]
14. Bastakoti, B.P.; Hsu, Y.C.; Liao, S.H.; Wu, K.C.W.; Inoue, M.; Yusa, S.I.; Nakashima, K.; Yamauchi, Y. Inorganic-Organic Hybrid Nanoparticles with Biocompatible Calcium Phosphate Thin Shells for Fluorescence Enhancement. *Chem. Asian J.* **2013**, *8*, 1301–1305. [CrossRef] [PubMed]
15. Shinto, Y.; Uchida, A.; Korkusuz, F.; Araki, N.; Ono, K. Calcium hydroxyapatite ceramic used as a delivery system for antibiotics. *J. Bone Jt. Surg. Br.* **1992**, *74*, 600–604. [CrossRef]
16. Zhao, X.Y.; Zhu, Y.J.; Chen, F.; Lu, B.Q.; Qi, C.; Zhao, J.; Wu, J. Calcium Phosphate Hybrid Nanoparticles: Self-Assembly Formation, Characterization, and Application as an Anticancer Drug Nanocarrier. *Chem. Asian J.* **2013**, *8*, 1306–1312. [CrossRef] [PubMed]
17. Mukesh, U.; Kulkarni, V.; Tushar, R.; Murthy, R.S.R. Methotrexate Loaded Self Stabilized Calcium Phosphate Nanoparticles: A Novel Inorganic Carrier for Intracellular Drug Delivery. *J. Biomed. Nanotechnol.* **2009**, *5*, 99–105. [CrossRef] [PubMed]
18. Liang, P.; Zhao, D.; Wang, C.Q.; Zong, J.Y.; Zhuo, R.X.; Cheng, S.X. Facile preparation of heparin/$CaCO_3$/CaP hybrid nano-carriers with controllable size for anticancer drug delivery. *Colloids Surf. B Biointerfaces* **2013**, *102*, 783–788. [CrossRef] [PubMed]
19. Rout, S.R.; Behera, B.; Maiti, T.K.; Mohapatra, S. Multifunctional magnetic calcium phosphate nanoparticles for targeted platin delivery. *Dalton Trans.* **2012**, *41*, 10777–10783. [CrossRef]
20. Meihua, Y.; Siddharth, J.; Peter, T.; Jiezhong, C.; Wenyi, G.; Chengzhong, Y. Hyaluronic acid modified mesoporous silica nanoparticles for targeted drug delivery to CD44-overexpressing cancer cells. *Nanomaterials* **2013**, *5*, 178–183.
21. Tasis, D.; Tagmatarchis, N.; Bianco, A.; Prato, M. Chemistry of Carbon Nanotubes. *Chem. Rev.* **2006**, *106*, 1105–1136. [CrossRef]
22. Zhang, R.F.; Zhang, Y.Y.; Zhang, Q.; Xie, H.H.; Qian, W.Z.; Wei, F. Growth of Half-Meter Long Carbon Nanotubes Based on Schulz–Flory Distribution. *ACS Nano* **2013**, *7*, 6156–6161. [CrossRef]
23. Im, O.; Li, J.; Wang, M.; Zhang, L.G.; Keidar, M. Biomimetic three-dimensional nanocrystalline hydroxyapatite and magnetically synthesized single-walled carbon nanotube chitosan nanocomposite for bone regeneration. *Int. J. Nanomed.* **2012**, *7*, 2087–2099.
24. Pastorin, G.; Wu, W.; Wieckowski, S.; Briand, J.P.; Kostarelos, K.; Prato, M.; Bianco, A. Double functionalisation of carbon nanotubes for multimodel drug delivery. *Chem. Commun.* **2006**, *11*, 1182–1184. [CrossRef] [PubMed]
25. Khoee, S.; Bafkary, R.; Fayyazi, F. DOX delivery based on chitosan-capped graphene oxide mesoporous silica nanohybride as pH-responsive nanocarriers. *J. Sol. Gel. Sci. Technol.* **2017**, *81*, 493–504. [CrossRef]
26. Ren, J.; Shen, S.; Wang, D.; Xi, Z.; Guo, L.; Pang, Z.; Qian, Y.; Sun, X.; Jiang, X. The Targeted Delivery of Anticancer Drugs to Brain Glioma by PEGylated Oxidized Multi-Walled Carbon Nanotubes Modified with Angiopep-2. *Biomaterials* **2012**, *33*, 3324–3333. [CrossRef] [PubMed]

27. Wu, P.; Chen, X.; Hu, N.; Tam, U.C.; Blixt, O.; Zettl, A.; Bertozzi, C.R. Biocompatible Carbon Nanotubes Generated by Functionalization with Glycodendrimers. *Angew. Chem. Int. Ed. Engl.* **2008**, *47*, 5022–5025. [CrossRef]
28. Bhatt, A.; Jain, A.; Gurnany, E.; Jain, R.; Modi, A.; Jain, A. Carbon Nanotubes: A Promising Carrier for Drug Delivery and Targeting. In *Nanoarchitectonics for Smart Delivery and Drug Targeting*, 1st ed.; William Andrew: Norwich, NY, USA, 2016; pp. 465–501.
29. Chauhan, G.; Chopra, V.; Tyagi, A.; Rath, G.; Sharma, K.; Goyal, K. Gold nanoparticles composite-Folic Acid Conjugated Graphene Oxide Nanohybrids for Targeted Chemo-Thermal Cancer Ablation: In-vitro Screening and In-vivo Studies. *Eur. J. Pharm. Sci.* **2017**, *1*, 351–361. [CrossRef] [PubMed]
30. Yang, S.T.; Cao, L.; Luo, P.G.; Lu, F.; Wang, X.; Wang, H.; Meziani, M.J.; Liu, Y.; Qi, G.; Sun, Y.P. Carbon Dots for Optical Imaging in Vivo. *J. Am. Chem. Soc.* **2009**, *131*, 11308–11309. [CrossRef]
31. Huang, Y.S.; Lu, Y.J.; Chen, J.P. Magnetic graphene oxide as a carrier for targeted delivery of chemotherapy drugs in cancer therapy. *J. Magn. Magn. Mater.* **2017**, *427*, 34–40. [CrossRef]
32. Barahuie, F.; Saifullah, B.; Dorniani, D.; Fakurazi, S.; Karthivashan, G.; Hussein, M.Z.; Elfghi, F.M. Graphene oxide as a nanocarrier for controlled release and targeted delivery of an anticancer active agent, chlorogenic acid. *Mater. Sci. Eng. C* **2107**, *74*, 177–185. [CrossRef]
33. Zhang, W.; Guo, Z.; Huang, D.; Liu, Z.; Guo, X.; Zhong, H. Synergistic effect of chemo-photothermal therapy using PEGylated graphene oxide. *Biomaterials* **2011**, *32*, 8555–8561. [CrossRef]
34. Qin, X.; Guo, Z.; Liu, Z.; Zhang, W.; Wan, M.; Yang, B. Folic acid-conjugated graphene oxide for cancer targeted chemo-photothermal therapy. *J. Photochem. Photobiol. B Biol.* **2013**, *120*, 156–162. [CrossRef] [PubMed]
35. Zhang, L.; Xia, J.; Zhao, Q.; Liu, L.; Zhang, Z. Functional Graphene Oxide as a Nanocarrier for Controlled Loading and Targeted Delivery of Mixed Anticancer Drugs. *Small* **2010**, *6*, 537–544. [CrossRef] [PubMed]
36. Zhao, X.; Wei, Z.; Zhao, Z.; Miao, Y.; Qiu, Y.; Yang, W.; Jia, X.; Liu, Z.; Hou, H. Design and Development of Graphene Oxide Nanoparticle/Chitosan Hybrids Showing pH-Sensitive Surface Charge-Reversible Ability for Efficient Intracellular Doxorubicin Delivery. *ACS Appl. Mater. Interfaces* **2018**, *10*, 6608–6617. [CrossRef]
37. Liu, Z.; Robinson, J.T.; Sun, X.; Dai, H. PEGylated Nano-Graphene Oxide for Delivery of Water Insoluble Cancer Drugs. *J. Am. Chem. Soc.* **2008**, *130*, 10876–10877. [CrossRef] [PubMed]
38. Masoudipour, E.; Kashanian, S.; Maleki, N. A targeted drug delivery system based on dopamine functionalized nano graphene oxide. *Chem. Phys. Lett.* **2017**, *668*, 56–63. [CrossRef]
39. Khalid, M.E. Nanodiamond as a drug delivery system: Applications and prospective. *J. Appl. Pharm. Sci.* **2011**, *1*, 29–39.
40. Mochalin, V.; Olga, S.; Dean, H.; Yury, G. The properties and applications of nanodiamonds. *Nat. Nanotechnol.* **2012**, *7*, 11–23. [CrossRef] [PubMed]
41. Yu, S.J.; Kang, M.W.; Chang, H.C.; Chen, K.M.; Yu, Y.C. Bright fluorescent NDs: No photobleaching and low cytotoxicity. *J. Am. Chem. Soc.* **2005**, *127*, 17604–17605. [CrossRef] [PubMed]
42. Chen, M.; Pierstorff, E.D.; Lam, R.; Li, S.Y.; Huang, H.; Osawa, E.; Ho, D. Nanodiamond-Mediated Delivery of Water-Insoluble Therapeutics. *ACS Nano* **2009**, *3*, 2016–2022. [CrossRef] [PubMed]
43. Amanda, M.; Suzanne, A.; Ciftan, H.; Olga, A. Nanodiamond Particles: Properties and Perspectives for Bioapplications. *Crit. Rev. Solid State Mater. Sci.* **2009**, *34*, 18–74.
44. Upal, R.; Vadym, D.; Andriy, D.; Jesse, R.; Paul, B.; Venkata, A.; Liu, X.; Thomas, G.; Surendra, S.; Madhavan, N. Characterization of Nanodiamond based anti-HIV drug Delivery to the Brain. *Sci. Rep.* **2018**, *8*, 1603–1615.
45. Slowing, I.; Viveroescoto, J.; Wu, C.; Lin, V. Mesoporous silica nanoparticles as controlled release drug delivery and gene transfection carriers. *Adv. Drug Deliv. Rev.* **2008**, *60*, 1278–1288. [CrossRef] [PubMed]
46. Sharma, K.; Kaushik, A.; Jayant, R.; Nair, M. Nanomaterials for Drug Delivery. In *Advances in Personalized Nanotherapeutics*; Springer: Berlin, Germany, 2017; pp. 57–77.
47. Slowing, I.; Trewyn, B.; Victor, S.Y. Effect of Surface Functionalization of MCM-41-Type Mesoporous Silicaon the Endocytosis by Human Cancer Cells. *J. Am. Chem. Soc.* **2006**, *128*, 14792–14793. [CrossRef] [PubMed]
48. Sabu, T.; Yves, G.; Neethu, N.; Parvathy, R.C.; Reny, T.T. Nanotechnology Applications for Tissue Engineering; gold nanoparticles in cancer drug delivery. *Elsevier Inc.* **2015**, *13*, 5637–5655.
49. Gibson, J.D.; Khanal, B.P.; Zubarev, E.R. Paclitaxel-Functionalized Gold Nanoparticles. *J. Am. Chem. Soc.* **2007**, *129*, 11653–11661. [CrossRef] [PubMed]

50. Dhar, S.; Daniel, W.L.; Giljohann, D.A.; Mirkin, C.A.; Lippard, S.J. Polyvalent Oligonucleotide Gold Nanoparticle Conjugates as Delivery Vehicles for Platinum(IV) Warheads. *J. Am. Chem. Soc.* **2009**, *131*, 14652–14653. [CrossRef] [PubMed]
51. Brown, S.D.; Nativo, P.; Smith, J.A.; Stirling, D.; Edwards, P.R.; Venugopal, B.; Flint, D.J.; Plumb, J.A.; Graham, D.; Wheate, N.J. Gold Nanoparticles for the Improved Anticancer Drug Delivery of the Active Component of Oxaliplatin. *J. Am. Chem. Soc.* **2010**, *132*, 4678–4684. [CrossRef] [PubMed]
52. Chen, Y.H.; Tsai, C.Y.; Huang, P.Y.; Chang, M.Y.; Cheng, P.C.; Chou, C.H.; Chen, D.H.; Wang, C.R.; Shiau, A.L.; Wu, C.L. Methotrexate Conjugated to Gold Nanoparticles Inhibits Tumor Growth in a Syngeneic Lung Tumor Model. *Mol. Pharm.* **2007**, *4*, 713–722. [CrossRef] [PubMed]
53. Kim, K.; Oh, K.S.; Park, D.Y.; Lee, J.Y.; Lee, B.S.; Kim, I.S.; Kim, K.; Kwon, I.C.; Sang, Y.K.; Yuk, S.H. Doxorubicin/gold-loaded core/shell nanoparticles for combination therapy to treat cancer through the enhanced tumor targeting. *J. Control. Release* **2016**, *228*, 141–149. [CrossRef]
54. Manivasagan, P.; Bharathiraja, S.; Bui, N.Q.; Jang, B.; Oh, Y.O.; Lim, I.G.; Oh, J. Doxorubicin-loaded fucoidan capped gold nanoparticles for drug delivery and photoacoustic imaging. *Int. J. Biol. Macromol.* **2016**, *91*, 578–588. [CrossRef]
55. Dhamechaa, D.; Jalalpure, S.; Jadhav, K.; Jagwani, S.; Chavan, R. Doxorubicin loaded gold nanoparticles: Implication of passive targeting on anticancer efficacy. *Pharmacol. Res.* **2016**, *113*, 547–556. [CrossRef] [PubMed]
56. Ravera, M.; Perin, E.; Gabano, E.; Zanellato, I.; Panzarasa, G.; Sparnacci, K.; Laus, M.; Osella, D. Functional fluorescent nonporous silica nanoparticles as carriers for Pt(IV) anticancer prodrugs. *J. Inorg. Biochem.* **2015**, *151*, 132–142. [CrossRef] [PubMed]
57. Al-Ajmi, M.F.; Hussain, A.; Ahmed, F. Novel synthesis of ZnO nanoparticles and their enhanced anticancer activity: Role of ZnO as a drug carrier. *Ceram. Int.* **2016**, *42*, 4462–4469. [CrossRef]
58. Vajtai, R. *Springer Handbook of Nanomaterials*; Springer: Berlin, Germany, 2013.
59. Sasaki, N.; Sudoh, Y. X-ray pole figure analysis of apatite crystals and collagen molecules in bone. *Calcif. Tissue Int.* **1997**, *60*, 361–367. [CrossRef] [PubMed]
60. Balasundaram, G.; Webster, T.J. Nanotechnology and biomaterials for orthopaedic medical applications. *Nanomedicine* **2006**, *1*, 169–176. [CrossRef] [PubMed]
61. Webster, T.J.; Ahn, E.S. Nanostructured biomaterials for tissue engineering bone. *Adv. Biochem. Eng. Biotechnol.* **2007**, *103*, 275–308.
62. Zhang, Z.G.; Li, Z.H.; Mao, X.Z.; Wang, W.C. Advances in bone repair with nanobiomaterials: Mini-review. *Cytotechnology* **2011**, *63*, 437–443. [CrossRef] [PubMed]
63. Tran, N.; Webster, T.J. Nanotechnology for bone materials. *WIREs Nanomed. Nanobiotech.* **2009**, *1*, 336–351. [CrossRef] [PubMed]
64. Marquis, M.E.; Lord, E.; Bergeron, E.; Drevelle, O.; Park, H.; Cabana, F.; Senta, H.; Faucheux, N. Bone cells-biomaterials interactions. *Front. Biosci.* **2009**, *14*, 1023–1067. [CrossRef]
65. Scheller, E.L.; Krebsbach, P.H.; Kohn, D.H. Tissue engineering: State of the art in oral rehabilitation. *J. Oral Rehabil.* **2009**, *36*, 368–389. [CrossRef]
66. Yao, C.; Slamovich, E.B.; Webster, T.J. Enhanced osteoblast functions on anodized titanium with nanotube-like structures. *J. Biomed. Mater. Res. A* **2008**, *85*, 157–166. [CrossRef] [PubMed]
67. Zhang, B.; Kwok, C.T. Hydroxyapatite-anatase-carbon nanotube nanocomposite coatings fabricated by electrophoretic codeposition for biomedical applications. *J. Mater. Sci. Mater. Med.* **2011**, *22*, 2249–2259. [CrossRef] [PubMed]
68. Kealley, C.; Elcombe, M.; Van Riessen, A.; Ben-Nissan, B. Development of carbon nanotube-reinforced hydroxyapatite bioceramics. *Phys. B Condens. Matter* **2006**, *385*, 496–498. [CrossRef]
69. Sirivisoot, S.; Yao, C.; Xiao, X.; Sheldon, B.W.; Webster, T.J. Greater osteoblast functions on multiwalled carbon nanotubes grown from anodized nanotubular titanium for orthopedic applications. *Nanotechnology* **2007**, *18*, 365102. [CrossRef]
70. Abrishamchian, A.; Hooshmand, T.; Mohammadi, M.; Najafi, F. Preparation and characterization of multi-walled carbon nanotube/hydroxyapatite nanocomposite film dip coated on Ti–6Al–4V by sol–gel method for biomedical applications: An in vitro study. *Mater. Sci. Eng. C* **2013**, *33*, 2002–2010. [CrossRef] [PubMed]

71. Facca, S.; Lahiri, D.; Fioretti, F.; Messadeq, N.; Mainard, D.; Benkirane-Jessel, N.; Agarwal, A. In Vivo Osseointegration of Nano-Designed Composite Coatings on Titanium Implants. *ACS Nano* **2011**, *5*, 4790–4799. [CrossRef] [PubMed]
72. Balani, K.; Agarwal, A. Wetting of carbon nanotubes by aluminum oxide. *Nanotechnology* **2008**, *19*, 165701. [CrossRef]
73. Correa-Duarte, M.A.; Wagner, N.; Rojas Chapana, J.; Morsczeck, C.; Thie, M.; Giersig, M. Fabrication and Biocompatibility of Carbon Nanotube-Based 3D Networks as Scaffolds for Cell Seeding and Growth. *Nano Lett.* **2004**, *4*, 2233–2236. [CrossRef]
74. Abarrategi, A.; Gutierrez, M.C.; Moreno Vicente, C.; Hortiguela, M.J.; Ramos, V.; Lopez Lacomba, J.L.; Ferrer, M.L.; Del Monte, F. Multiwall carbon nanotube scaffolds for tissue engineering purposes. *Biomaterials* **2008**, *29*, 94–102. [CrossRef]
75. Huisheng, P.; Qingwen, L.; Tao, C. *Industrial Applications of Carbon Nanotubes*; Xue, Y., Ed.; Elsevier Inc.: Amsterdam, The Netherlands, 2017.
76. Keefer, E.W.; Botterman, B.R.; Romero, M.I.; Rossi, A.F.; Gross, G.W. Carbon nanotube coating improves neuronal recordings. *Nat. Nanotechnol.* **2008**, *3*, 434–439. [CrossRef]
77. Webster, T.J.; Siegel, R.W.; Bizios, R. Osteoblast adhesion on nanophase ceramics. *Biomaterials.* **1999**, *20*, 1221–1227. [CrossRef]
78. Li, H.; Khor, K.; Cheang, P.; Khor, K. Impact formation and microstructure characterization of thermal sprayed hydroxyapatite/titania composite coatings. *Biomaterials* **2003**, *24*, 949–957. [CrossRef]
79. Siddharthan, A.; Sampath, K.S.; Seshadri, S.K. In situ composite coating of titania hydroxyapatite on commercially pure titanium by microwave processing. *Surf. Coat. Technol.* **2010**, *204*, 1755–1763. [CrossRef]
80. Kuwabara, A.; Hori, N.; Sawada, T.; Hoshi, N.; Watazu, A.; Kimoto, K. Enhanced biological responses of a hydroxyapatite/TiO_2 hybrid structure when surface electric charge is controlled using radiofrequency sputtering. *Dent. Mater. J.* **2012**, *31*, 368–376. [CrossRef] [PubMed]
81. Ahmed, H.H.; Aglan, H.A.; Mabrouk, M.; Abd Rabou, A.A.; Beherei, H.H. Enhanced mesenchymal stem cell proliferation through complexation of selenium/titanium nanocomposites. *J. Mater. Sci. Mater. Med.* **2019**, *30*, 24. [CrossRef] [PubMed]
82. Tavakol, S.; Nikpour, M.R.; Amani, A.; Soltani, M.; Rabiee, S.M.; Rezayat, S.M.; Chen, P.; Jahanshahi, M. Bone regeneration based on nanohydroxyapatite and hydroxyapatite/chitosan nanocomposites: An in-vitro and in-vivo comparative study. *J. Nanopart. Res.* **2013**, *15*, 1373–1389. [CrossRef]
83. Si, Y. Fluorescent Nanomaterials for Bioimaging and Biosensing: Application on E.coli Bacteria. PhD Thesis, École normale supérieure de Cachan—ENS Cachan, Paris, France, 2015.
84. Cai, W.; Shin, D.W.; Chen, K.; Gheysens, O.; Cao, Q.; Wang, S.X.; Gambhir, S.S.; Chen, X. Peptide-Labeled Near-Infrared Quantum Dots for Imaging Tumor Vasculature in Living Subjects. *Nano Lett.* **2006**, *6*, 669–676. [CrossRef]
85. Keren, S.; Zavaleta, C.; Cheng, Z.; De La Zerda, A.; Gheysens, O.; Gambhir, S.S. Noninvasive molecular imaging of small living subjects using Raman spectroscopy. *Proc. Natl. Acad. Sci. USA* **2008**, *105*, 5844–5849. [CrossRef] [PubMed]
86. Mulder, W.J.; Castermans, K.; Van Beijnum, J.R.; Oude Egbrink, M.G.; Chin, P.T.; Fayad, Z.A.; Löwik, C.W.; Kaijzel, E.L.; Que, I.; Storm, G.; et al. Molecular imaging of tumor angiogenesis using alphavbeta3-integrin targeted multimodal quantum dots. *Angiogenesis* **2009**, *12*, 17–24. [CrossRef] [PubMed]
87. Chen, K.; Li, Z.B.; Wang, H.; Cai, W.; Chen, X. Dual-modality optical and positron emission tomography imaging of vascular endothelial growth factor receptor on tumor vasculature using quantum dots. *Eur. J. Nucl. Med. Mol. Imaging* **2008**, *35*, 2235–2244. [CrossRef]
88. Gao, X.; Cui, Y.; Levenson, R.M.; Chung, L.W.K.; Nie, S. In vivo cancer targeting and imaging with semiconductor quantum dots. *Nat. Biotechnol.* **2004**, *22*, 969–976. [CrossRef] [PubMed]
89. Yong, K.T.; Hong, D.; Indrajit, R.; Wing Cheung, L.; Earl, J.; Anirban, M.; Paras, N. Imaging Pancreatic Cancer Using BioconjugatedInP Quantum Dots. *ACS Nano* **2009**, *3*, 502–510. [CrossRef] [PubMed]
90. Cai, W.B.; Chen, X.Y. Preparation of peptide-conjugated quantum dots for tumor vasculature-targeted imaging. *Nat. Protoc.* **2008**, *3*, 89–96. [CrossRef] [PubMed]

91. Li, K.; Cao, X.; Hou, W.; Zheng, L.; Peng, C.; Xiao, T.; Guo, R.; Shen, M.; Zhang, G.; Shi, X. Facile formation of dendrimer-stabilized gold nanoparticles modified with diatrizoic acid for enhanced computed tomography imaging applications. *Nanoscale* **2012**, *4*, 6768.
92. Kravets, V.; Almemar, Z.; Jiang, K.; Culhane, K.; Machado, R.; Hagen, G.; Kotko, A.; Dmytruk, I.; Spendier, K.; Pinchuk, A. Imaging of Biological Cells Using Luminescent Silver Nanoparticles. *Nanoscale Res. Lett.* **2016**, *11*, 30. [CrossRef] [PubMed]
93. Oostendorp, M.; Douma, K.; Hackeng, T.M.; Dirksen, A.; Post, M.J.; Van Zandvoort, M.A.; Backes, W.H. Quantitative Molecular Magnetic Resonance Imaging of Tumor Angiogenesis Using cNGR-Labeled Paramagnetic Quantum Dots. *Cancer Res.* **2008**, *68*, 7676–7683. [CrossRef] [PubMed]
94. Zhang, C.; Jugold, M.; Woenne, E.C.; Lammers, T.; Morgenstern, B.; Mueller, M.M.; Zentgraf, H.; Bock, M.; Eisenhut, M.; Semmler, W.; et al. Specific targeting of tumor angiogenesis by RGD-conjugated ultra small superparamagnetic iron oxide particles using a clinical 1.5-T magnetic resonance scanner. *Cancer Res.* **2007**, *15*, 1555–1562. [CrossRef] [PubMed]
95. CaSeo, W.S.; Lee, J.H.; Sun, X.; Suzuki, Y.; Mann, D.; Liu, Z.; Terashima, M.; Yang, P.C.; McConnell, M.V.; Nishimura, D.G.; et al. FeCo/graphitic-shell nanocrystals as advanced magnetic-resonance-imaging and near-infrared agents. *Nat. Mater.* **2007**, *5*, 971–976.
96. Liu, Z.; Li, X.; Tabakman, S.M.; Jiang, K.; Fan, S.; Dai, H. Multiplexed Multi-Color Raman Imaging of Live Cells with Isotopically Modified Single Walled Carbon Nanotubes. *J. Am. Chem. Soc.* **2008**, *130*, 13540–13541. [CrossRef]
97. Smith, B.R.; Cheng, Z.; De, A.; Koh, A.L.; Sinclair, R.; Gambhir, S.S. Real-Time Intravital Imaging of RGD–Quantum Dot Binding to Luminal Endothelium in Mouse Tumor Neovasculature. *Nano Lett.* **2008**, *8*, 2599–2606. [CrossRef]
98. Cai, W.; Chen, K.; Li, Z.B.; Gambhir, S.S.; Chen, X. Dual-Function Probe for PET and Near-Infrared Fluorescence Imaging of Tumor Vasculature. *J. Nucl. Med.* **2007**, *48*, 1862–1870. [CrossRef] [PubMed]
99. Zhuang, L.; Rui, P. Inorganic nanomaterials for tumor angiogenesis. *Eur. J. Nucl. Med. Mol. Imaging* **2010**, *37*, 147–163.
100. Liu, Z.; Chen, K.; Davis, C.; Sherlock, S.; Cao, Q.; Chen, X.; Dai, H. Drug Delivery with Carbon Nanotubes for In vivo Cancer Treatment. *Cancer Res.* **2008**, *68*, 6652–6660. [CrossRef] [PubMed]
101. Sharifi, S.; Seyednejad, H.; Laurent, S.; Atyabi, F.; Saei, A.A.; Mahmoudi, M. Superparamagnetic iron oxide nanoparticles for in vivo molecular and cellular imaging. *Contrast Media Mol. Imaging* **2015**, *10*, 329–355. [CrossRef]
102. Afsaneh, L.; Saeed, S.; Sophie, L.; Saeed, S. Dual nano-sized contrast agents in PET/MRI: A systematic review. *Contrast Media Mol. Imaging* **2016**, *11*, 428–447.
103. Li, C.; Tao, C.; Ismail, O.; Guizhi, Z.; Emir, Y.; Mingxu, Y.; Cuichen, W.; Jing, Z.; Erqun, S.; Cheng, Z.; et al. Gold-Coated Fe_3O_4 Nanoroses with Five Unique Functions for Cancer Cell Targeting, Imaging, and Therapy. *Adv. Funct. Mater.* **2014**, *26*, 1772–1780. [CrossRef]
104. Zhang, Y.; Wen, S.; Zhao, L.; Li, D.; Liu, C.; Jiang, W.; Gao, X.; Gu, W.; Ma, N.; Zhao, J.; et al. Ultra stable polyethyleneimine-stabilized gold nanoparticles modified with polyethylene glycol for blood pool, lymph node and tumor CT imaging. *Nanoscale* **2015**, *8*, 5567–5577. [CrossRef] [PubMed]
105. Yigit, M.V.; Zhu, L.; Ifediba, M.A.; Zhang, Y.; Carr, K.; Moore, A.; Medarova, Z. Noninvasive MRI-SERS Imaging in Living Mice Using an Innately Bimodal Nanomaterial. *ACS Nano* **2011**, *5*, 1056–1066. [CrossRef]
106. Wisniewski, N.; Reichert, M. Methods for reducing biosensor membrane biofouling. *Colloids Surf. B Biointerfaces* **2000**, *18*, 197–219. [CrossRef]
107. Desai, T.A.; Hansford, D.; Ferrari, M. Characterization of micromachined silicon membranes for immunoisolation and bioseparation applications. *J. Memb. Sci.* **1999**, *159*, 221–231. [CrossRef]
108. Max, L.G.Q.; Song, Z.X. *Nanoporous Materials: Science and Engineering*; World Scientific: Singapore, 2004; Volume 4, ISBN 178326179X.
109. Langley, P.J.; Hulliger, J. Nanoporous and mesoporous organic structures: New openings for materials research. *Chem. Soc. Rev.* **1999**, *28*, 279–291. [CrossRef]
110. Beherei, H.H.; Shaltout, A.A.; Mabrouk, M.; Abdelwahed, N.A.M.; Das, D.B. Influence of niobium pentoxide particulates on the properties of brushite/gelatin/alginate membranes. *J. Pharm. Sci.* **2018**, *107*, 1361–1371. [CrossRef] [PubMed]

111. Trautmann, C.; Brüchle, W.; Spohr, R.; Vetter, J.; Angert, N. Pore geometry of etched ion tracks in polyimide. *Nucl. Instrum. Methods Phys. Res. Sect. B Beam Interact. Mater. Atoms.* **1996**, *111*, 70–74. [CrossRef]
112. Tsuru, T. Inorganic porous membranes for liquid phase separation. *Sep. Purif. Methods* **2001**, *30*, 191–220. [CrossRef]
113. Metz, S.; Trautmann, C.; Bertsch, A.; Renaud, P. Polyimide microfluidic devices with integrated nanoporous filtration areas manufactured by micromachining and ion track technology. *J. Micro. Microeng.* **2003**, *14*, 324. [CrossRef]
114. Desai, T.A.; Chu, W.H.; Tu, J.K.; Beattie, G.M.; Hayek, A.; Ferrari, M. Microfabricated immunoisolatingbiocapsules. *Biotechnol. Bioeng.* **1998**, *57*, 118–120. [CrossRef]
115. Desai, T.; Bhatia, S. *Therapeutic Micro/Nano Technology*; Springer: New York, NY, USA, 2006; Volume 1, ISBN 9780387255651.
116. Tong, H.D.; Jansen, H.V.; Gadgil, V.J.; Bostan, C.G.; Berenschot, E.; van Rijn, C.J.M.; Elwenspoek, M. Silicon nitride nanosieve membrane. *Nano Lett.* **2004**, *4*, 283–287. [CrossRef]
117. Masuda, H.; Fukuda, K. Ordered Metal Nanohole Arrays Made by a Two-Step Replication of Honeycomb Structures of Anodic Alumina. *Science* **1995**, *268*, 1466–1468. [CrossRef] [PubMed]
118. Masuda, H.; Hasegwa, F.; Ono, S. Self-ordering of cell arrangement of anodic porous alumina formed in sulfuric acid solution. *J. Electrochem. Soc.* **1997**, *144*, L127–L130. [CrossRef]
119. Foll, H.; Christophersen, M.; Carstensen, J.; Hasse, G. Formation and application of porous silicon. *Mater. Sci. Eng. R* **2002**, *39*, 93–141. [CrossRef]
120. Lewis, A.L. Phosphorylcholine-based polymers and their use in the prevention of biofouling. *Colloids Surf. B Biointerfaces* **2000**, *18*, 261–275. [CrossRef]
121. Harrison, D.J.; Turner, R.F.B.; Baltes, H.P. Characterization of Perfluorosulfonic Acid Polymer Coated Enzyme Electrodes and a Miniaturized Integrated Potentiostat for Glucose Analysis in Whole Blood. *Anal. Chem.* **1988**, *60*, 2002–2007. [CrossRef] [PubMed]
122. Langer, R.; Tirrell, D.A. Designing materials for biology and medicine. *Nature* **2004**, *428*, 487–492. [CrossRef] [PubMed]
123. Jackson, E.A.; Hillmyer, M.A. Nanoporous membranes derived from block copolymers: From drug delivery to water filtration. *ACS Nano* **2010**, *4*, 3548–3553. [CrossRef] [PubMed]
124. Jeon, G.; Yang, S.Y.; Kim, J.K. Functional nanoporous membranes for drug delivery. *J. Mater. Chem.* **2012**, *22*, 14814–14834. [CrossRef]
125. Ishihara, K.; Nomura, H.; Mihara, T.; Kurita, K.; Iwasaki, Y.; Nakabayashi, N. Why do phospholipid polymers reduce protein adsorption? *J. Biomed. Mater. Res.* **1998**, *39*, 323–330. [CrossRef]
126. Lee, J.H.; Kopecek, J.; Andrade, J.D. Protein-resistant surfaces prepared by PEO-containing block copolymer surfactants. *J. Biomed. Mater. Res.* **1989**, *23*, 351–368. [CrossRef] [PubMed]
127. Lindner, E.; Cosofret, V.V.; Ufer, S.; Buck, R.P.; Kao, W.J.; Neuman, M.R.; Anderson, J.M. Ion-selective membranes with low plasticizer content: Electroanalytical characterization and biocompatibility studies. *J. Biomed. Mater. Res.* **1994**, *28*, 591–601. [CrossRef]
128. Morra, M. On the molecular basis of fouling resistance. *J. Biomater. Sci. Polym. Ed.* **2000**, *11*, 547–569. [CrossRef]
129. McHugh, A.J. The role of polymer membrane formation in sustained release drug delivery systems. *J. Control. Release* **2005**, *109*, 211–221. [CrossRef]
130. Stamatialis, D.F.; Papenburg, B.J.; Gironés, M.; Saiful, S.; Bettahalli, S.N.M.; Schmitmeier, S.; Wessling, M. Medical applications of membranes: Drug delivery, artificial organs and tissue engineering. *J. Memb. Sci.* **2008**, *308*, 1–34. [CrossRef]
131. Banerjee, P.; Das, R.; Das, P.; Mukhopadhyay, A. *Membrane Technology*; Carbon Nanostructures, Wiley: Hoboken, NJ, USA, 2018; ISBN 9780080479385.
132. Wick, S.M. Transdermal Nitroglycerin Delivery System. U.S. Patent 2017, 4,751,087, 14 June 1988.
133. Merus Labs Luxco II S.à R.L. Estraderm MX 50 summary of product characteristics. 2019. Available online: https://www.medicines.org.uk/emc/product/5839/smpc (accessed on 27 January 2019).
134. Pharmaceuticals, J. Duragesic (Fentanyl Transdermal System) for transdermal administration, summary of product charcteristics. 2016. Available online: https://www.fda.gov/.../fentanyl-transdermal-system-marketed-duragesic-information (accessed on 28 January 2019).

135. Pharma, T. Androderm Transdermal Patch (New Zealand data sheet) 2018. Available online: https://www.nps.org.au/medicine-finder/androderm-transdermal-patch (accessed on 28 January 2019).
136. Orosz, K.E.; Gupta, S.; Hassink, M.; Abdel-Rahman, M.; Moldovan, L.; Davidorf, F.H.; Moldovan, N.I. Delivery of antiangiogenic and antioxidant drugs of ophthalmic interest through a nanoporous inorganic filter. *Mol. Vis.* **2004**, *10*, 555–565. [PubMed]
137. Pharriss, B.B.; Erickson, R.; Bashaw, J.; Hoff, S.; Place, V.A.; Zaffaroni, A. Progestasert: A Uterine Therapeutic System for Long-term Contraception: I. Philosophy and Clinical Efficacy. In Proceedings of the 29th Annual Meeting of The American Fertility Society, San Francisco, CA, USA, 5–7 April 1974.
138. Pharmaceuticals, B.H. Levonorgestrel-releasing intrauterine system (Mirena). Data sheet 2008, Bayer Heal. Available online: https://www.mirena-us.com/about-miren (accessed on 27 January 2019).
139. Chien, Y.W.; Lin, S. Drug Delivery: Controlled Release. In *Encycl. Pharm. Technol*, 3rd ed.; CRC Press: Boca Raton, FL, USA, 2007; pp. 1082–1103.
140. Transdermal Infusion System summary of product characteristics. 2014. Available online: www.merck.com/product/patent/home.html%0ACopyright (accessed on 27 January 2019).
141. Eiyo, T. Frandol tape Drug Information Sheet 2016. Available online: https://bciq.biocentury.com/products/frandol_tape_isosorbide_dinitrate (accessed on 27 January 2019).
142. Gupta, S.K.; Okerholm, R.A.; Eller, M.; Wei, G.; Rolf, C.N.; Gorsline, J. Comparison of the pharmacokinetics of two nicotine transdermal systems: Nicoderm and habitrol. *J. Clin. Pharmacol.* **1995**, *35*, 493–498. [CrossRef] [PubMed]
143. Valeant MINITRAN- nitroglycerin patch summary of product characteristics, Dly. med 2019. Available online: https://dailymed.nlm.nih.gov/dailymed/drugInfo.cfm?setid=15b3ea32-3422-4aa8-b4c2-41ee6d1aa566 (accessed on 28 January 2019).
144. Place, V.A.; Nichols, K.C. Transdermal Delivery of Testosterone with TESTODERMTM to Provide a Normal Circadian Pattern of Testosterone. *Ann. N. Y. Acad. Sci.* **1991**, *618*, 441–449. [CrossRef] [PubMed]
145. Ajithkumar, T.V.; Hatcher, H.M. Breast cancer. *Spec. Train. Oncol.* **2017**, *5*, 115–133.
146. Preeti, K.; Jain, R.; Choukse, R.; Dubey, P.K.; Agrawal, S. Ocusert as A Novel Drug Delivery System. *Int. J. Pharm. Biol. Arch.* **2013**, *4*, 614–619.
147. Ferguson, T.H.; Needham, G.F.; Wagner, J.F. Compudose: An implant system for growth promotion and feed efficiency in cattle. *J. Control. Release* **1988**, *8*, 45–54. [CrossRef]
148. Labs, M.; Luxco, S. Deponit 5 summary of product characteristics. eMC 2019. Available online: http://www.datapharm.org.uk (accessed on 28 January 2019).
149. Yang, W.W.; Pierstorff, E. Reservoir-based polymer drug delivery systems. *J. Lab. Autom.* **2012**, *17*, 50–58. [CrossRef]
150. Maitz, M.F. Applications of synthetic polymers in clinical medicine. *BiosurfaceBiotribol.* **2015**, *1*, 161–176. [CrossRef]
151. Macleod, A.M.; Campbell, M.; Cody, J.D.; Daly, C.; Donaldson, C.; Grant, A.; Khan, I.; Rabindranath, K.S.; Vale, L.; Wallace, S.; et al. Cellulose, modified cellulose and synthetic membranes in the haemodialysis of patients with end-stage renal disease. *Cochrane Database Syst. Rev.* **2005**, *20*, 20–42. [CrossRef]
152. Pruitt, L.; Furmanski, J. Polymeric biomaterials for load-bearing medical devices. *JOM* **2009**, *61*, 1–24. [CrossRef]
153. Jeon, G.; Yang, S.Y.; Byun, J.; Kim, J.K. Electrically actuatable smart nanoporous membrane for pulsatile drug release. *Nano Lett.* **2011**, *11*, 1284–1288. [CrossRef] [PubMed]
154. Ulbricht, M. Advanced functional polymer membranes. *Polymer* **2006**, *47*, 2217–2262. [CrossRef]
155. Kambe, T.N. Microbial Degradation of Polyurethane, Polyester Polyurethanes and Polyether Polyurethanes. *Appl. Microbiol. Biotechnol.* **1999**, *51*, 134–140. [CrossRef]
156. Shenoy, D.; Little, S.; Langer, R.; Amiji, M. Poly (ethylene oxide)-modified poly (β-amino ester) nanoparticles as a pH-sensitive system for tumor-targeted delivery of hydrophobic drugs In vitro evaluations. *Mol. Pharm.* **2005**, *2*, 357–366. [CrossRef] [PubMed]
157. Foll, H.; Christophersen, M.; Carstensen, J.; Hasse, G. Formation and application of porous silicon. *Mater. Sci. Eng. R* **2002**, *280*, 1–49. [CrossRef]
158. Yang, Q.; Adrus, N.; Tomicki, F.; Ulbricht, M. Composites of functional polymeric hydrogels and porous membranes. *J. Mater. Chem.* **2011**, *21*, 2783–2811. [CrossRef]

159. Groth, T.; Seifert, B.; Albrecht, W.; Malsch, G.; Gross, U.; Fey-Lamprecht, F.; Michanetzis, G.; Missirlis, Y.; Engbers, G. Development of polymer membranes with improved haemocompatibility for biohybrid organ technology. *Clin. Hemorheol. Microcirc.* **2005**, *32*, 129.
160. Park, Y.J.; Nam, K.H.; Ha, S.J.; Pai, C.M.; Chung, C.P.; Lee, S.J. Porous poly(l-lactide) membranes for guided tissue regeneration and controlled drug delivery: Membrane fabrication and characterization. *J. Control. Release* **1997**, *43*, 151–160. [CrossRef]
161. Causa, F.; Netti, P.A.; Ambrosio, L.; Ciapetti, G.; Baldini, N.; Pagani, S.; Martini, D.; Giunti, A. Poly-caprolactone/hydroxyapatite composites for bone regeneration: In vitro characterization and human osteoblast response. *J. Biomed. Mater. Res. Part. A* **2006**, *76*, 151–162. [CrossRef]
162. Kim, S.Y.; Kanamori, T.; Noumi, Y.; Wang, O.C.; Shinbo, T. Preparation of porous poly(d,L-lactide) and poly(d,L-lactide-co-glycolide) membranes by a phase inversion process and investigation of their morphological changes as cell culture scaffolds. *J. Appl. Polym. Sci.* **2004**, *92*, 2082–2092. [CrossRef]
163. Vienken, J.; Diamantoglou, M.; Hahn, C.; Kamusewitz, H.; Paul, D. Considerations on developmental aspects of biocompatible dialysis membranes. *Artif. Organs* **1995**, *19*, 398–406. [CrossRef] [PubMed]
164. Iwasaki, Y.; Uchiyama, S.; Kurita, K.; Morimoto, N.; Nakabayashi, N. A nonthrombogenic gas-permeable membrane composed of a phospholipid polymer skin film adhered to a polyethylene porous membrane. *Biomaterials* **2002**, *23*, 3421. [CrossRef]
165. Queiroz, D.P.; Norberta de Pinho, M. Structural characteristics and gas permeation properties of polydimethylsiloxane/poly(propylene oxide) urethane/urea bi-soft segment membranes. *Polymer* **2005**, *46*, 2346. [CrossRef]

Article

In Vitro and In Silico Analyses of Nicotine Release from a Gelisphere-Loaded Compressed Polymeric Matrix for Potential Parkinson's Disease Interventions

Pradeep Kumar, Yahya E. Choonara, Lisa C. du Toit, Neha Singh and Viness Pillay *

Wits Advanced Drug Delivery Platform Research Unit, Department of Pharmacy and Pharmacology, School of Therapeutic Sciences, Faculty of Health Sciences, University of the Witwatersrand, Johannesburg, 7 York Road, Parktown 2193, South Africa; pradeep.kumar@wits.ac.za (P.K.); yahya.choonara@wits.ac.za (Y.E.C.); lisa.dutoit@wits.ac.za (L.C.d.T); neni2707@gmail.com (N.S.)

* Correspondence: viness.pillay@wits.ac.za; Tel.: +27-11-717-2274

Received: 13 October 2018; Accepted: 12 November 2018; Published: 15 November 2018

Abstract: This study aimed to develop a prolonged-release device for the potential site-specific delivery of a neuroprotective agent (nicotine). The device was formulated as a novel reinforced crosslinked composite polymeric system with the potential for intrastriatal implantation in Parkinson's disease interventions. Polymers with biocompatible and bioerodible characteristics were selected to incorporate nicotine within electrolyte-crosslinked alginate-hydroxyethylcellulose gelispheres compressed within a release rate-modulating external polymeric matrix, comprising either hydroxypropylmethylcellulose (HPMC), polyethylene oxide (PEO), or poly(lactic-*co*-glycolic) acid (PLGA) to prolong nicotine release. The degradation and erosion studies showed that the produced device had desirable robustness with the essential attributes for entrapping drug molecules and retarding their release. Zero-order drug release was observed over 50 days from the device comprising PLGA as the external matrix. Furthermore, the alginate-nicotine interaction, the effects of crosslinking on the alginate-hydroxyethycellulose (HEC) blend, and the effects of blending PLGA, HPMC, and PEO on device performance were mechanistically elucidated using molecular modelling simulations of the 3D structure of the respective molecular complexes to predict the molecular interactions and possible geometrical orientation of the polymer morphologies affecting the geometrical preferences. The compressed polymeric matrices successfully retarded the release of nicotine over several days. PLGA matrices offered minimal rates of matrix degradation and successfully retarded nicotine release, leading to the achieved zero-order release for 50 days following exposure to simulated cerebrospinal fluid (CSF).

Keywords: alginate gelispheres; textural analysis; crosslinked matrices; PLGA discs; prolonged release; powder flow properties

1. Introduction

Parkinson's disease (PD), a central nervous system disorder, is associated with difficulties with respect to movement, typically outlined by the patient's shuffling gait due to unstable posture, bradykinesia, tremor, and rigidity. The progressive neuronal loss in PD is prolonged, requiring clinical neuroprotection and/or disease modification as therapeutic strategies for effectively reducing PD-related disability. The potential application of nicotine as a neuroprotectant in PD was demonstrated as far back as 1926 [1]. Nicotine has previously been shown to protect against the degeneration of nigrostriatal neurons via evoking the release of dopamine from the striatum [2–5]. PD is still an incurable disease. The available therapeutic options only offer symptomatic relief and primarily aim to improve the functionality of the patient with no intervention towards the progression of the associated neurodegeneration. Additionally, a major challenge is to improve patient compliance to therapy by reducing the side-effects associated with continuous multiple dosing via the oral route subsequently leading to erratic plasma drug levels.

Generally, prolonged parenteral zero-order drug delivery systems have the capacity to minimize dose-related side-effects owing to a constant and sustained drug release profile. This is applicable specifically to drugs with narrow therapeutic indices, wherein such systems can provide a reduction in administered dose, avoidance of fluctuations in plasma drug levels, reduced frequency of administration, and hence the enhancement of patient compliance [6,7]. The use of polymers is a popular means of achieving controlled drug release due to the simplicity, cost-effectiveness, ease of manufacturing, versatility, and the ability to deliver compounds with a wide range of solubilities [8].

Hydroxypropylmethylcellulose (HPMC) and polyethylene oxide (PEO) have been used extensively to formulate controlled-release drug delivery systems. Both polymers are hydrophilic and undergo simultaneous swelling and erosion when exposed to hydration. Depending on the solubility of incorporated drugs within monolithic matrices of such polymers, the actives diffuse out into the bulk medium preceding or following erosion of the polymer [9,10]. The mechanisms relating to the swelling and hydration of hydrophilic polymeric matrices may include, but are not limited to, polymeric chain extension, disentanglement, and solvent accommodation to macroscopic characteristics such as drug release [11,12]. In addition, formulation factors such as the quantity of drug loading, drug-polymer ratio, drug particle size and molecular mass, polymer viscosity and molecular weight, and presence of formulation excipients and release modulators significantly affect the drug release rate from hydro-swelling matrices [13,14]. Poly(lactic-*co*-glycolic acid) (PLGA), a polyhydroxy acid polymeric carrier, is capable of delivering drugs as a controlled-release site-specific system and has the advantage of being moderately aqueous-soluble for controlled drug diffusion [15]. Biocompatibility studies have indicated that PLGA is biodegradable and considerably well tolerated when implanted into the CNS [16]. Consequently, no surgical intervention or procedures are required for removal of implantable devices prepared with PLGA. Fournier and co-workers (2003) established that regardless of the implantation site, PLGA devices initiate only a moderate and non-specific inflammatory reaction, mainly due to mechanical insult during the implantation procedure [16].

The purpose of this study was to incorporate nicotine-loaded alginate-hydroxyethycellulose (HEC) gelispheres into an external polymeric matrix with the aim of modulating drug release to achieve prolonged zero-order release. The alginate-HEC gelispheres were previously developed by the authors and were physically crosslinked using divalent ions [17]. HPMC, PEO, and PLGA were selected as the polymers for formulating the device due to the desirable inherent characteristics of the polymers for implantation into the CNS. The study was further aimed to elucidate the molecular mechanisms inherent to the drug-polymer interactions, formation of the gelispheres, and finally the fabrication and performance of various composite polymeric matrices using static lattice atomistic simulations.

2. Materials and Methods

2.1. Materials

Nicotine was obtained from Sigma-Aldrich, St. Louis, MO, USA. The polymers employed for gelisphere preparation included sodium alginate (FMC BioPolymer, Drammen, Norway) and HEC (Hercules Inc., Natrosol® Pharm, Wilmington, DE, USA). The crosslinking reagents employed were calcium chloride (CaCl2) and barium chloride (BaCl2) (UniLab®, Saarchem (Pty) Ltd., Krugersdorp, South Africa). PLGA (Resomer® L503, Evonik Nutrition & Care GmbH, Essen, Germany), HPMC (Methocel™ K4M, Dow Chemical Company, Pittsburg, CA, USA) and PEO (PolyoxTM WSR 303; Union Carbide, Bound Brook, NJ, USA) were employed as the external matrices.

2.2. Methods

2.2.1. Formulation of the Reinforced Alginate-HEC Gelispheres

The alginate-HEC gelispheres were formulated employing ionotropic gelation as described elsewhere by the authors [17]. Briefly, an alginate (2% *w*/*v*) and HEC (1% *w*/*v*; the reinforcing agent) dispersion was prepared in double-deionised water and subsequently homogenized. Thereafter, nicotine (1% *w*/*v*) was added to the polymeric dispersion through titration at a rate of 5 mL/min. The resulting solution was immediately added into the crosslinking solution comprising $BaCl_2$ and $CaCl_2$ (2% *w*/*v*). The crosslinked alginate-HEC (ALG-HEC) gelispheres were crosslinked for 30 min and thereafter washed with deionized water. Formulations were then exposed to 2 M HCl subsequent to curing in order to precipitate alginic acid in the crosslinked system and retard the hydration and swelling dynamics of the gelispheres.

2.2.2. Formulation of the Gelisphere-Loaded External Polymeric Matrices

Briefly, 500 mg of a dry polymer mixture comprising either HPMC, PEO, PLGA, or various binary combinations of the polymers and gelispheres equivalent to 20 mg of entrapped nicotine (approximately 100 mg gelispheres) were compressed into 10-mm flat-faced discs (Beckman® Hydraulic Press; Beckman Instruments Inc., Fullerton, CA, USA), applying a compression force of 8 tons. Table 1 summarizes the compositions of the various formulations investigated.

Table 1. Polymeric compositions of the various external matrix formulations. Friability results for polymeric matrices expressed as % weight loss after 4 (% WL_4) and 20 (% WL_{20}) minutes in a friabilator at 25 rpm. HPMC: hydroxypropylmethylcellulose; PEO: polyethylene oxide; PLGA: poly(lactic-*co*-glycolic) acid.

Formulation Code	HPMC (mg)	PEO (mg)	PLGA (mg)	% WL_4	% WL_{20}
HPMC	500	-	-	0.081	0.307
PEO	-	500	-	0.101	0.133
PLGA	-	-	500	0.000	0.026
HPMC-PEO	250	250	-	0.050	0.107
HPMC-PLGA	250	-	250	0.051	0.064
PEO-PLGA	-	250	250	0.038	0.102
HPMC-PEO-PLGA	166	166	166	0.025	0.051

2.2.3. Evaluation of the Flowability and Friability of the Polymeric Material

To evaluate the flowability of the powders the angle of repose (φ) method was employed. A conical funnel was attached to a retort stand with the bottom orifice (diameter 7 mm) at a 10-cm height above the horizontal surface. Samples of 3 g of powder were placed into the funnel while keeping the orifice closed. The center-point and a graduated set of axes was marked on a piece of graphical paper and placed beneath the funnel. The contents of the funnel were allowed to flow through the orifice and the angle of repose (φ) was calculated employing Equation (1):

$$\tan\varphi = 2h/D \tag{1}$$

where h is the powder height at center-point (mm) and D is the diameter of powder bed formed (mm).

Friability analysis evaluated the ability of the compressed gelisphere-loaded matrices to withstand mechanical damage due to the variation in particle size. The matrix friability was measured using a Roche® Friabilator (Hoffman la Roche, Basel, Switzerland). Samples were accurately weighed and placed into the friabilator (N = 3). Rotation times of 4 min and 20 min at 25 rpm were used. Matrices were then removed and the surface was lightly dabbed to remove any loose particles and re-weighed. The mass loss was calculated employing Equation (2).

$$WL_t = (W_0 - W_t)/W_0 \times 100 \tag{2}$$

where WL_t is the mass loss at time t (%), W_t is the mean mass at t; and W_0 is the mean mass at time t_0.

2.2.4. Determination of the Brinell Hardness Number of the Compressed Polymeric Matrices

The change in physicomechanical properties of the matrices following exposure to simulated cerebrospinal fluid (CSF) was assessed by measuring the Brinell Hardness Number (BHN). The BHN is an indication of the force required to indent the surface of the gelisphere-loaded matrices and represents the surface hardness of the compressed polymeric matrices. A TA.XTplus Texture Analyser (StableMicrosystems®, Surrey, UK) was used to generate force-distance profiles for matrices exposed to simulated CSF over predetermined time intervals (pH 7.4; 37 °C). The input parameters employed with a 5-kg load cell were pre-test speed (1.0 mm/s), test speed (0.5 mm/s), post-test speed (1.0 mm/s), and trigger force (0.5 N). The BHN (N/mm^2) was calculated employing Equation (3).

$$BHN = \frac{2F}{\pi D\left[D - (D^2 - d^2)^{\frac{1}{2}}\right]} \tag{3}$$

where D is the diameter of the ball probe (3.175 mm), d is the indentation distance (0.25 mm), and F is the force (N) generated.

2.2.5. Evaluation of Conductivity Changes Following Polymeric Matrix Gelation

An OAKTON TDS Test™ Kit (Model WD-35661-70; OAKTON Instruments, Solon, OH, USA) was employed to measure the conductivity of the PBS (0.1 M; pH 7.4) solution containing the polymeric discs at predetermined intervals of time over a period of 30 days. Samples were extracted from the solutions and diluted 1:10 in double deionized water. The electrode was immersed in the diluted polymer-PBS solution and conductivity (μSiemens) was read.

2.2.6. Evaluation of the Matrix Erosion upon Hydration

Pre-weighed hydrated discs were removed from simulated CSF at predetermined intervals and dried under ambient conditions (21 °C) for 7 days. The dried discs were re-weighed and the percentage weight loss (% WL) calculated.

2.2.7. Evaluation of Surface Morphology of the Matrices

Scanning electron microscopy (SEM) was employed to assess the changes in the porosity of the polymer matrices following hydration. Dried samples of the discs were cross-sectioned and coated with a thin layer of colloidal graphite. Thereafter they were mounted onto aluminum stubs and coated with a thin layer of gold–palladium under an electrical potential of 20 keV. Samples were viewed in the FE-SEM (JEOL JSM-840, Tokyo, Japan) at a magnification of 65 and probe current of 30 nanoamps. SEM images were processed using Mathematica™ 8.0 (Wolfram Research, Champaign, IL, USA) as described in the supplementary data. Initially, images were cropped to restrict image content to that of the scaffold.

2.2.8. Textural Analysis of the Hydrated and Unhydrated Gelisphere-Loaded External Polymeric Matrices

The Texture Analyzer (TAXT.plus, Stable Micro Systems®, Surrey, UK) was employed to conduct textural analysis of the gelisphere-loaded polymeric matrices (N = 5). The texture analyzer was fitted with a 10-mm cylindrical steel probe and a 40-kg load cell was used. Matrix resilience, deformability modulus, and fracture energy of the crosslinked gelispheres were assessed according to their G and M content in both the unhydrated and hydrated states. For the later test, gelispheres were immersed in 15 mL double deionised water in glass vials and agitated at 50 rpm in a shaker bath (Labline® Instruments, Melrose Park, IL, USA) maintained at 37 °C. Samples (N = 10) were removed and analyzed for four 24-h intervals and evaluated for the above stated entities. The association of the monomers, i.e., GG, GM, and MM blocks and their interaction with the crosslinking cations on the mechanical properties was also examined. Samples with an identical G/M ratio were compared according to differences in viscosity. The input parameters employed with a 5-kg load cell (50% strain) were pre-test speed (1.0 mm/s), test speed (0.5 mm/s), post-test speed (1.0 mm/s), and trigger force (0.5 N). The force-time profile was employed for the determination of matrix resilience. This refers to the ability of the gelisphere to recover upon application of deforming stress to its original state of equilibrium. This was calculated as the ratio of the areas under the compression and decompression curve. A forcedistance profile generated for the gelispheres was employed to obtain the area under the curve (AUC), which enabled the evaluation of the energy needed to rupture the gelisphere matrix, i.e., the fracture energy (Nm) while the gradient to the peak force, known as the deformability modulus (N/mm), corresponds to the deformability modulus of the gelisphere matrix [18].

2.2.9. In Vitro Drug Release Studies of the Gelisphere-Loaded External Polymeric Matrices

In vitro drug release studies were performed on the nicotine-loaded matrices using simulated cerebrospinal fluid (CSF). Gelisphere-loaded external polymeric matrices were immersed in 100 mL of 0.1 M phosphate-buffered saline (PBS) (pH 7.4; 37 °C) in sealed 150-mL glass jars that were placed in a shaker bath (Labex, Stuart SBS40 South Africa) and agitated at 50 rpm. At predetermined intervals, 5-mL samples were removed and the samples were analyzed by UV spectroscopy (245 nm; Specord® 40, Analytik Jena UV-VIS, Aargau, Switzerland). To maintain sink conditions; 5 mL of nicotine-free PBS were replaced into the glass jars.

2.2.10. Static Lattice Atomistic Molecular Structural Simulations

To elucidate the interaction profile of individual molecules within the alginate-drug complex, alginate-HEC gelisphere matrices, and the external matrices, molecular mechanics computations were conducted in a vacuum using the HyperChem™ 8.0.8 Molecular Modeling System (Hypercube Inc., Gainesville, FL, USA) and ChemBio3D Ultra 11.0 (CambridgeSoft Corporation, Cambridge, UK). The polymers (PLGA and PEO) were drawn in their syndiotactic stereochemistry as 3D models; the carbohydrate (alginate) and cellulose derivatives (HEC and HPMC) were built from standard bond lengths and angles using a sugar builder module; and the structure of nicotine was drawn with natural bond angles. The structures so-obtained were energetically optimized using a progressive minimization paradigm employing MM+ and the Amber 3 (Assisted Model Building and Energy Refinements) force field. The conformer having the lowest energy was used to create the polymer-polymer complexes. A complex of one molecule with another was assembled by disposing them in a parallel way, and the same procedure of energy minimization was repeated to generate the final models: alginate-nicotine (ALG–NCT), ALG–HEC, ALG–HEC–Ca^{2+}, ALG–HEC–Ba^{2+}, PLGA–HPMC, HPMC–PEO, PLGA–PEO, and PLGA–HPMC–PEO. Full geometry optimization was carried out in vacuum employing the Polak–Ribiere conjugate gradient algorithm until an RMS gradient of 0.001 kcal/mol was reached. For molecular mechanics calculations in vacuum, the force fields were utilized with a distance-dependent dielectric constant scaled by a factor of 1. The 1–4 scale factors were electrostatic = 0.5 and van der Waals = 0.5 [19].

3. Results and Discussion

3.1. Flowability and Friability of the Polymeric Matrices

The flowability results indicated that PLGA had the greatest degree of inter-particle interaction (cohesion). Hence, its high angle of repose (AR) (40.40°) indicated that it had poor flow properties. When combined with HPMC, their mutually attractive interaction was indicated by the highest AR of 52.41°, which implies that the combination had the potential for producing highly dense compressed discs. PEO on the other hand indicated excellent flow properties, as it demonstrated a lower AR of 13.49°. This can be explained in terms of the fact that it was predominately composed of spherical granular particles. These particles had a smaller surface area per particle to interact with each other, as compared to PLGA and HPMC. Its incorporation with PLGA and HPMC resulted in a significant decline in AR (11.46° and 12.74°, respectively) and improved overall flowability, and thus it negatively impacted on the cohesive tendency of the particles, which implied that matrices generated with the aforementioned combinations would generate less compact matrices. From a manufacturing process point of view, improved flowability permits easier processing conditions. However, this implies that additional substances (such as binders) may need to be added to enhance powder compressibility to generate discs. According to the United States Pharmacopoeia (USP 23), conventional compressed tablets that incur a loss in weight after 100 revolutions (which is the equivalent of 4 min in the friabilator at 25 rpm) in a friabilator of less than 0.5–1.0% are considered acceptable. The results indicated that the discs generated were well within the predetermined limits. PLGA discs were exceedingly well compressed and underwent negligible loss in weight following 500 revolutions (i.e., 20 min at 25 rpm) (Table 1). Furthermore, the incorporation of PLGA with HPMC and PEO into the disc resulted in an enhanced robustness imparted into the matrix, while the addition of PEO has the reverse effect, indicated by the greater friability.

3.2. Changes in the Brinell Hardness Number (BHN) of Compressed External Polymeric Matrices Following Hydration

Evaluating changes in the Brinell Hardness Number (BHN) of the discs allowed us to establish the rate at which water penetrated the surface of the discs and lead to their subsequent hydration and erosion. The faster this occurred, the more quickly the incorporated gelispheres became exposed to the surrounding aqueous medium and released the entrapped nicotine. The results demonstrated that the most significant changes in BHN occurred within the first day of exposure to simulated CSF (Figure 1a). Discs composed of PEO in particular experienced a significant decline in BHN value within the first 24 h (Figure 1a). Both PEO and HPMC hydrogels imbibed and retained a substantial amount of water from the surrounding buffer media to become hydrated, and subsequently underwent polymeric relaxation, generating a gel front on the surface of the disc. Subsequent to the initial 24-h decline, the discs underwent gelation to the core of their compressed matrices rapidly. Furthermore, the presence of PEO impeded cohesion between the HPMC and PLGA powder particles consequently generated matrices that were more prone to degradation upon exposure to hydration and mechanical stress. Due to the more complex nature of the HPMC polymer in comparison to PEO, it did not undergo polymeric relaxation as rapidly. Thus, the BHN value for HPMC declined most significantly between days 1 and 6. PLGA proved to be the exception to this rule. It demonstrated a steady decline in BHN over the 30 days. The combination of PLGA–HPMC also depicted a gradual, albeit more rapid decline in BHN, indicating that the presence of PLGA with the HPMC generated sufficient cohesion between the polymers to restrict the diffusion of water molecules into the compressed polymeric matrix.

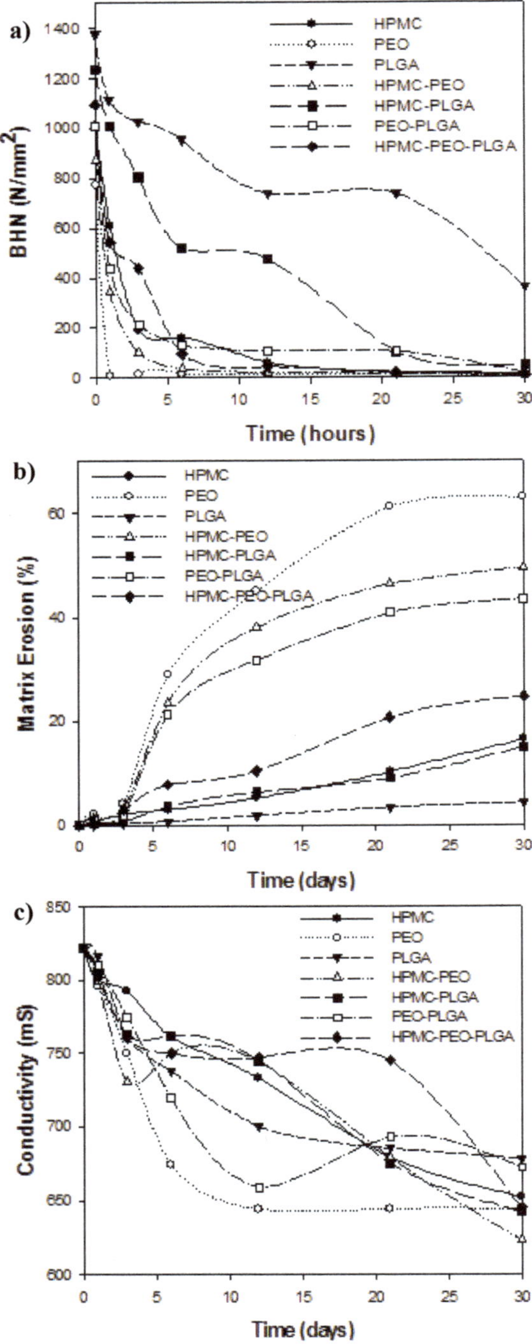

Figure 1. Physical properties of compressed polymeric discs following exposure to simulated CSF over a period of 30 days: (**a**) Changes in the Brinell Hardness Number (N/mm^2; N = 3; SD < 23.67 N/mm^2); (**b**) Matrix erosion (%; N = 3; SD < 0.05%); and (**c**) Changes in conductivity (μSiemens; N = 3; SD < 16.34 μS).

3.3. *Evaluation of Erosion of Compressed External Polymeric Matrix*

In contact with dissolution medium (in this case simulated CSF), hydrophilic polymers such as PEO and HPMC either swell into a hydro-gel layer or undergo erosion or both. Such swelling behavior is proportional to the rate of hydration and along with relative mobilities of dissolution medium and drug it dictates the kinetics as well as mechanism of drug release [20]. As portrayed in the previous results, PEO discs demonstrated lower gel strength, greater water uptake, and therefore demonstrated greater matrix erosion as depicted in Figure 1b. HPMC discs on the other hand demonstrated greater matrix swelling with negligible matrix erosion (Figure 1b). However, from the results it was observed that the rate of drug release by far superseded the rate of matrix degradation; hence it was concluded that the rate of drug release in this case was predominately a result of the diffusion of the drug from the matrices as opposed to matrix degradation.

As mentioned previously, PLGA is a hydrophobic bioerodible polymer. Thus, the mechanics of polymeric degradation and matrix erosion therefore occur via two main mechanisms, namely bulk erosion and surface erosion [21]. The degradation of such polymers is based on the conversion of the macromolecule into its water-soluble monomers via the hydrolytic cleavage of the ester bonds that constitute the backbone of the polymer. While bulk erosion does occur, a second mechanism which is surface erosion also takes place concurrently. This refers to the occurrence of hydrolytic breakdown of the polymer confined to the surface of the polymeric matrix. This was reflected by the steady decline in its BHN value over time following hydration (Figure 1b). Drug release therefore involved complex combination of both the drug diffusion from the surface as a result of bulk erosion as explained further in the manuscript.

3.4. *Changes in Conductivity of Compressed External Polymeric Matrices Following Hydration*

These results inversely correlated with those observed for matrix erosion i.e., the greater the matrix erosion, the lower the observed conductivity of the solution. This indicated a greater degree of interaction between the untangling polymeric chains and the ions in solution. As greater quantities of the polymer unraveled from the disc, i.e., the matrix eroded, it ionized and interacted with ions in the surrounding PBS. Hence, we observed a decline in the conductivity of the solution over the 30 days (Figure 1c). The greatest decline was observed for PEO. This is not surprising as it undergoes the most rapid rates of matrix degradation. Since the most significant phase of degradation of the PEO discs occurred within the first 12 days of exposure to simulated CSF, conductivity declined exponentially until this point, after which equilibrium was attained. HPMC demonstrated a steady, almost linear decline in conductivity until day 21, after which a significant amount of the matrix had degraded to interact with surrounding counter-ionic species. As expected, PLGA expressed the least change in electrical conductivity, since it demonstrated the minimal change in weight (matrix erosion) over time following hydration.

3.5. *Evaluation of Porosity Changes upon Hydration*

Scanning electron micrographs of the discs demonstrated the progression of pore formation and enlargement, formation of a trabeculated network of channels within the structure followed by erosion, and loss of integrity (Figure 2). PLGA discs demonstrated the most intact matrices in comparison to other systems. Figure 2a indicates the formation of a porous structure within the matrix at day 21 following exposure to simulated CSF. Larger pores (diameter of approximately 300 μm) were limited to the periphery of the discs, while the core of the matrix was relatively intact (where pores have a diameter of approximately 50 μm).

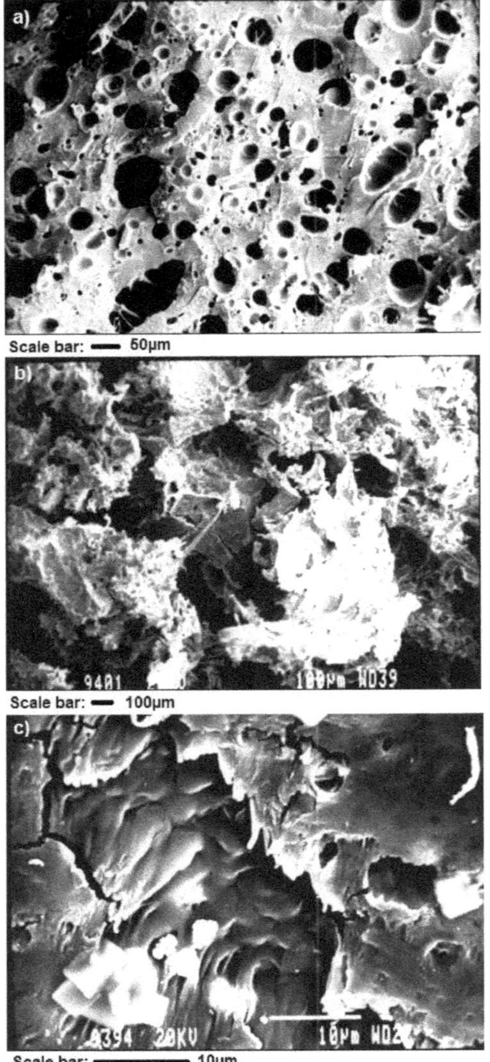

Figure 2. Scanning electron micrographs (Magnification 65×) of compressed: (**a**) PLGA, (**b**) HPMC–PLGA, and (**c**) PEO-PLGA discs following exposure to simulated cerebrospinal fluid (CSF) after 21 days.

This matrix did not expand significantly, nor did it undergo any significant erosion. In contrast by day 21, HPMC–PLGA discs (Figure 2b) had a significantly more patent network, with pore sizes in the core of the matrix having comparatively larger diameters. The disc expanded in width and cross-sectional diameter due to the tendency of HPMC to 'swell'. PEO–PLGA discs (Figure 2c) demonstrated a significant degree of erosion within the matrix while simultaneously indicating a greater degree of swelling and erosion. By day 21, the entire network, including the core of the matrix, was extremely patent and highly trabeculated. The majority of the matrix appeared to be composed of

PLGA. A detailed image processing analysis of the PLGA matrix is provided in the supplementary data (Figure S1).

3.6. In Vitro Drug Release from Nicotine-Loaded Compressed Gelisphere Polymeric Matrices

In vitro drug release studies in simulated CSF indicated that reinforced alginate gelispheres released 100% of entrapped drug within 50 h through a biphasic first- and zero-order release profile (Figure 3a). It was observed that reinforced alginate gelispheres incorporated into PLGA discs demonstrated zero-order release kinetics that were able to provide controlled and prolonged release for approximately 50 days (Figure 3b). Drug release kinetics indicated a rapid burst in drug release after day 35, where after a first-order release profile was obtained. An additional 40% of entrapped nicotine was released between days 35 and 40. In retrospect it was observed that PLGA had the maximum cohesive tendency between its particles and has the ability to be most compressible. Being a hydrophobic polymer, its main degradative mechanisms are through the hydrolytic cleavage of its polymeric backbone. Consequently, these discs underwent minimal matrix degradation.

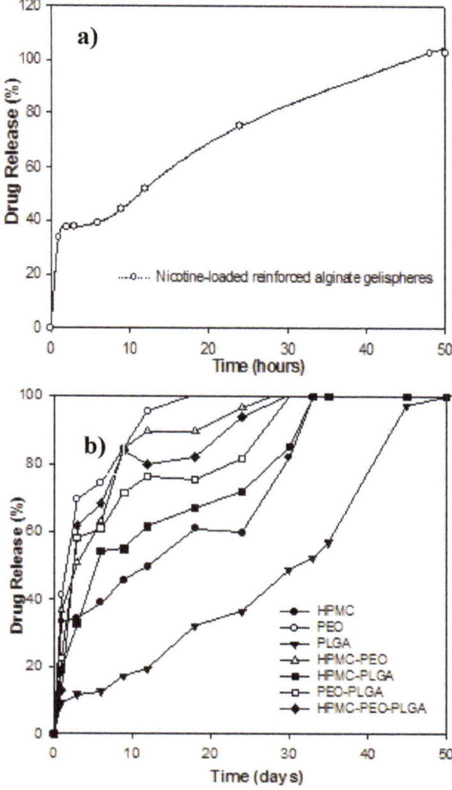

Figure 3. Drug released (%) from: (**a**) nicotine-loaded reinforced alginate gelispheres, and (**b**) nicotine-loaded reinforced alginate gelispheres compressed into polymeric discs following exposure to simulated cerebrospinal fluid (CSF) over a period of 50 hours ($N = 3$; SD < 0.53) and 50 days ($N = 3$; SD < 0.59%) respectively.

The rapid release rates observed from PEO and HPMC matrices (Figure 3b) allowed us to conclude that drug release occurred primarily by Fickian diffusion. The presence of PEO in matrices enhanced the rate of drug release from the compressed discs. Its hydrophilic character allowed rapid disentanglement and degradation of the compressed matrices. Its tendency was therefore to undergo comparatively rapid bulk erosion when it came in contact with the alkaline hydrating medium (Figure 3b). HPMC is a hydrogel; hence, it has the capacity to imbibe a significant amount of water before beginning to degrade. Furthermore, as observed from previous results, its particles were sufficiently cohesive to generate compact discs that are able to provide zero-order release following the achievement of an equilibrium hydrated state (discussed under later sections). Between days 3 and 18, a linear drug release profile was observed from HPMC discs (Figure 3b). Prior to this, a burst release was observed, whereby approximately 33.25% of the entrapped nicotine was released within the first 24 h. Combining HPMC and PLGA had a unique effect on drug release. While in this case zero-order release was also observed, they exceeded those observed with the individual polymers. A burst effect was observed until day 6, by which time 54.31% of the drug had been released (Figure 3b). Thereafter, drug release followed a linear zero-order release profile until day 32. Release kinetics for all other combinations exhibited rapid burst rates of drug release and predominately first-order release (Figure 3b).

3.7. Molecular Mechanics Energy Relationship (MMER) Analysis

Molecular mechanics energy relationship (MMER), analysis, a method for analytico-mathematical representation of potential energy surfaces, was used to provide information about the contributions of valence terms, noncovalent Coulombic terms, and noncovalent van der Waals interactions for the drug-polysaccharide complex, crosslinked polysaccharide morphologies, and the polymer-polymer composites. The MMER model for the potential/steric energy factors in various molecular complexes can be written as:

$$E_{\text{molecule/complex}} = V_{\sum} = V_{\text{b}} + V_{\theta} + V_{\varphi} + V_{\text{ij}} + V_{\text{hb}} + V_{\text{el}} \quad (4)$$

where $V_{\sum}$ is related to total steric energy for an optimized structure, V_{b} corresponds to bond stretching contributions, V_{θ} denotes bond angle contributions, V_{φ} represents the torsional contribution from dihedral angles, V_{ij} incorporates van der Waals interactions due to non-bonded interatomic distances, V_{hb} symbolizes hydrogen bond energy function, and V_{el} stands for electrostatic energy.

3.7.1. Energy Minimizations Involving Drug-Polymer Morphologies

The energy-minimized and geometrically optimized conformation depicting the alginate–nicotine (ALG–NCT) complex is presented in Figure 4. The component binding energies and intrinsic molecular attributes are listed in Equations (5)–(7) and Table 2, respectively. It is evident from Figure 4 that NCT and ALG show close interaction via formation of H-bonds between C–O–C...N–H and C=O...N–H group of ALG and NCT, respectively. The final energy of the drug–polymer complex was stabilized by all three non-bonding interactions (van der Waals interactions, H-bonding and electrostatic interactions) between the constituent molecules ALG and NCT. The surface-to-volume ratio (SVR) of the complex ALG–NCT (SVR = 0.451) displayed lower SVR than the individual molecules (Table 2) further confirming the stability of the complex—the lower the SVR, the more stable the structure. However, the finally globally minimized energy values suggest that the ALG–NCT complex is destabilized by a binding energy of ~4 kcal/mol (Equations (5)–(7)) due to exceptionally high torsional contribution arising due to deviations from optimum dihedral angles (Equation (7)). Such interactions may induce dipoles in the molecular complex due to the torsional strains experienced around dihedral angles.

$$E_{\text{ALG}} = -23.369V_{\sum} = 2.148V_{\text{b}} + 26.810V_{\theta} + 23.631V_{\varphi} + 8.649V_{\text{ij}} - 4.500V_{\text{hb}} - 80.110V_{\text{el}} \quad (5)$$

$$E_{\text{NCT}} = 13.283V_{\sum} = 0.388V_{\text{b}} + 3.672V_{\theta} + 6.198V_{\varphi} + 3.023V_{\text{ij}} \quad (6)$$

$$E_{\text{ALG-NCT2}} = 7.204V_{\sum} = 3.403V_{\text{b}} + 34.575V_{\theta} + 54.275V_{\varphi} + 8.182V_{\text{ij}} - 5.407V_{\text{hb}} - 87.824V_{\text{el}} \quad (7)$$

$$[\Delta E_{\text{BINDING}} = 4.007 \text{ kcal/mol}]$$

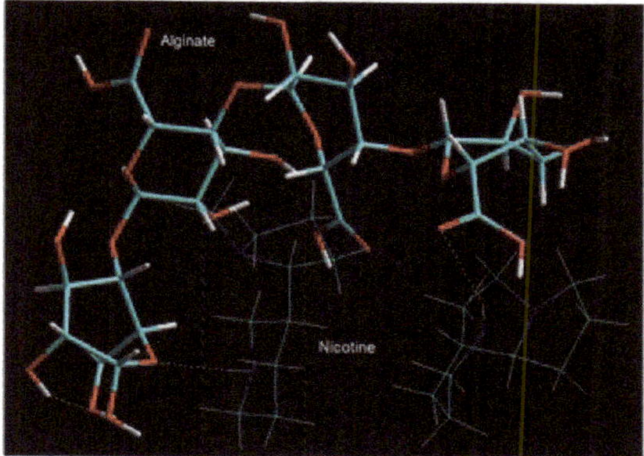

Figure 4. Visualization of geometrical preferences of nicotine in complexation with alginate after molecular mechanics simulations. Color codes: C (cyan), O (red), N (blue), and H (white).

Table 2. Calculated molecular attributes of the complexes involving alginate (ALG) and nicotine (NCT).

Structure	Molecular Attributes		
	Surface Area (grid)	Volume	Surface-To-Volume Ratio
ALG	828.99	1518.13	0.546
NCT[2] #	603.16	1088.42	0.554
ALG–NCT[2] #	1026.16	2273.41	0.451

ΔRef = $Ref_{(HostGuest)} - Ref_{(Host)} - Ref_{(Guest)}$. # Data represent two molecules of nicotine.

It can be deduced from the above discussion that although nicotine formed a polyelectrolyte complex in association with alginate, the complex being unstable as it was might have led to release of nicotine upon hydration of the matrix. The above findings corroborated with the need of strategies for controlling the release of nicotine from the alginate matrix. We propose that this may be achieved by three different approaches applied together. Firstly, this may be achieved by inclusion of another polymer to stabilize the alginate matrix. Secondly, the strengthening of the alginate matrix can be achieved by crosslinking the hydrogel matrix using divalent ions such as Ca^{2+} or Ba^{2+}. The third approach is by incorporating the crosslinked blend hydrogel–matrix in a release rate-modulating external polymeric matrix.

3.7.2. Mechanistic Elucidation of Crosslinked Polymer Conformations

The general molecular mechanics program was used in this study to compute the energy of the polymer–saccharide (ALG–HEC) and the cation-polymer-saccharide conformations. The geometrical conformation of the ALG–HEC after SLAS are shown in Figure 5a and the corresponding energy attributes are depicted in Equations (5), (8) and (9).

$$E_{HEC} = 130.862V_{\sum} = 4.017V_b + 80.816V_\theta + 35.115V_\varphi + 10.28V_{ij} - 0.231V_{hb} + 0.863V_{el} \quad (8)$$

$$E_{ALG\text{-}HEC} = -4.459V_{\sum} = 6.685V_b + 44.109V_\theta + 77.607V_\varphi + 9.085V_{ij} - 4.062V_{hb} - 137.886V_{el} \quad (9)$$

$$[\Delta E_{BINDING} = -111.952 \text{ kcal/mol}]$$

$$E_{ALG\text{-}HEC\text{-}Ca2+} = -10.907V_{\Sigma} = 6.666V_b + 44.137V_\theta + 77.733V_\varphi + 2.794V_{ij} - 4.018V_{hb} - 138.22V_{el} \quad (10)$$

$$[\Delta E_{BINDING} = -118.4 \text{ kcal/mol}]$$

$$E_{ALG\text{-}HEC\text{-}Ba2+} = -13.641V_{\Sigma} = 6.700V_b + 44.245V_\theta + 77.981V_\varphi + 0.116V_{ij} - 4.060V_{hb} - 138.625V_{el} \quad (11)$$

$$[\Delta E_{BINDING} = -121.134 \text{ kcal/mol}]$$

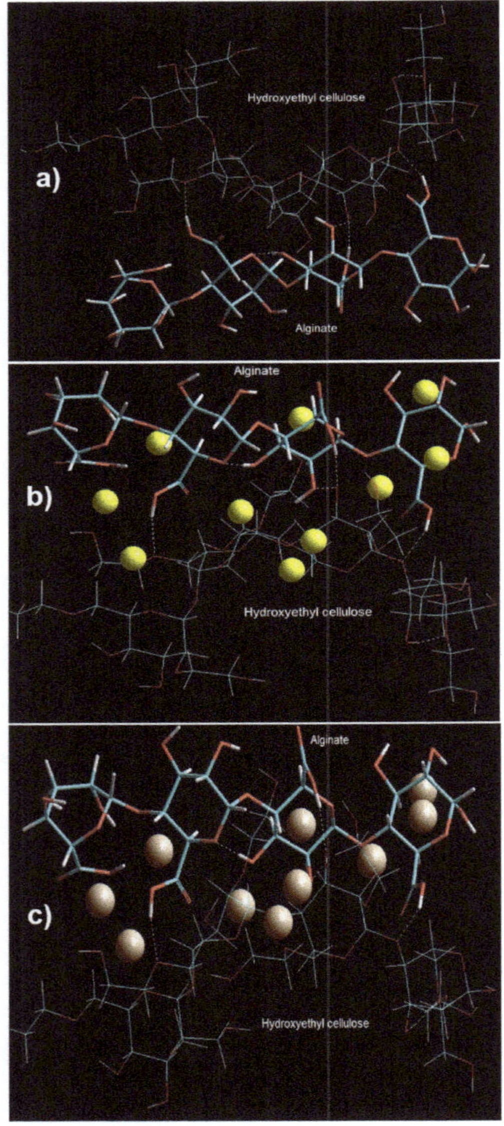

Figure 5. Visualization of geometrical preference of (**a**) Hydroxyethyl cellulose in complexation with alginate; (**b**) oligosaccharide–Ca^{2+} complex, and (**c**) oligosaccharide–Ba^{2+} complex derived after molecular mechanics simulations. Color codes: C (cyan), O (red), N (blue), Ca (yellow), Ba (brown), and H (white).

The final minimized energy values for the ALG-HEC complex demonstrated a stabilization of binding energy by 111.952 kcal/mol and was significantly supported by the van der Waals interactions due to the presence of ethyl aliphatic groups in the constituent polymers. The interaction led to the formation of both intermolecular H-bonds between the sugar moieties (Figure 5a). These underlying weak chemical interactions may not cause a structural change in the polymers but may initiate aggregation of the aliphatic chains creating dense localized regions not characteristic of the bulk polymers. In addition, ALG-induced intramolecular hydrogen bonding in HEC and vice versa may influence the hydration process of the polymer matrix. This energy stabilization along with the matrix entanglement may provide a prolonged nicotine release along with reduced burst release as compared to alginate alone.

We further screened ALG–HEC interactions for two divalent cations (Ca^{2+} and Ba^{2+}) using SLAS, wherein the ion probes were placed within the van der Waals surface of the polymeric chains as shown in Figure 5a,b. The component binding energies listed in Equations (5), (8), (10) and (11) confirmed that the ALG–HEC–Ba^{2+} formed a more thermodynamically stable conformer than ALG–HEC–Ca^{2+}. Interestingly, the polymer chains formed an ordered secondary structure outlined by the presence of cation binding sites as ordered arrays and within the polysaccharide fragments. These energy optimizations were supported by the electrostatic forces while the van der Waals interactions and the torsion angle energies destabilized the cation-crosslinked molecules. The geometrical conformation of the polymer chains and the steric organization of the cations conformed the role of stereospecificity, ligand spacing and position of the coordination shell for the cation interaction display (Figure 5). A close look at Figure 5 revealed that Ba^{2+} provided significantly close packed 3D architecture as compared to Ca^{2+} and hence increased the tendency of the polymer network to form intraglycosidic hydrogen bonds and an increase in hydration energy. In essence, the torsional strains, cavity formation contribution and hydration factors together contributed to complexation energies. This behavior of the Ca^{2+} and Ba^{2+} was also in agreement with experimental data in the case of ALG–HEC gelispheres where the corrugated polysaccharide chains might "offer several oxygen atoms whose stereochemical arrangement fits into the coordination sphere" of the cations resulting in prolonged release of nicotine from gelisphere matrices.

3.7.3. 3D Computational Modeling for Polymer–Polymer Complexes

The measured differences in the flowability and friability properties and texture profile of external matrices were elucidated by randomly disposing PLGA or PEO or PLGA/PEO around HPMC to form HPMC–PLGA, HPMC–PEO and PLGA/HPMC/PEO.

$$E_{\text{HPMC}} = 49.713V_{\Sigma} = 2.089V_b + 18.821V_\theta + 22.360V_\varphi + 6.776V_{ij} - 0.335V_{hb} \quad (12)$$

$$E_{\text{PLGA}} = 2.021V_{\Sigma} = 0.522V_b + 4.905V_\theta + 2.575V_\varphi - 5.982V_{ij} \quad (13)$$

$$E_{\text{PEO}} = 29.483V_{\Sigma} = 0.643V_b + 6.092V_\theta + 15.675V_\varphi + 7.072V_{ij} \quad (14)$$

$$E_{\text{PLGA-HPMC}} = 32.098V_{\Sigma} = 2.366V_b + 21.460V_\theta + 26.393V_\varphi - 17.375V_{ij} - 0.747V_{hb} \quad (15)$$

$$[\Delta E_{\text{BINDING}} = -19.636 \text{ kcal/mol}]$$

$$E_{\text{PLGA-PEO}} = 11.572V_{\Sigma} = 1.240V_b + 8.231V_\theta + 18.895V_\varphi - 16.794V_{ij} \quad (16)$$

$$[\Delta E_{\text{BINDING}} = -19.932 \text{ kcal/mol}]$$

$$E_{\text{HPMC-PEO}} = 53.607V_{\Sigma} = 2.540V_b + 17.290V_\theta + 42.725V_\varphi - 7.945V_{ij} - 1.003V_{hb} \quad (17)$$

$$[\Delta E_{\text{BINDING}} = -25.236 \text{ kcal/mol}]$$

$$E_{\text{PLGA-HPMC-PEO}} = 40.662V_{\Sigma} = 3.043V_b + 25.015V_\theta + 40.473V_\varphi - 26.637V_{ij} - 1.233V_{hb} \quad (18)$$

$$[\Delta E_{\text{BINDING}} = -40.555 \text{ kcal/mol}]$$

It is evident from Equations (12)–(18) that the formation of HPMC–PLGA, PLGA–PEO, HPMC–PEO and PLGA/HPMC/PEO (in vacuum) were accompanied by energy stabilization of −19.636 kcal/mol, −19.932 kcal/mol, −25.236 kcal/mol and −40.555 kcal/mol, respectively. Since all four molecular complexes displayed –ve energy of binding, it can be estimated that the polymers demonstrated compatibility and stability to form a polymeric blend in a dried state [22].

The chemical interactions within the polymeric blends were primarily represented by non-bonding interactions (van der Waals forces) and were accountable for the for mechanical strength and flow characteristics of the composites. This energy minimization involving HPMC demonstrated extensive rotations among the monosaccharide residues leading to the formation of a strained architecture which was eventually relieved by angle adjustments and bond length, resulting in H-bonding within the PLGA–HPMC and PEO–HPMC composites (Figure 6). As reported earlier in this paper, PLGA–HPMC–PEO produced highly dense compressed discs due to high angle of repose and flowability. This could be due to the interactions of aliphatic groups of PLGA/PEO with pendent groups of HPMC → formation of unfavorable regions → large steric modulations and overcoming of torsional barriers → large accessible potential energy surface → formation of a polymeric matrix. This became even more applicable when the matrices were hydrated. Given the high torsional strain within the HPMC matrices, it may induce degradation of the polymeric matrix in order to acquire energy stabilization upon hydration. The movement of water molecules inside the torsional strains may in turn lead to an early release of drug molecules. This corroborates with the experimental results wherein rapid release rates were observed from PEO and HPMC matrices (Figure 3). This torsional restraint relaxation during hydration also led to the decrease in BHN values in HPMC and PEO matrices as compared to PLGA matrices (Figure 1).

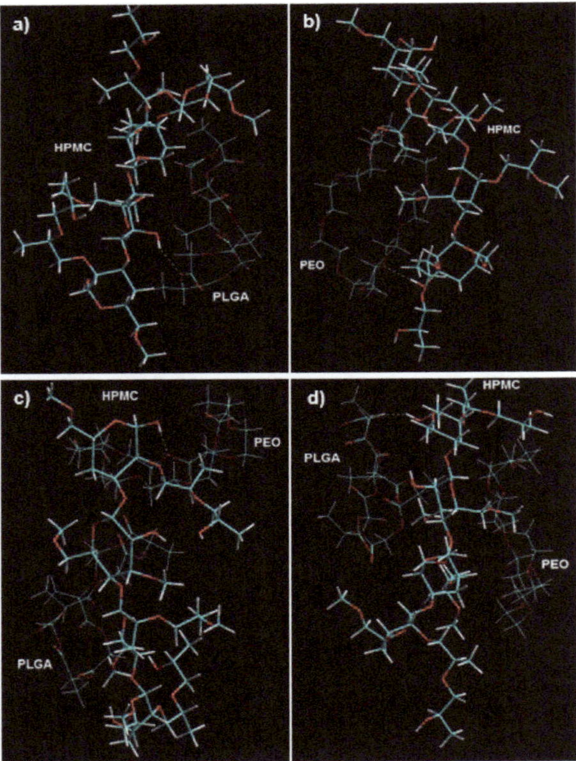

Figure 6. Visualization of geometrical preference of (**a**) PLGA in complexation with HPMC; (**b**) PEO in complexation with HPMC; (**c**) PLGA–HPMC–PEO tripolymeric comples with PEO H-bonded to HPMC; and (**d**) PLGA–HPMC–PEO tripolymeric complex with PLGA H-bonded to HPMC after molecular mechanics simulations. Color codes: C (cyan), O (red), N (blue), and H (white).

4. Conclusions

The intended drug delivery system that this study sought to develop was one that could provide sustained zero-order release of nicotine from a crosslinked polyspheric system. This would allow for the attainment of constant drug levels at the site of action and minimize adverse effects associated with fluctuating drug levels, as well as optimizing patient therapy and compliance. PLGA matrices incorporating calcium barium-crosslinked alginate-hydroxyethylcellulose gelispheres offered minimal rates of matrix degradation and successfully retarded nicotine release, leading to the achieved zero-order release for 50 days following exposure to simulated CSF. The computational modeling methods that were used for the prediction of preferred molecular conformations of the drug–polymer and polymer-polymer complexes using force field minimizations and the modes of interaction corroborated well the experimental results. The developed drug delivery system displays a great potential for use in the treatment of Parkinson's disease.

Supplementary Materials: The following are available online at http://www.mdpi.com/1999-4923/10/4/233/s1, Figure S1: Image processing of micrograph of a PLGA scaffold: (**a**) BLURimage: blurred SEMimage; (**b**) CQimage: colorquantized image of BLURimage; (**c**) UniHisto: uniform histogram of CQimage; (**d**) UniHistoSep: separated uniform histogram of CQimage after image processing; (**e**) NonUniHisto: non-uniform histogram of SEMimage before image processing; and (**f**) NonUniHistoSep: separated non-uniform histogram of SEMimage before image processing. Note that each color channel is only capable of highlighting most of the pores, but not all.

Author Contributions: Conceptualization, Y.E.C. and V.P.; Methodology, P.K., L.D.d.T. and N.S.; Software, P.K.; Formal Analysis, P.K. and N.S.; Resources, V.P.; Writing—Original Draft Preparation, P.K.; Writing—Review & Editing, all authors; Supervision, Y.E.C. and V.P.; Funding Acquisition, V.P.

Funding: This work was funded by the National Research Foundation (NRF) of South Africa.

Conflicts of Interest: The authors declare no conflict of interest. The funders had no role in the design of the study; in the collection, analyses, or interpretation of data; in the writing of the manuscript, or in the decision to publish the results.

References

1. Moll, H. The treatment of post encephalitic Parkinsonism by nicotine. *Br. Med. J.* **1926**, *1*, 1079–1081. [CrossRef] [PubMed]
2. Quik, M.; Zhang, D.; McGregor, M.; Bordia, T. Alpha7 nicotinic receptors as therapeutic targets for Parkinson's disease. *Biochem. Pharmacol.* **2015**, *97*, 399–407. [CrossRef] [PubMed]
3. Barreto, G.E.; Iarkov, A.; Moran, V.E. Beneficial effects of nicotine, cotinine and its metabolites as potential agents for Parkinson's disease. *Front. Aging Neurosci.* **2014**, *6*, 340. [CrossRef] [PubMed]
4. Mouhape, C.; Costa, G.; Ferreira, M.; Abin-Carriquiry, J.A.; Dajas, F.; Prunell, G. Nicotine-induced neuroprotection in rotenone in vivo and in vitro models of Parkinson's disease: Evidences for the involvement of the labile iron pool level as the underlying mechanism. *Neurotox. Res.* **2018**, in press. [CrossRef] [PubMed]
5. Villafane, G.; Thiriez, C.; Audureau, E.; Straczek, C.; Kerschen, P.; Cormier-Dequaire, F.; Van Der Gucht, A.; Gurruchaga, J.-M.; Quéré-Carne, M.; Paul, M.; et al. High-dose transdermal nicotine in Parkinson's disease patients: A randomized, open-label, blinded-endpoint evaluation phase 2 study. *Eur. J. Neurol.* **2018**, *25*, 120–127. [CrossRef] [PubMed]
6. Oliveira, S.S.M.; Oliveira, F.S.; Gaitani, C.M. Microparticles as a strategy for low-molecular-weight heparin delivery. *J. Pharm. Sci.* **2011**, *100*, 1783–1792. [CrossRef] [PubMed]
7. Pahwa, R.; Saini, N.; Kumar, V.; Kohli, K. Chitosan-based gastroretentive floating drug delivery technology: An updated review. *Exp. Opin. Drug Deliv.* **2012**, *9*, 525–539. [CrossRef] [PubMed]
8. Kamaly, N.; Yameen, B.; Wu, J.; Farokhzad, O.C. Degradable controlled-release polymers and polymeric nanoparticles: Mechanisms of controlling drug release. *Chem. Rev.* **2016**, *116*, 2602–2663. [CrossRef] [PubMed]
9. Hu, A.; Chen, C.; Mantle, M.D.; Wolf, B.; Gladden, L.F.; Rajabi-Siahboomi, A.; Missaghi, S.; Mason, L.; Melia, C.D. The properties of HPMC:PEO extended release hydrophilic matrices and their response to ionic environments. *Pharm. Res.* **2017**, *34*, 941–956. [CrossRef] [PubMed]

10. Zhang, Q.; Fassihi, M.A.; Fassihi, R. Delivery considerations of highly viscous polymeric fluids mimicking concentrated biopharmaceuticals: Assessment of injectability via measurement of total work done "WT". *AAPS PharmSciTech* **2018**, *19*, 1520–1528. [CrossRef] [PubMed]
11. Kumar, P.; Bhatia, M. Functionalization of chitosan/methylcellulose interpenetrating polymer network microspheres for gastroretentive application using central composite design. *PDA J. Pharm. Sci. Technol.* **2010**, *64*, 497–506. [PubMed]
12. Colombo, P.; Bettini, R.; Peppas, N.A. Observation of swelling process and diffusion front position during swelling in hydroxypropyl methyl cellulose (HPMC) matrices containing a soluble drug. *J. Control. Release* **1999**, *61*, 83–91. [CrossRef]
13. Jamzad, S.; Tutunji, L.; Fassihi, R. Analysis of macromolecular changes and drug release from hydrophilic matrix systems. *Int. J. Pharm.* **2005**, *292*, 75–85. [CrossRef] [PubMed]
14. Mansour, H.M.; Sohn, M.; Al-Ghananeem, A. Materials for pharmaceutical dosage forms: Molecular pharmaceutics and controlled release drug delivery aspects. *Int. J. Mol. Sci.* **2010**, *11*, 3298–3322. [CrossRef] [PubMed]
15. Kapoor, D.N.; Bhatia, A.; Kaur, R.; Sharma, R.; Kaur, G.; Dhawan, S. PLGA: A unique polymer for drug delivery. *Ther. Deliv.* **2015**, *6*, 41–58. [CrossRef] [PubMed]
16. Fournier, E.; Passirani, C.; Montero-Menei, C.N.; Benoit, J.P. Biocompatibility of implantable synthetic polymeric drug carriers: Focus on brain biocompatibility. *Biomaterials* **2003**, *24*, 3311–3331. [CrossRef]
17. Choonara, Y.E.; Pillay, V.; Khan, R.A.; Singh, N.; Du Toit, L.C. Mechanistic evaluation of alginate-HEC gelisphere compacts for controlled intrastriatal nicotine release in Parkinson's disease. *J. Pharm. Sci.* **2009**, *98*, 2059–2072. [CrossRef] [PubMed]
18. Bawa, P.; Pillay, V.; Choonara, Y.E.; du Toit, L.C.; Ndesendo, V.M.K.; Kumar, P. A composite polyelectrolytic matrix for controlled oral drug delivery. *AAPS PharmSciTech* **2011**, *12*, 227–238. [CrossRef] [PubMed]
19. Kumar, P.; Choonara, Y.E.; Pillay, V. In silico analytico-mathematical interpretation of biopolymeric assemblies: Quantification of energy surfaces and molecular attributes via atomistic simulations. *Bioeng. Transl. Med.* **2018**, *3*, 222–231. [CrossRef] [PubMed]
20. Caccavo, D.; Cascone, S.; Lamberti, G.; Barba, A.A. Controlled drug release from hydrogel-based matrices: Experiments and modeling. *Int. J. Pharm.* **2015**, *486*, 144–152. [CrossRef] [PubMed]
21. Rafiei, P.; Haddadi, A. Pharmacokinetic Consequences of PLGA nanoparticles in docetaxel drug delivery. *Pharm. Nanotechnol.* **2017**, *5*, 3–23. [CrossRef] [PubMed]
22. Kumar, P.; Choonara, Y.E.; Toit, L.C.; Modi, G.; Naidoo, D.; Pillay, V. Novel high-viscosity polyacrylamidated chitosan for neural tissue engineering: Fabrication of anisotropic neurodurable scaffold via molecular disposition of persulfate-mediated polymer slicing and complexation. *Int. J. Mol. Sci.* **2012**, *13*, 13966–13984. [CrossRef] [PubMed]

Article

Cell Internalization in Fluidic Culture Conditions Is Improved When Microparticles Are Specifically Targeted to the Human Epidermal Growth Factor Receptor 2 (HER2)

Inmaculada Mora-Espí [1], Elena Ibáñez [1], Jorge Soriano [1], Carme Nogués [1], Thorarinn Gudjonsson [2,3] and Leonardo Barrios [1,*]

1 Unitat de Biologia Cel·lular, Departament de Biologia Cel·lular, Fisiologia i Immunologia, Facultat de Biociències, Universitat Autònoma de Barcelona, Bellaterra, 08193 Barcelona, Spain; xapaxin@gmail.com (I.M.-E.); Elena.ibanez@uab.cat (E.I.); jorge.soriano@uab.cat (J.S.); carme.nogues@uab.cat (C.N.)

2 Biomedical Center, University of Iceland, 101 Reykjavík, Iceland; tgudjons@hi.is

3 Department of Anatomy, Faculty of Medicine, and Department of Laboratory Hematology, University Hospital, 101 Reykjavik, Iceland

* Correspondence: lleonard.barrios@uab.cat; Tel.: +34-93-581-2776; Fax: +34-93-581-2295

Received: 21 February 2019; Accepted: 8 April 2019; Published: 11 April 2019

Abstract: Purpose: To determine if the specific targeting of microparticles improves their internalization by cells under fluidic conditions. Methods: Two isogenic breast epithelial cell lines, one overexpressing the Human Epidermal Growth Factor Receptor 2 (HER2) oncogene (D492HER2) and highly tumorigenic and the other expressing HER2 at much lower levels and non-tumorigenic (D492), were cultured in the presence of polystyrene microparticles of 1 μm in diameter, biofunctionalized with either a specific anti-HER2 antibody or a non-specific secondary antibody. Mono- and cocultures of both cell lines in static and fluidic conditions were performed, and the cells with internalized microparticles were scored. Results: Globally, the D492 cell line showed a higher endocytic capacity than the D492HER2 cell line. Microparticles that were functionalized with the anti-HER2 antibody were internalized by a higher percentage of cells than microparticles functionalized with the non-specific secondary antibody. Although internalization was reduced in fluidic culture conditions in comparison with static conditions, the increase in the internalization of microparticles biofunctionalized with the anti-HER2 antibody was higher for the cell line overexpressing HER2. Conclusion: The biofunctionalization of microparticles with a specific targeting molecule remarkably increases their internalization by cells in fluidic culture conditions (simulating the blood stream). This result emphasizes the importance of targeting for future in vivo delivery of drugs and bioactive molecules through microparticles.

Keywords: microfluidics; coculture; HER2; polystyrene μPs; biofunctionalization

1. Introduction

Drug targeting has the potential to improve the therapeutic efficacy and mitigate the non-specific effects of many drugs. In the last years, several types of drug delivery vehicles have been developed, including monoclonal antibodies [1], peptides [2], proteins [3], lipoproteins [4], carbohydrates [5], and polymeric nanoparticles [6,7]. Compared with the number of studies in which nanoparticles (NPs) are used [8–12], only a small number of studies involve the use of microparticles (μPs) [13–15]. It has been reported that small sizes and positive charges favor NP intake by cells [14–16], but in some cases, larger particle sizes could be advantageous for preventing non-specific interactions and internalization into normal non-phagocytic cells or for optimal tissue entrapment and transient retention [17]. Moreover, to target cancer cells, NPs and μPs surfaces can be modified to increase the interaction with plasma membrane-specific markers like the transferrin receptor [18], the folate receptor [19], or the human epidermal growth factor receptor 2 (HER2, also known as ERBB2) [20,21]. HER2 is a receptor tyrosine kinase which is overexpressed by some types of cancer cells and is considered a marker of poor clinical outcome in breast and ovarian cancer [22,23]. Some treatments directed to this target have already been approved and are clinically used, such as the anti-HER2 monoclonal antibody trastuzumab, alone or in combination with emtansine (T-DM1) [24], and the HER2 tyrosine kinase activity inhibitor lapatinib.

Traditionally, in vitro studies on drug carriers and drug release have been performed in static monolayer cell cultures. However, studies in microfluidic environments, mimicking the circulatory system, are currently gaining interest [12]. Compared with static cultures, microfluidic studies allow for better predictions about how a drug or a drug carrier running in a circulating flow, will interact with cells [25–27]. On the other hand, because normal and tumoral cells are intermingled in vivo, cocultures of normal and tumoral cells can better simulate tissue conditions than monocultures [28,29].

The overall objective of the present study was to evaluate the efficiency of targeting μPs to cells in physiological-like conditions (fluidic culture conditions). Two isogenic breast epithelial cell lines were used, one normal (D492) and the other overexpressing HER2 (D492HER2). Moreover, polystyrene μPs of 1 μm in diameter were biofunctionalized with a specific targeting protein, an anti-HER2 antibody, or with a non-specific secondary antibody. The specific objectives of the study were to evaluate and compare the cell internalization of these μPs in different culture conditions: monoculture versus coculture conditions, and static versus fluidic culture conditions.

2. Material and Methods

2.1. Biofunctionalization of Polystyrene μPs

Carboxylate polystyrene μPs of 1 μm in diameter (Polybead® Carboxylate Microspheres. Polysciences, Inc., Warrington, PA, USA) were biofunctionalized with two different targeting molecules: (1) mouse anti-c-ERBB2/c-Neu (Ab-5) clone TA-1 (Millipore, Darmstadt, Germany), herein referred to as antiH, and (2) goat anti-mouse IgG2a secondary antibody Alexa Fluor® 647 conjugate (Life Technologies, Carlsbad, CA, USA), herein referred to as secAb. Biofunctionalization was carried out using the PolyLink Protein Coupling Kit for COOH Microspheres (Polysciences) according to the manufacturer's instructions. The size of the μPs before and after biofunctionalization was analyzed by transmission electronic microscopy (TEM) (JEOL, JEM 2011). Biofunctionalization was evaluated under a fluorescence inverted microscope (Olympus IX71, Olympus, Hamburg, Germany) and by the change in the ζ-potential.

Microscopically, biofunctionalization of μPs with secAb (μP-secAb) was evaluated directly on the basis of their far-red fluorescence emission. On the other hand, μPs biofunctionalized with antiH (μP-antiH) were incubated for 5 min with chicken anti-mouse IgG (H+L) secondary antibody Alexa Fluor® 488 conjugate (1:500. Life Technologies) before the evaluation of green fluorescence emission.

Biofunctionalized and non-biofunctionalized μPs were separately resuspended in H14 culture medium [30] and sonicated for 5 min (Fisherbrand FB15047, Fisher Scientific, Germany) to achieve a monodispersed sample. Their ζ-potential was then measured with a Zetasizer Nano ZS (Malvern Instruments, Malvern, UK).

2.2. Cell Lines

Two isogenic breast epithelial cell lines, D492 and D492HER2, were used in the study. D492 is a non-tumorigenic cell line with stem cell properties that expresses low levels of the HER2 oncogene [30,31]. D492HER2 was generated by overexpressing the HER2 oncogene in D492 and is highly tumorigenic [32]. Both cell lines constitutively express green fluorescent protein (GFP).

The cells were cultured in serum-free H14 culture medium [30,31] at 37 °C and 5% CO_2 (standard conditions). As explained below, cell culture was performed as follows: mono- or cocultures in static conditions, and cocultures in fluidic conditions.

2.3. Cell Cultures in Static Conditions

For monoculture experiments, cells were seeded at a density of 60,000 cells/well in 24-well plates (μ-Plate 24-Well ibiTreat: #1.5 polymer coverslip. ibidi, Martinsried, Germany). For coculture experiments, 30,000 cells of each cell line (D492 and D492HER2) were seeded together in each well. In both cases, the cells were maintained for 24 h in standard culture conditions prior to performing any experiments.

To analyze μP internalization, μPs (μP-antiH or μP-secAb) were sonicated for 5 min, diluted (1:100), counted with a hemocytometer, and then added at a proportion of 45 μP/cell to the cell cultures and incubated for further 24 h in standard culture conditions.

2.4. Cell Cultures in Fluidic Conditions

For these experiments, only cocultures were performed. Before seeding, channel slides (μ-Slides I 0.8 mm ibiTreat, ibidi) were coated with bovine collagen type I (Advanced Biomatrix, San Diego, CA, USA) to enhance cell adhesion. Then, 1.5×10^5 cells of each cell line (D492 and D492HER2) were seeded in H14 medium containing 2% penicillin/streptomycin (Biowest, Nuaillé, France) and incubated in standard culture conditions. After 24 h, the slides were connected to a microfluidic system consisting of a perfusion set (Perfusion Set Red ID 1.6 mm, ibidi) filled with fresh H14 medium containing the μPs (μP-secAb or μP-antiH) and a Fluidic Unit connected to an ibidi Pump, controlled by a pump-control software (ibidi). The microfluidic system was kept at 37 °C and 5% CO_2. The cultures were maintained for 24 h under a unidirectional flow rate fixed at 4.32 mL/min with a shear stress of 1.50 dyn/cm^2 and a pressure of 7.9 mbar, as recommended by the manufacturer.

2.5. Evaluation of Microparticles Internalization

After being cultured in either static or fluidic conditions, the cells were washed three times with phosphate buffer saline (PBS) at room temperature (RT), fixed for 15 min with 4% paraformaldehyde (Sigma-Aldrich, St Louis, MO, USA) in PBS, and washed again with PBS (three times). Next, the fixed cells were permeabilized with 0.1% Triton X-100 (Sigma-Aldrich) in PBS for 10 min at RT, washed with PBS (three times), and blocked with 3% bovine serum albumin (BSA) (Sigma-Aldrich) in PBS for 40 min. PBS with 3% BSA was also employed to dilute the antibodies used in this work.

The cells from the monocultures were incubated with Alexa Fluor® 546 Phalloidin (1:40, Life Technologies) to label actin microfilaments and, in the case of samples with μP-antiH, also with goat anti-mouse IgG1 secondary antibody Alexa Fluor® 647 conjugate (1:150. Life Technologies) for 1 h at RT to detect μP-antiH.

To distinguish between D492 and D492HER2 in cocultures, the cells were first incubated with rabbit anti-HER2 monoclonal antibody (1:200, Cell Signaling, Danvers, MA, USA) overnight at 4 °C. Then, the samples were washed three times with PBS and incubated for 2.5 h at RT with Alexa Fluor® 546 Phalloidin to label actin microfilaments to visualize the cell limit and chicken anti-rabbit IgG (H+L) Alexa Fluor® 405 conjugate secondary antibody (1:150, Life Technologies) to label HER2 in the plasma membrane; for cells incubated with μP-antiH, goat anti-mouse IgG1 secondary antibody Alexa Fluor® 647 conjugate was also used.

Finally, the cells were washed three times with PBS and maintained at 4 °C in PBS until evaluation under a confocal laser scanning microscope (CLSM. Olympus, Tokyo, Japan). Orthogonal projections of z-stacks of at least 100 cells for each cell line were evaluated in each replicate. The xyz sequentially acquired images allowed for assessment of whether the particles were inside the cells or attached to their surfaces.

2.6. Statistical Analyses

At least three replicates of each experiment were performed. To compare μP internalization in each experimental condition and for each cell line, ANOVA with post-hoc Tukey HSD test was used; $p < 0.05$ was considered statistically significant.

3. Results

3.1. Microparticles Characterization after Biofunctionalization

Biofunctionalization did not affect the size of μPs, as can be observed in TEM images (Figure 1A). Moreover, biofunctionalization was confirmed in two ways, microscopically and by analyzing the ζ-potential. As expected, under fluorescence microscopy, μP-secAb emitted far-red fluorescence, and μP-antiH emitted green fluorescence after incubation with an Alexa® 488-conjugated secondary antibody (Figure 1B). Biofunctionalization was also confirmed by changes in the μP surface charge. Non-biofunctionalized polystyrene carboxylate μPs (μP-COOH) showed a ζ-potential value of −32.3 mV, whereas μP-secAb and μP-antiH increased their ζ-potential to smaller negative values of −11.23 mV and −11.5 mV, respectively (Figure 1C).

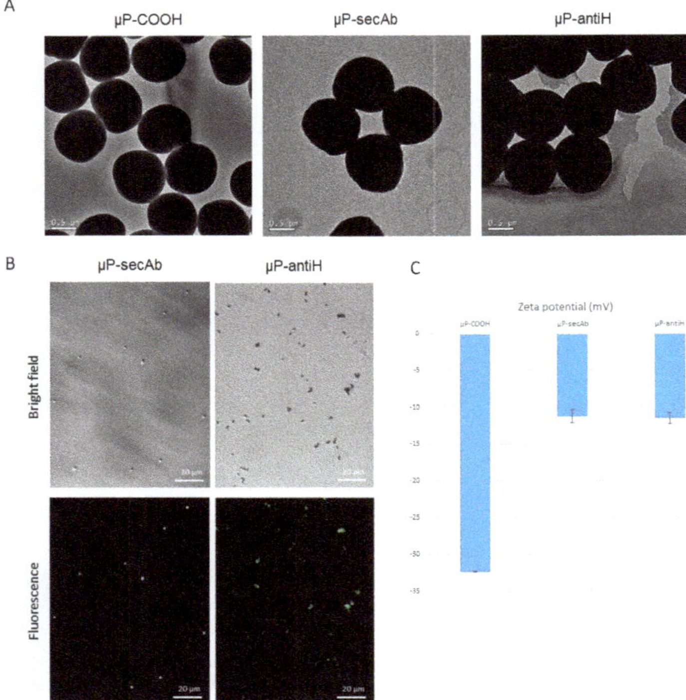

Figure 1. Characterization of microparticles (μP) biofunctionalization. (**A**) Transmission electronic microscopy (TEM) images of microparticles before (COOH) and after biofunctionalization (μP-secAb and μP-antiH). (**B**) Images of microparticles biofunctionalized with a secondary antibody (μP-secAb) or an anti-HER2 antibody (μP-antiH) in bright-field (upper panels) and fluorescence (lower panels) microscopy. (**C**) Zeta potential before (COOH) and after biofunctionalization (μP-secAb and μP-antiH).

3.2. Microparticles Internalization by Cells

The internalization of μPs by cells was evaluated through the orthogonal images captured by a CLSM in both static and fluidic culture conditions (examples in Figures 2 and 3, respectively). Staining of actin filaments was useful to visualize the cell perimeter and, together with the orthogonal projections, allowed us to clearly distinguish between internalized and non-internalized μPs. From these images, the number of cells with at least one internalized μP was scored.

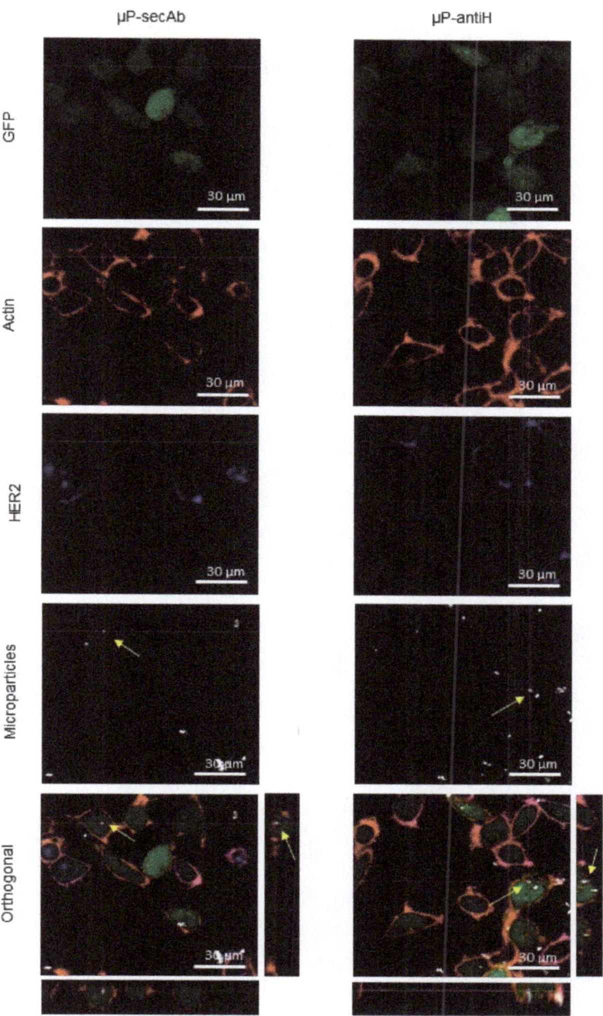

Figure 2. Immunofluorescence analysis by confocal laser scanning microscope (CLSM) of cells cultured in static conditions. Confocal images of D492 and D492HER2 cells cocultured in static conditions and incubated with microparticles biofunctionalized with a non-specific secondary antibody (μP-secAb) or a specific anti-HER2 antibody (μP-antiH). Cells, constitutively expressing green fluorescent protein (GFP, green), were incubated with Alexa Fluor® 546 Phalloidin (red) to label actin microfilaments and Alexa Fluor® 405 conjugate secondary antibody (blue) to label HER2 in the plasma membrane. The arrows point to some examples of μPs located inside the cells.

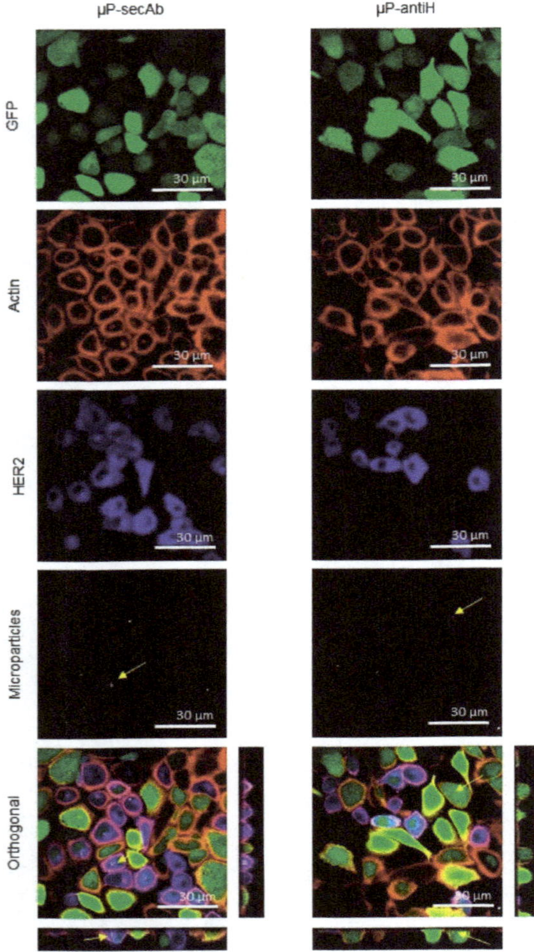

Figure 3. Immunofluorescence analysis by CLSM of cells cultured in fluidic conditions. Confocal images of D492 and D492HER2 cells cocultured in fluidic conditions and incubated with microparticles biofunctionalized with a non-specific secondary antibody (µP-secAb) or a specific anti-HER2 antibody (µP-antiH). The cells, constitutively expressing GFP (green), were incubated with Alexa Fluor® 546 Phalloidin (red) to label actin microfilaments and Alexa Fluor® 405 conjugate secondary antibody (blue) to label HER2 in the plasma membrane. The arrows point to some examples of µPs located inside the cells.

As can be seen in Figure 4, in all conditions, the percentage of D492 cells with internalized microparticles was always higher than that of D492HER2 cells, indicating that D492 cells have an inherent superior capacity to internalize microparticles. Regarding the importance of specific biofunctionalization in µPs recognition and intake by the cells, the internalization related to non-specific binding due to the intrinsic cell endocytic capacity, was represented by the percentage of cells with internalized µPs biofunctionalized with the non-specific antibody (µPs-secAb). In contrast, the internalization related to the specific recognition of µPs by the cells was represented by the increase in the percentage of cells with internalized µPs when these were specifically functionalized (μP-antiH) to recognize a cell membrane receptor (HER2). For both cell lines, the biofunctionalization with a specific targeting antibody (antiH) resulted in higher internalization percentages than the biofunctionalization with a non-specific targeting antibody (secAb). The differences between static monoculture and coculture conditions were not significant. As expected, fluidic culture conditions globally decreased internalization, but again internalization was higher for microparticles which were specifically biofunctionalized (μP-antiH) than for those that were biofunctionalized with a non-specific antibody (μP-secAb). The increase in the percentage of cells with internalized μP-antiH was also higher for cells with HER2 overexpression (D492HER2) than for cells without HER2 overexpression (D492) (Figure 4). Remarkably, the increase in the percentage of cells with internalized μP-antiH with regard to μP-secAb internalization was higher in fluidic conditions than in static conditions, especially in D492HER2 cells (164% versus 77–99% in D492HER2 cells, and 100% versus 22–35% in D492 cells).

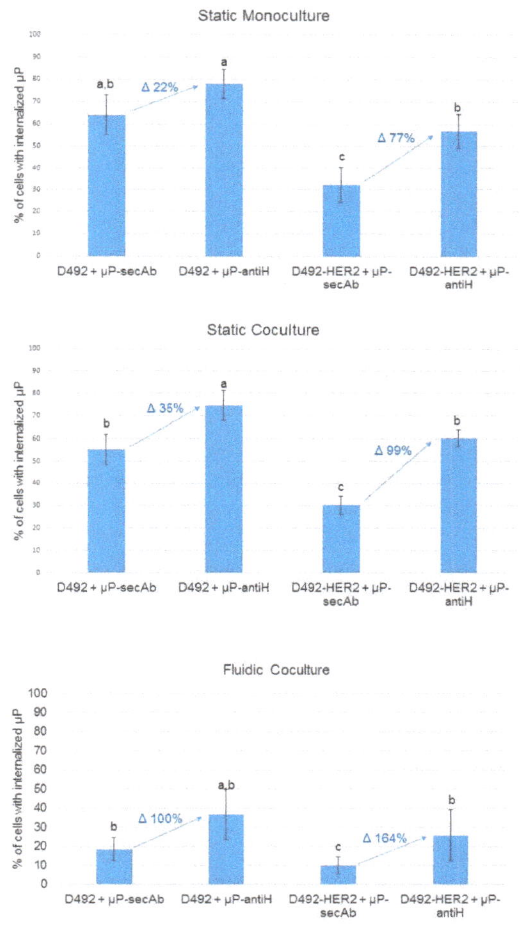

Figure 4. Microparticles internalization by monocultured or cocultured D492 and D492HER2 cells in static and fluidic conditions. Percentages of cells with internalized microparticles biofunctionalized with a non-specific secondary antibody (μP-secAb) or a specific anti-HER2 antibody (μP-antiH). Statistically significant differences are indicated with different letters on top of the bars. An increase of μPs internalization, as a percentage, is indicated in blue together with blue arrows.

4. Discussion

Polystyrene μPs were successfully biofunctionalized with an anti-HER2 antibody or with a secondary antibody, as demonstrated by the detection of fluorescence under a microscope and by changes in their ζ-potential. The reduction in μPs electronegativity could help in their interaction with the plasma membrane, which contains negatively charged saccharides. In fact, it has been described that positively or slightly negatively charged particle surfaces favor cell intake [14,33–35].

The cell lines used in this study are isogenic breast epithelial cell lines. D492 was established by isolating suprabasal cells from a reduction mammoplasty and subsequently immortalizing them using the *E6* and *E7* oncogenes from the human papilloma virus [30,31]. D492 has stem cell properties evaluated by the ability to generate luminal and myoepithelial cells and, in 3D culture, to form branching ductal–alveolar-like structures. D492HER2 was generated by overexpressing the HER2 oncogene in D492 [32]. Interestingly, D492 showed higher percentages of μPs internalization than D492HER2. A possible explanation could be the differences in phenotype between these cells. D492HER2 cells undergo an epithelial to mesenchymal transition (EMT) and express mesenchymal markers, which could reduce their ability to internalize μPs [31,32]. Higher rates of μP internalization have also been described in MCF10A breast epithelial normal cells than in SKBR3 breast epithelial tumoral cells, which also overexpress HER2 [14].

The molecules used to bio-functionalize the μPs also influenced the internalization rates. For both cell lines, the percentage of cells with internalized μP-antiH was higher than that of cells with internalized μP-secAb. This result was expected, since we hypothesized that a specific recognition between the cells and the μPs, such as that afforded by antiH biofunctionalization, could result in an increase in μP internalization. Also as expected, the increase in the percentages of μPs internalization obtained with antiH biofunctionalization was higher for D492HER2 cells, which overexpress HER2, than for D492 cells, which express HER2 at much lower levels [32]. These results clearly indicate that when μPs are biofunctionalized with molecules that specifically interact with plasma membrane receptors, their internalization can be improved, and the level of improvement is related to the number of receptors that are present on the cell surface.

Finally, in relation to the culture conditions, our results indicate that the percentage of cells with internalized μPs clearly decreased in fluidic conditions compared with static conditions. This was probably because the establishment of prior transient contacts between the plasma membrane and the μPs may be more difficult in fluidic conditions or in high shear stress conditions [36,37]. One can hypothesize that in the case of dynamic cultures, specific targeting may be helpful to improve the internalization rates, as it may facilitate the frequency and strength of μPs-cell contacts. Our results agree with this hypothesis because, for both cell lines, the increase of μPs-antiH internalization was higher in fluidic than in static cocultures when compared with μPs-secAb internalization, which opens the door to scale-up studies.

The present results, together with previous research, emphasize the relevance of cell type, specific targeting, and culture conditions such as mono- or cocultures and static or fluidic, to the potential application of NPs and μPs for drug delivery to cancer cells.

5. Conclusions

The present study indicates that, for all culture conditions, μPs biofunctionalized with anti-HER2 antibodies were more efficiently internalized by cells expressing HER2 than μPs biofunctionalized with a non-specific antibody. Moreover, in fluidic culture conditions (simulating the blood stream), the specificity of targeting was still more useful for μP internalization, because the increase in internalization was higher for cells overexpressing HER2. Overall, the results presented emphasize the importance of targeting not only for directing μPs to the appropriate cells but also for achieving reasonable internalization rates, thereby opening the door to the use of microparticles as carriers for drug delivery.

Author Contributions: Conceptualization: E.I., C.N., T.G., and L.B.; Data curation: I.M.-E., J.S., and L.B.; Formal analysis: I.M.-E., J.S., and C.N.; Funding acquisition: C.N.; Investigation: I.M.-E.; Project administration: C.N.; Supervision: L.B.; Validation: E.I., C.N., T.G., and L.B.; Writing—original draft: I.M.-E. and L.B.; Writing—review & editing: I.M.-E., E.I., J.S., C.N., T.G., and L.B.

Funding: This research was funded by the Spanish Ministerio de Ciencia e Innovación (MAT2014-57960-C3-3-R and MAT2017-86357-C3-3-R) and the Generalitat de Catalunya (2017-SGR-503). I.M.E. Thanks to the Spanish Ministerio de Ciencia e Innovación for a predoctoral grant.

Acknowledgments: The authors wish to thank the Servei de Microscòpia at the Universitat Autònoma de Barcelona.

Conflicts of Interest: The authors declare no conflict of interest.

References

1. Sato, K.; Nagaya, T.; Choyke, P.L.; Kobayashi, H. Near infrared photoimmunotherapy in the treatment of pleural disseminated NSCLC: Preclinical experience. *Theranostics* **2015**, *5*, 698–709. [CrossRef]
2. You, H.; Yoon, H.E.; Jeong, P.H.; Ko, H.; Yoon, J.H.; Kim, Y.C. Pheophorbide-a conjugates with cancer-targeting moieties for targeted photodynamic cancer therapy. *Bioorg. Med. Chem.* **2015**, *23*, 1453–1462. [CrossRef] [PubMed]
3. Hamblin, M.R.; Newman, E.L. Photosensitizer targeting in photodynamic therapy I. Conjugates of haematoporphyrin with albumin and transferrin. *J. Photochem. Photobiol. B Biol.* **1994**, *26*, 45–56. [CrossRef]
4. Hamblin, M.R.; Newman, E.L. Photosensitizer targeting in photodynamic therapy II. Conjugates of haematoporphyrin with serum lipoproteins. *J. Photochem. Photobiol. B Biol.* **1994**, *26*, 147–157. [CrossRef]
5. Park, S.Y.; Baik, H.J.; Oh, Y.T.; Oh, K.T.; Youn, Y.S.; Lee, E.S. A smart polysaccharide/drug conjugate for photodynamic therapy. *Angew. Chem. Int. Ed.* **2011**, *50*, 1644–1647. [CrossRef] [PubMed]
6. Koopaei, M.N.; Dinarvand, R.; Amini, M.; Rabbani, H.; Emami, S.; Ostad, S.N.; Atyabi, F. Docetaxel immunonanocarriers as targeted delivery systems for HER 2-positive tumor cells: preparation, characterization, and cytotoxicity studies. *Int. J. Nanomed.* **2011**, *6*, 1903–1912.
7. Zheng, Y.; Yu, B.; Weecharangsan, W.; Piao, L.; Darby, M.; Mao, Y.; Koynova, R.; Yang, X.; Li, H.; Xu, S.; et al. Transferrin-conjugated lipid-coated PLGA nanoparticles for targeted delivery of aromatase inhibitor 7alpha-APTADD to breast cancer cells. *Int. J. Pharm.* **2010**, *390*, 234–241. [CrossRef]
8. Calero, M.; Gutiérrez, L.; Salas, G.; Luengo, Y.; Lázaro, A.; Acedo, P.; Morales, M.P.; Miranda, R.; Villanueva, A. Efficient and safe internalization of magnetic iron oxide nanoparticles: Two fundamental requirements for biomedical applications. *Nanomed. Nanotechnol. Biol. Med.* **2014**, *10*, 733–743. [CrossRef]
9. Yang, Q.; Li, L.; Sun, W.; Zhou, Z.; Huang, Y. Dual stimuli-responsive hybrid polymeric nanoparticles self-assembled from POSS-Based starlike copolymer-drug conjugates for efficient intracellular delivery of hydrophobic drugs. *ACS Appl. Mater. Interfaces* **2016**, *8*, 13251–13261. [CrossRef] [PubMed]
10. Liu, X.; Wu, F.; Tian, Y.; Wu, M.; Zhou, Q.; Jiang, S.; Niu, Z. Size Dependent cellular uptake of Rod-like bionanoparticles with different aspect ratios. *Sci. Rep.* **2016**, *6*, 24567. [CrossRef] [PubMed]
11. Chatterjee, D.K.; Fong, L.S.; Zhang, Y. Nanoparticles in photodynamic therapy: An emerging paradigm. *Adv. Drug Deliv. Rev.* **2008**, *60*, 1627–1637. [CrossRef] [PubMed]
12. Chang, J.-Y.; Wang, S.; Allen, J.S.; Lee, S.H.; Chang, S.T.; Choi, Y.-K.; Friedrich, C.; Choi, C.K. A novel miniature dynamic microfluidic cell culture platform using electro-osmosis diode pumping. *Biomicrofluidics* **2014**, *8*, 044116.
13. Patiño, T.; Nogués, C.; Ibáñez, E.; Barrios, L. Enhancing microparticle internalization by nonphagocytic cells through the use of noncovalently conjugated polyethyleneimine. *Int. J. Nanomed.* **2012**, *7*, 5671.
14. Patiño, T.; Soriano, J.; Barrios, L.; Ibáñez, E.; Nogués, C. Surface modification of microparticles causes differential uptake responses in normal and tumoral human breast epithelial cells. *Sci. Rep.* **2015**, *5*, 11371. [CrossRef] [PubMed]
15. Zauner, W.; Farrow, N.A.; Haines, A.M. In vitro uptake of polystyrene microspheres: Effect of particle size, cell line and cell density. *J. Control. Release* **2001**, *71*, 39–51. [CrossRef]
16. Gratton, S.E.A.; Ropp, P.A.; Pohlhaus, P.D.; Luft, J.C.; Madden, V.J.; Napier, M.E.; DeSimone, J.M. The effect of particle design on cellular internalization pathways. *Proc. Natl. Acad. Sci. USA* **2008**, *105*, 11613–11618. [CrossRef]

17. Kutscher, H.L.; Chao, P.; Deshmukh, M.; Sundara Rajan, S.; Singh, Y.; Hu, P.; Joseph, L.B.; Stein, S.; Laskin, D.L.; Sinko, P.J. Enhanced passive pulmonary targeting and retention of PEGylated rigid microparticles in rats. *Int. J. Pharm.* **2010**, *402*, 64–71. [CrossRef]
18. Tros de Ilarduya, C.; Düzgüneş, N. Delivery of therapeutic nucleic acids via transferrin and transferrin receptors: Lipoplexes and other carriers. *Expert Opin. Drug Deliv.* **2013**, *10*, 1583–1591. [CrossRef]
19. Wong, P.T.; Choi, S.K. Mechanisms and implications of dual-acting methotrexate in folate-targeted nanotherapeutic delivery. *Int. J. Mol. Sci.* **2015**, *16*, 1772–1790. [CrossRef]
20. Stuchinskaya, T.; Moreno, M.; Cook, M.J.; Edwards, D.R.; Russell, D. A Targeted photodynamic therapy of breast cancer cells using antibody-phthalocyanine-gold nanoparticle conjugates. *Photochem. Photobiol. Sci.* **2011**, *10*, 822–831. [CrossRef]
21. Awada, G.; Gombos, A.; Aftimos, P.; Awada, A. Emerging drugs targeting human epidermal growth factor receptor 2 (Her2) in the treatment of breast cancer. *Expert Opin. Emerg. Drugs* **2016**, *21*, 91–101. [CrossRef] [PubMed]
22. Slamon, D.; Clark, G.; Wong, S.; Levin, W.; Ullrich, A.; McGuire, W. Human breast cancer: Correlation of relapse and survival with amplification of the HER-2/neu oncogene. *Science* **1987**, *235*, 177–182. [CrossRef] [PubMed]
23. Slamon, D.; Godolphin, W.; Jones, L.; Holt, J.; Wong, S.; Keith, D.; Levin, W.; Stuart, S.; Udove, J.; Ullrich, A.; et al. Studies of the HER-2/neu proto-oncogene in human breast and ovarian cancer. *Science* **1989**, *244*, 707–712. [CrossRef] [PubMed]
24. Lewis Phillips, G.D.; Li, G.; Dugger, D.L.; Crocker, L.M.; Parsons, K.L.; Mai, E.; Blättler, W.A.; Lambert, J.M.; Chari, R.V.J.; Lutz, R.J.; et al. Targeting HER2-positive breast cancer with trastuzumab-DM1, an antibody-cytotoxic drug conjugate. *Cancer Res.* **2008**, *68*, 9280–9290. [CrossRef] [PubMed]
25. Calibasi Kocal, G.; Güven, S.; Foygel, K.; Goldman, A.; Chen, P.; Sengupta, S.; Paulmurugan, R.; Baskin, Y.; Demirci, U. Dynamic Microenvironment induces phenotypic plasticity of esophageal cancer cells under flow. *Sci. Rep.* **2016**, *6*, 38221. [CrossRef] [PubMed]
26. Carvalho, R.M.; Maia, F.R.; Silva-Correia, J.; Costa, B.M.; Reis, R.L.; Oliveira, J.M. A semiautomated microfluidic platform for real-time investigation of nanoparticles' cellular uptake and cancer cells' tracking. *Nanomedicine* **2017**, *12*, 581. [CrossRef] [PubMed]
27. Damiati, S.; Kompella, U.B.; Damiati, S.A.; Kodzius, R. Microfluidic devices for drug delivery systems and drug screening. *Genes* **2018**, *9*, 103. [CrossRef] [PubMed]
28. Tobar, N.; Guerrero, J.; Smith, P.C.; Martínez, J. NOX4-dependent ROS production by stromal mammary cells modulates epithelial MCF-7 cell migration. *Br. J. Cancer* **2010**, *103*, 1040–1047. [CrossRef]
29. Arrigoni, C.; Bersini, S.; Gilardi, M.; Moretti, M. In vitro co-culture models of breast cancer metastatic progression towards bone. *Int. J. Mol. Sci.* **2016**, *17*, 1405. [CrossRef]
30. Gudjonsson, T.; Nielsen, H.L.; Rønnov-jessen, L.; Bissell, M.J.; Petersen, O.W. Isolation, immortalization and characterization of a human breast epithelial cell line with stem cell properties. *Genes Dev.* **2002**, 693–706. [CrossRef]
31. Sigurdsson, V.; Hilmarsdottir, B.; Sigmundsdottir, H.; Fridriksdottir, A.J.R.; Ringnér, M.; Villadsen, R.; Borg, A.; Agnarsson, B.A.; Petersen, O.W.; Magnusson, M.K.; et al. Endothelial induced EMT in breast epithelial cells with stem cell properties. *PLoS ONE* **2011**, *6*, e23833. [CrossRef]
32. Ingthorsson, S.; Andersen, K.; Hilmarsdottir, B.; Maelandsmo, G.M.; Magnusson, M.K.; Gudjonsson, T. HER2 induced EMT and tumorigenicity in breast epithelial progenitor cells is inhibited by coexpression of EGFR. *Oncogene* **2016**, *35*, 4244–4255. [CrossRef]
33. Chithrani, B.D.; Chan, W.C.W. Elucidating the mechanism of cellular uptake and removal of protein-coated gold nanoparticles of different sizes and shapes. *Nano Lett.* **2007**, *7*, 1542–1550. [CrossRef] [PubMed]
34. Dausend, J.; Musyanovych, A.; Dass, M.; Walther, P.; Schrezenmeier, H.; Landfester, K.; Mailänder, V. Uptake mechanism of oppositely charged fluorescent nanoparticles in HeLa Cells. *Macromol. Biosci.* **2008**, *8*, 1135–1143. [CrossRef] [PubMed]
35. Fröhlich, E. The role of surface charge in cellular uptake and cytotoxicity of medical nanoparticles. *Int. J. Nanomed.* **2012**, *7*, 5577. [CrossRef] [PubMed]

36. Farokhzad, O.C.; Khademhosseini, A.; Jon, S.; Hermmann, A.; Cheng, J.; Chin, C.; Kiselyuk, A.; Teply, B.; Eng, G.; Langer, R. Microfluidic System for studying the interaction of nanoparticles and microparticles with cells. *Anal. Chem.* **2005**, *77*, 5453–5459. [CrossRef]
37. Strobl, F.G.; Breyer, D.; Link, P.; Torrano, A.A.; Bräuchle, C.; Schneider, M.F.; Wixforth, A. A surface acoustic wave-driven micropump for particle uptake investigation under physiological conditions in very small volumes. *Beilstein J. Nanothechnol.* **2015**, *6*, 414–419. [CrossRef] [PubMed]

Article

The Process–Property–Performance Relationship of Medicated Nanoparticles Prepared by Modified Coaxial Electrospraying

Weidong Huang [1,2], Yuan Hou [3], Xinyi Lu [3], Ziyun Gong [3], Yaoyao Yang [3], Xiao-Ju Lu [1,*], Xian-Li Liu [2,*] and Deng-Guang Yu [3,*]

1 School of Chemistry and Chemical Engineering, Hubei Polytechnic University, Huangshi 435003, China; neweydong@hbpu.edu.cn

2 Hubei Key Laboratory of Mine Environmental Pollution Control and Remediation, School of Environmental Science and Engineering, Hubei Polytechnic University, Huangshi 435003, China

3 School of Materials Science and Engineering, University of Shanghai for Science and Technology, Shanghai 200093, China; 1626410104@st.usst.edu.cn (Y.H.); 1626418101@st.usst.edu.cn (X.L.); 1626410115@st.usst.edu.cn (Z.G.); yyyang@usst.edu.cn (Y.Y.)

* Correspondence: luxiaoju@hbpu.edu.cn (X.-J.L.); liuxianli@hbpu.edu.cn (X.-L.L.); ydg017@usst.edu.cn (D.-G.Y.); Tel.: +86-714-634-8814 (X.-J.L.); +86-714-636-8937 (X.-L.L.); +86-21-5527-0632 (D.-G.Y.)

Received: 5 April 2019; Accepted: 7 May 2019; Published: 10 May 2019

Abstract: In pharmaceutical nanotechnology, the intentional manipulation of working processes to fabricate nanoproducts with suitable properties for achieving the desired functional performances is highly sought after. The following paper aims to detail how a modified coaxial electrospraying has been developed to create ibuprofen-loaded hydroxypropyl methylcellulose nanoparticles for improving the drug dissolution rate. During the working processes, a key parameter, i.e., the spreading angle of atomization region (θ, °), could provide a linkage among the working process, the property of generated nanoparticles and their functional performance. Compared with the applied voltage (V, kV; $D = 2713 - 82V$ with ${R_{\theta V}}^2 = 0.9623$), θ could provide a better correlation with the diameter of resultant nanoparticles (D, nm; $D = 1096 - 5\theta$ with ${R_{D\theta}}^2 = 0.9905$), suggesting a usefulness of accurately predicting the nanoparticle diameter. The drug released from the electrosprayed nanoparticles involved both erosion and diffusion mechanisms. A univariate quadratic equation between the time of releasing 95% of the loaded drug (t, min) and D ($t = 38.7 + 0.097D - 4.838 \times 10^5 D^2$ with a R^2 value of 0.9976) suggests that the nanoparticle diameter has a profound influence on the drug release performance. The clear process-property-performance relationship should be useful for optimizing the electrospraying process, and in turn for achieving the desired medicated nanoparticles.

Keywords: coaxial electrospraying; polymeric nanoparticles; spreading angle; process-property-performance relationship

1. Introduction

Today, nanomaterials play one of the most important roles in the research and development of modern pharmaceutics [1–3]. New material processing procedures [4–8], combined with different kinds of raw materials [9–13] and novel innovative strategies for constructing functional products [14–18], are frequently introduced into this application field for providing efficacious drug delivery and enhancing the therapeutic effects of active pharmaceutical ingredients (APIs). Among them, electrohydrodynamic atomization (EHDA) is a popular technique for creating nanoproducts, which mainly includes electrospraying and electrospinning. These new methods explore electrical energy to atomize the working fluid for producing solid products at micro or nano scale [19–23].

The past two decades have witnessed the rapid progress of electrosprayed nanoparticles being utilized as functional products in a wide variety of fields [24–30]. In pharmaceutics, further developments of medicated electrosprayed nanoparticles include creating complex nanostructures (just as its counterpart electrospinning [31–36]), production on large scales, potential clinical applications, and commercial products [37,38]. However, the electro–hydro–dynamic working process is still far from being understood, due to the overlap of several disciplines such as fluid mechanics, electric dynamics, and polymer rheology during the extremely fast drying processes of electrospraying [39,40]. Even a purposeful and conscious manipulation of the electrosprayed nanoparticle's diameter is very hard to realize.

Shown in Figure 1 is a diagram about the single-fluid electrospraying process and the possible experimental parameters that can exert significant influences on the diameters of resultant nanoparticles. An electrospraying apparatus brings together the working fluid and electrostatic energy at the convergent point, i.e., the nozzle of spraying head. Between the two electrodes consisting of spraying head and collector, the working fluids are atomized and solidified into particles within several decades of microseconds. Based on this, all the experimental parameters can be divided into three categories which are concluded in Figure 1. Correspondingly, the resultant nanoparticles' diameter (D) can be a function of working fluid's property (w), operation conditions (o), and environmental parameters (e), i.e., $D = f$ (w,o,e).

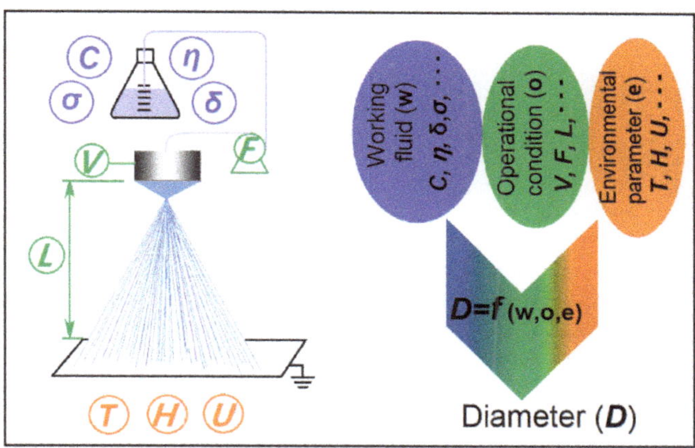

Figure 1. A diagram of the single-fluid electrospraying process and the experimental parameters exerting influence on the diameter of resultant nanoparticle.

During the past several decades, numerous publications have investigated the influence of particular parameters on electrosprayed products. These articles disclosed the process-property relationship for intentionally manipulating the particles' diameters, morphology, and surface smoothness [41–43]. However, there are too many parameters that can exert significant influence on the final products during the electrospraying processes [44,45]. For example, the properties of working fluid include polymer concentration (C), viscosity (η), surface tension (δ), and also conductivity (σ). The operational parameters include the applied voltage (V), the fluid flow rate (F), and also the particle collected distance (L). The environmental conditions include temperature (T), humidity (H), possible vacuum (U), and sometime with hot air blowing.

Thus, it is difficult to manipulate the diameter of final nanoparticles accurately through particular experimental parameters. In contrast, the parameters of the electrospraying itself seem to be neglected. Compared with the experimental parameters that can be controlled directly by researchers, very few publications have reported uncontrollable parameters in relation to working processes, such as the size and angle of a Taylor cone, the length of straight fluid jets, and the spreading angle of the atomization region.

Based on the above-mentioned knowledge, here for the first time, we have investigated the influence of spreading an angle of the atomization region on the diameter of resultant nanoparticles. Meanwhile, the influence of applied voltage on the spreading angle and nanoparticles' diameters, and the size of medicated nanoparticle on the drug fast release performance were also studied. Thus, an example about how to disclose the process-property-performance relationship of medicated nanoparticles prepared by modified coaxial electrospraying is showed. In the experiments, ibuprofen (IBU) and hydroxypropyl methylcellulose (HPMC) were selected as the model drug and polymer matrix, respectively. IBU, a typical nonsteroidal anti-inflammatory drug, is broadly exploited to treat pain, fever and inflammation. However, its poor water solubility always limits its fast action for achieving a desired therapeutic effect [46–48]. HPMC is commonly utilized as an excipient in oral tablet, eye drops, and capsule formations. It can be used both as a delaying agent for controlled release, as well as an enhancer to improve the soluble rate of a soluble drug [49,50].

2. Materials and Methods

2.1. Materials

IBU was provided by Wuhan Anruike Biological Pharmaceutical Co., Ltd. (Wuhan, China). HPMC powders (2910 5cps, M_n = 428,000 g/mol, methoxy content = 28.0–30.0%, hydroxypropoxy content = 7.5–12%) were obtained from Shandong Fine Chemical Co., Ltd. (Jinan, China). Ethanol and dichloromethane (DCM) were purchased from Shanghai Lingfeng Chemical Testing Co. Ltd. (Shanghai, China). All other chemicals are analytical reagents, and water was distilled twice before use.

2.2. Modified Coaxial Electrospraying

A solidifiable solution consisting of 7% (*w*/*v*) HPMC and 3% (*w*/*v*) IBU in a mixture of ethanol and DCM (1:1, *v*:*v*) was prepared and utilized as the core fluid. Pure solvent ethanol was used as the shell fluid. Four nanoparticles referred to as P1, P2, P3, and P4 were prepared at an applied voltage of 16, 17, 18, and 19 kV, respectively. For all preparations, the particle-collected distance was fixed at 15 cm. The shell and core fluid flow rates were 0.2 and 0.8 mL/h, respectively. The electrospraying processes were recorded using a digital camera (PowerShot A640, Tokyo, Japan).

2.3. Morphology of the Prepared Nanoparticles

The surface morphological characterization of the prepared nanoparticles was observed under scanning electron microscopy (SEM; Quanta FEG450, FEI Corporation, Hillsboro, OR, USA) at 20 kV of accelerated voltage. Before the observation, the samples were sputter-coated with gold under vacuum. The images were analyzed by ImageJ software with the measuring of over 100 different nanoparticles and their diameters presented as mean ± S.D.

2.4. Drug Fast Release Performance

An amount of 20 mg medicated nanoparticles was placed in 50 mL of phosphate buffer solution (PBS) with a pH value of 7.0. The buffer solutions including samples were incubated in a shaking bath at 37 ± 0.1 °C and an agitation speed of 50 rpm. Each type of sample was repeated 6 times. At predetermined time intervals, 1 mL of sample solutions were withdrawn and replaced with 1 mL fresh medium. The amount of IBU released from the nanofibers was measured using a UV-Vis Spectrophotometer (UV-2102PC, Unico Instrument Co. Ltd., Shanghai, China) by measuring the absorbance at 264 nm. The calibration curve was obtained, and all concentrations were evaluated in percentage as mean ± S.D. using Equation (1).

$$Accumulative\ release\ (\%) = \frac{Amount\ of\ drug\ release}{Amount\ of\ initial\ drug} \times 100 \quad (1)$$

3. Results and Discussion

3.1. Preparations of the Medicated Particles Using the Modified Coaxial Electrospraying

A diagram about the modified coaxial electrospraying process is shown in Figure 2a. Right from the first publication about coaxial electrospraying [25], all the shell fluids must be solidifiable to ensure the creation of core-shell nano-/micro-structures. However, this limitation was broken by Yu and his co-worker. They reported a modified coaxial process, in which fluids without solidifiable properties using a single-fluid electrospraying process can be introduced into the coaxial process and was explored as the shell working fluids [43–45]. This new process greatly expanded the capability of electrospraying as it was deemed capable of generating additional kinds of nanostructures, including core-shell solid structure, nanocoating and homogeneous particles with a high quality (Figure 2a).

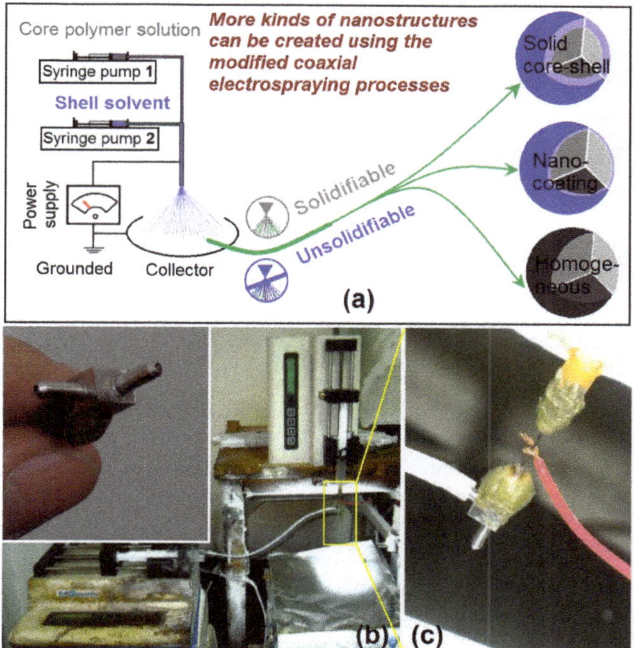

Figure 2. The modified coaxial electrospraying: (**a**) A diagram about the process, by which many kinds of nanostructures can be created through the unsolidifiable shell fluids; (**b**) implementation of the modified coaxial electrospraying, the upper-left inset shows the home-made concentric spraying head; (**c**) the connection of power supply and working fluid with the spraying head.

The organization of electrospraying systems and the implementation of an electrospraying process are exhibited in Figure 2b. A home-made concentric spraying head was exploited to implement the modified coaxial electrospraying (the upper-left inset of Figure 2b), which can also be utilized to conduct coaxial electrospinning [49]. The inner metal capillary has an inner and outer diameter of 0.3 mm and 0.6 mm, respectively. The orifice of the outer capillary has a diameter of 1.2 mm.

Two syringe pumps (KDS100 and KDS 200, Cole-Parmer®, Vernon Hills, IL, USA) were employed to drive the core and shell liquids to the spraying head (Figure 2b). A copper line was directly attached on the concentric spraying head to convey the electrostatic energy to the working fluids (Figure 2c).

During the EHDA process, the applied voltage is one of the most important parameters that have significant influence on the final products. As for the working fluids systems, there are often suitable working ranges for almost all the experimental parameters. In the present systems, when the applied voltages were changed from 16 kV to 19 kV, all the coaxial electrospraying processes were run continuously, smoothly and robustly. With an enlarged shooting, the images of Taylor cones are shown in Figure 3a–d. The estimated spreading angles of three measurements are (42 ± 3)°, (70 ± 6)°, (83 ± 5)°, and (92 ± 4)°, for an applied voltage of 16, 17, 18 and 19 kV, respectively. As the voltage elevated, the spreading angles increased correspondingly. For reference, to achieve the clearest images these photos were taken under a certain tilted angle, however the camera lens should always keep the same height as the nozzle of a spraying head.

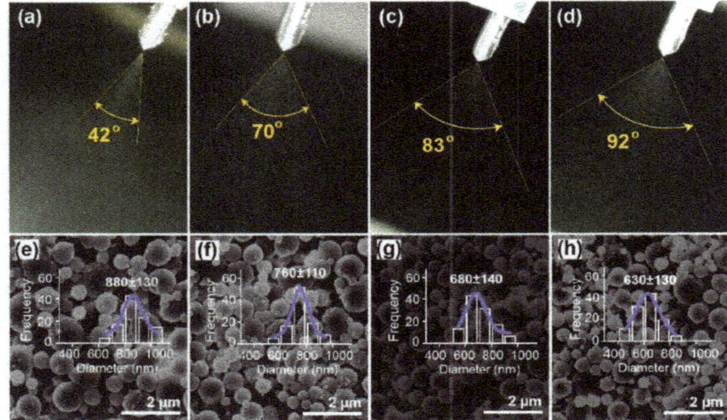

Figure 3. (**a–d**) The typical changes of spreading angles with the elevation of applied voltages (kV) from 16, to 17, 18, and 19, respectively ($n = 3$); (**e–h**) SEM images of the resultant nanoparticles and their diameter distributions (**e**) P1; (**f**) P2; (**g**) P3; (**h**) P4.

The SEM images of the prepared nanoparticles and their diameter distributions are shown in Figure 3e–h. Nanoparticles P1, P2, P3, and P4 have an estimated diameter of 880 ± 130, 760 ± 110, 680 ± 140, and 630 ± 130 nm, respectively. All the nanoparticles have a round surface with some satellites. As the applied voltages increased, the diameters decreased correspondingly.

3.2. The Process–Property Relationship and the Related Mechanism

Shown in Figure 4a is the influence of applied voltage on the spreading angle. The trend is clear that the spreading angle increased as the applied voltage elevated. A linear regression suggests these two parameters have a relationship of $\theta = 17V - 229$, with a correlation coefficient ${R_{\theta V}}^2 = 0.9623$. The applied voltage is an operational parameter that can be manipulated directly by researchers. The spreading angle is a process parameter, however it cannot be directly manipulated by certain operational parameters in relation to the working fluid's property. Although these two different kinds of parameters have a positive correlation, the spreading angle should also receive the influence of other parameters. It should be the normal oscillation of other operational parameters (such as the shell and core fluids' flow rates and the ambient conditions) that make the correlation coefficient vary from a number of one.

Figure 4b demonstrated the influence of applied voltage on the diameter of nanoparticles. A negative correlation is clear in that the nanoparticles' diameter decreased as the applied voltage elevated. A linear regression suggests these two parameters have a relationship of $D = 2713 - 82V$, with a correlation coefficient $R_{DV}{}^2 = 0.9624$. Diameter is one of the most important parameters of the medicated nanoparticles. This equation suggests that a certain relationship exists, which can then be explored for manipulating the size of final products. This strategy has been frequently demonstrated in literature [15]. However, as seen in the relationship between spreading angle and applied voltage, the relationship between nanoparticles' diameter and applied voltage is seemingly influenced by other parameters. Their normal oscillations would make the manipulation effect of final products' size using the applied voltage often unsatisfactory.

Although the applied voltage did not have a strong linear relationship with the spreading angle and the nanoparticles' diameter, the spreading angle had a better linear relationship with the size of nanoparticles, which is shown in Figure 4c. A linear equation can be regressed as $D = 1096 - 5\theta$, with a correlation coefficient $R_{D\theta}{}^2 = 0.9905$. The highly linear correlation between these two parameters is able to be anticipated. This is because any changes of almost all experimental parameters (including those about working fluids' property, those about operational conditions and those about the environment) will equally exert their influences on both the atomization processes and the successive solid particles' properties, represented by the spreading angle and the particles' diameter.

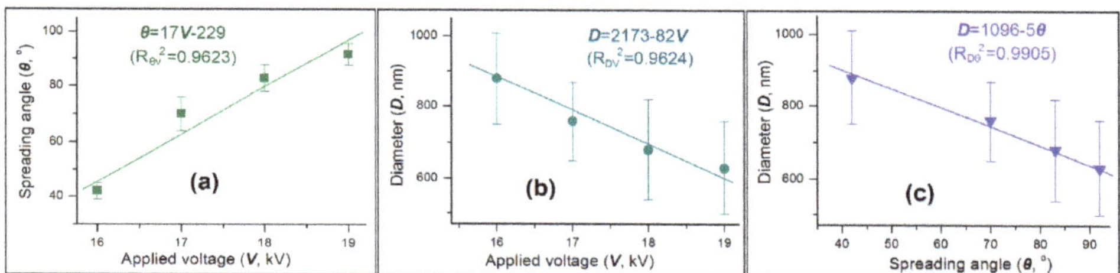

Figure 4. The applied voltage-spreading angle-nanoparticle diameter relationships: (**a**) The influence of applied voltage on the spreading angle; (**b**) the influence of applied voltage on the nanoparticles' diameter; (**c**) the accurate relationship between the spreading angle and the nanoparticles' diameter.

Despite the number of fluids an electrospraying process treats simultaneously, they will typically experience five steps, as originally demonstrated by single-fluid blending electrospraying. Shown in Figure 5, the five steps included; the fluid being charged, the formation of the Taylor cone, the convergent point of straight fluid jet, the atomization region, and the final collection of solid nanoparticles. The key intermediate steps comprise the three stages during which the working fluids are dried into solid particles. The third stage, i.e., the atomization region, often determines the solidification effects and the quality of the final solid products. Although many publications have investigated the formation of Taylor cones and the initiation of an EHDA process [49], little attention has been payed to the atomization process and which can be described by the spreading angle (θ).

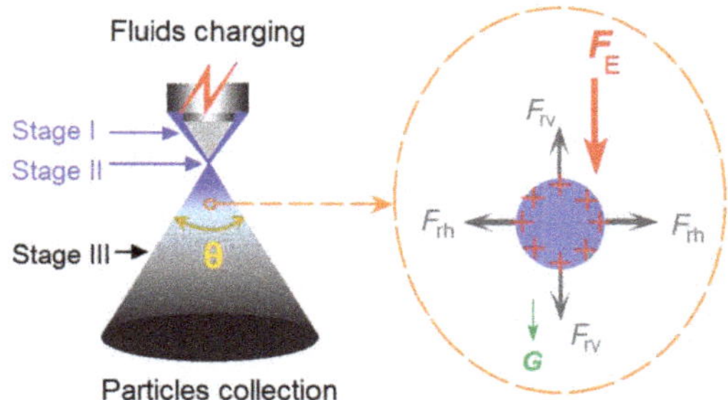

Figure 5. A diagram about the coaxial electrospraying and the force analysis of a charged droplet.

During the atomization process, the nascent mono-disperse fluid droplets formed by the Columbic explosion should be quickly split and shrunk owing to the accumulation of surface charges and the rapid evaporation of solvents resulting from the huge surface areas. For a certain droplet, the force analysis is shown in Figure 5. The droplet should have repelling forces from all sides, for example the vertical direction repelling forces F_{rv} and the horizontal direction repelling forces F_{rh}. Meanwhile, the droplet should receive the force (F_E) between the two electrodes and gravity (G), which pushed the droplets/nanoparticles from the nozzle of the spraying head to the collector during the solidification process. When the applied voltages elevated, the droplets should have more charges, and the repelling forces should be increased correspondingly. Within a fixed collected distance, the increase of horizontal direction repelling forces F_{rh} should expand the atomization region and increase the spreading angle automatically. Thus, the larger applied voltage provided, the bigger the spreading angle the atomization region had. Meanwhile, the increase of applied voltage should promote the Columbic explosion and the successive splitting of droplets, resulting in nanoparticles with smaller diameters. Both the spreading angle and the nanoparticles' diameter are the direct results of the applied voltage and are similarly influenced by the fluctuations of a series of other parameters. Thus, these two parameters have a very high correlation.

Additionally, from a standpoint of force analysis, the spreading angle is a combined effect of a series of forces under the electrical field, giving a hint that it can be a useful tool for accurately predicting the resultant nanoparticle's diameter. This discovery is very important because new methods can now be exploited to accurately predict the size of final products and give researchers more power in respect to manipulating the electrospraying process. As far as the measurement of spreading angle is concerned, a High Frequency Camera (HFC) can be utilized to record the working process of electrospraying and should be able to improve our present work.

3.3. The property-Performance Relationship and the Related Mechanism

Shown in Figure 6a is the drug in vitro release profiles from the four types of medicated HPMC nanoparticles. Although all of them were able to release over 50% of the contained drug at the first minute after they were placed into the dissolution media (52.7 ± 5.1%, 61.3 ± 4.7%, 72.1 ± 5.6%, and 82.4 ± 4.3% for nanoparticles P1, P2, P3, and P4, respectively), a trend soon became clear. The finer the particles were, the faster the loaded drug was exhausted from the nanoparticles. The times needed for releasing 95% of a drug were 8.82, 6.88, 4.45 and 3.04 min for nanoparticles P1, P2, P3, and P4, respectively.

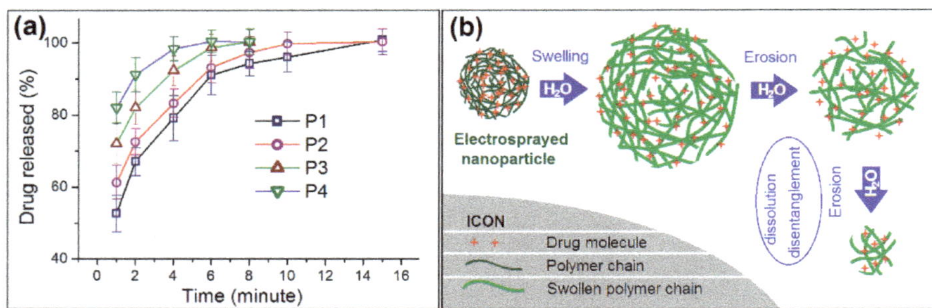

Figure 6. (**a**) The in vitro dissolution tests of the electrosprayed nanoparticles; (**b**) a schematic of the drug erosion mechanism from the medicated nanoparticles.

Improving the release rate and apparent solubility of poorly water-soluble drugs is always one of the major challenges in the fields of pharmaceutics and medicated nanomaterials. Traditionally, comminution of the drug powders with hydrophilic polymers such as HPMC is frequently performed to reduce the drug particle size. However, the drug fast release performance is often limited (such as

IBU-HPMC xerogel granule [51]) and the preparation process is time-consuming (such as IBU-HPMC nano-suspension [52]). In comparison, the present IBU-HPMC nanoparticles were able to provide a better performance about promoting the fast dissolution of IBU and could be generated using a simple and straightforward one-step process.

HPMC is a soluble and hydrophilic polymer. During the dissolution process, the increase of diameter has both positive and negative influence on the loaded drug molecules to free into the dissolution media. Shown as Figure 6b is a diagram about the drug release mechanism. In the electrosprayed nanoparticles, the drug molecules are homogeneously distributed all over them due to the extremely fast drying process of electrospraying. When these medicated nanoparticles are placed into water, they will absorb water to swell gradually. This is a process that the water molecules penetrate in the solid nanoparticles. Meanwhile, the drug molecules should leave the polymer chains and go into the penetrated water, and further diffuse outward to the bulk solution due to the concentration gradient. This is a typical drug diffusion mechanism. As the swelling goes forward, the outer layer HMPC molecules will free into the bulk solution themselves, together with the contained and penetrated drug molecules. Thus, the erosion mechanism also happens here.

As the increase of nanoparticles' diameters, on one hand, the penetration distance of water and diffusion distance of drug molecules should all increase correspondingly. This is to say the increase of diameter will prolong the drug release time period due to the diffusion mechanism. On the other hand, there is a competition between the swelling rate and the dissolution rate of HPMC in the gradually enlarged particles in water, the larger particles may accelerate the drug release through the easy erosion of the outer layer of the swollen particles to shorten the drug release time period. Thus, the two factors co-act on the drug release, and show a univariate quadratic equation relationship between the diameter of particle with release time apparently, as indicated in Figure 7. To fit the time needed to release 95% of the loaded drug (t, min) with the nanoparticles' diameter (D, nm), a relationship between them can be found, i.e., $t = 38.7 + 0.097D - 4.838 \times 10^5 D^2$ ($R^2 = 0.9976$). This univariate quadratic equation with a R^2 value of 0.9976 suggests that the diameter of nanoparticles has a profound influence on the drug release performance. As hinted by the primary power ($0.097D$), the increase of diameter will prolong the drug release time. However, the increase of diameter may also shorten the drug release time, as suggested by the quadratic term ($-4.838 \times 10^5 D^2$). This should have a close relationship with the property of drug loaded polymer HPMC and also the density of resultant nanoparticles.

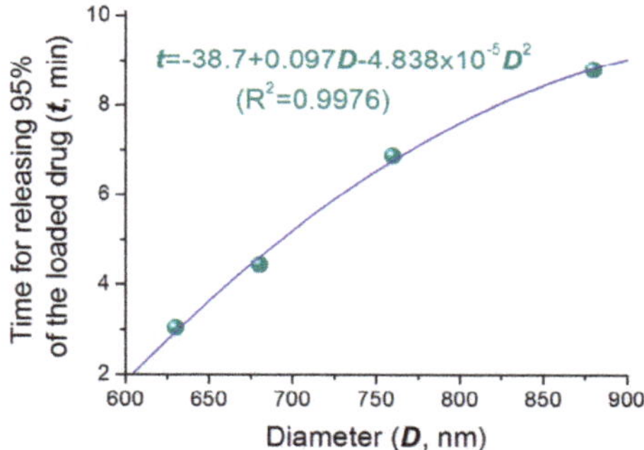

Figure 7. The relationship between the size of medicated HPMC nanoparticles and the time for releasing 95% of the loaded drug.

Poorly water-soluble drugs are one of the most difficult and long-existing issues in pharmaceutics [53–55]. Nanotechnologies have brought new lights on resolving this problem. However, how to take advantage of these advanced techniques comprises a challenge to the researchers. The present work shows a fine example to build clear process-property-performance relationship for exploring the modified coaxial electrospraying to create medicated nanoparticles. These nanoparticles can be further transferred into capsules or tablets for potential oral administration.

4. Conclusions and Perspectives

In the present work, the influences of spreading the angle of the atomization region (θ) during a modified coaxial electrospraying process on the resultant nanoparticles' diameter, and in turn on the drug dissolution rate from the prepared IBU-loaded HPMC nanoparticles were systematically investigated. With θ as a key parameter, a series of process-property-performance relationships were found. These relationships include $\theta = 17V - 229$ ($R_{\theta V}{}^2 = 0.9623$), $D = 2713 - 82V$ ($R_{DV}{}^2 = 0.9624$), $D = 1096 - 5\theta$ ($R_{D\theta}{}^2 = 0.9905$), and $t = 38.7 + 0.097D - 4.838 \times 10^5 D^2$ ($R^2 = 0.9976$). Compared with the applied voltage (V), θ could provide a better correlation with the diameter of resultant nanoparticles (D), suggesting its usefulness for accurately predicting the nanoproducts' size.

Today, electrospraying, is fast developing from treating a single fluid (mainly creating homogeneous particles [56,57]), to the treatment of double fluids (such as coaxial and side-by-side processes for generating core-shell and Janus particles [58–60]), and to the simultaneous treatment of three or even more working fluids [61,62]. In contrast, it is drawing increasing attention for producing particles on a larger-scale, just as its peer electrospinning [63,64]. Regardless of the different development directions of this advanced technique, how to build an accurate relationship between the experimental conditions and the final products' quality always poses a big challenge to the researchers. Here, a proof-of-concept method is shown that the spreading angle, as a process parameter, can be explored to predict the resultant nanoparticles' diameters in an accurate manner. Along this way, many new possibilities can be anticipated. For example, the precise control of the diameters of core-shell, Janus, and tri-layer nanoparticles and the size of their internal compartments. Particularly for fabrication of electrosprayed nanoparticles on a large scale, the process parameters (including those for characterizing Taylor cone and the atomization region) should be useful tools for elaborately manipulating the nanoproducts' quality and functional performances.

Author Contributions: Conceptualization, W.H., X.-J.L. and D.-G.Y.; experiments, W.H., X.L., Z.G., Y.H. and Y.Y.; writing—original draft preparation, W.H., and Y.Y.; writing—review and editing, all authors; supervision, D.-G.Y. and X.-L.L.; project administration, D.-G.Y. and X.-J.L.

Funding: The National Natural Science Foundation of China (No. 51803121 & 51402099), the Technical Innovation Project (Major Project) of Hubei, China (2016ACA176) and USST college student innovation projects (SH10252194 & 10252324/330) are appreciated.

Conflicts of Interest: The authors declare no conflicts of interest.

References

1. Hubbell, J.A.; Chikoti, A. Nanomaterials for Drug Delivery. *Science* **2012**, *337*, 303–305. [CrossRef] [PubMed]
2. Farokhzad, O.C. Nanotechnology for drug delivery: The perfect partnership. *Expert Opin. Drug Deliv.* **2008**, *5*, 927–929. [CrossRef] [PubMed]
3. Mitragotri, S.; Burke, P.; Langer, R. Overcoming the challenges in administering biopharmaceutical drugs: Formulation and delivery strategies. *Nat. Rev. Drug Discov.* **2014**, *13*, 655–672. [CrossRef] [PubMed]
4. Cheung, K.; Das, D.B. Microneedles for drug delivery: Trends and progress. *Drug Deliv.* **2016**, *23*, 2338–2354. [CrossRef]
5. Nayak, A.; Babla, H.; Han, T.; Das, D.B. Lidocaine carboxymethylcellulose with gelatine co-polymer hydrogel delivery by combined microneedle and ultrasound. *Drug Deliv.* **2016**, *23*, 658–669. [CrossRef]
6. Shen, Y.; Li, X.; Le, Y. Amorphous nanoparticulate formulation of sirolimus and its tablets. *Pharmaceutics* **2018**, *10*, 155. [CrossRef] [PubMed]
7. Yang, C.; Yu, D.G.; Pan, D.; Liu, X.K.; Wang, X.; Bligh, S.W.A.; Williams, G.R. Electrospun pH-sensitive core-shell polymer nanocomposites fabricated using a tri-axial processes. *Acta Biomater.* **2016**, *35*, 77–86. [CrossRef]

8. Mehta, P.; Haj-Ahmad, R.; Rasekh, M.; Arshad, M.S.; Smith, A.; van der Merwe, S.M.; Li, X.; Chang, M.W.; Ahmad, Z. Pharmaceutical biomaterial engineering via electrohydrodynamic atomization technologies. *Drug Discov. Today* **2017**, *22*, 157–165. [CrossRef]
9. Kumar, P.S.; Venkatesh, K.; Gui, E.L.; Jayaraman, S.; Singh, G.; Arthanareeswaran, G. Electrospun carbon nanofibers/TiO_2-PAN hybrid membranes for effective removal of metal ions and cationic dye. *Environ. Nanotech. Monit. Manag.* **2018**, *10*, 366–376. [CrossRef]
10. Mabrouk, M.; Chejara, D.R.; Mulla, J.A.S.; Badhe, R.V.; Choonara, Y.E.; Kumar, P.; du Toit, L.C.; Pillay, V. Design of a novel crosslinked HEC-PAA porous hydrogel composite for dissolution rate and solubility enhancement of efavirenz. *Int. J. Pharm.* **2015**, *490*, 429–437. [CrossRef] [PubMed]
11. Beherei, H.H.; Shaltout, A.A.; Mabrouk, M.; Abdelwahed, N.A.; Das, D.B. Influence of niobium pentoxide particulates on the properties of brushite/gelatin/alginate membranes. *J. Pharm. Sci.* **2018**, *107*, 1361–1371. [CrossRef] [PubMed]
12. Bareras-Urbina, C.G.; Ramírez-Wong, B.; López-Ahumada, G.A.; Burruel-Ibarra, S.E.; Martínez-Cruz, O.; Tapia-Hernández, J.A.; Rodriguez Felix, F. Nano-and micro-particles by nanoprecipitation: Possible application in the food and agricultural industries. *Int. J. Food Prop.* **2016**, *19*, 1912–1923. [CrossRef]
13. Chuysinuan, P.; Pengsuk, C.; Lirdprapamongkol, K.; Techasakul, S.; Svasti, J.; Nooeaid, P. Enhanced structural stability and controlled drug release of hydrophilic antibiotic-loaded alginate/soy protein isolate core-sheath fibers for tissue engineering applications. *Fiber. Polym.* **2019**, *20*, 1–10. [CrossRef]
14. Chakraborty, S.; Liao, I.C.; Adler, A.; Leong, K.W. Electrohydrodynamics: A facile technique to fabricate drug delivery systems. *Adv. Drug Deliver. Rev.* **2009**, *61*, 1043–1054. [CrossRef]
15. Bock, N.; Dargaville, T.R.; Woodruff, M.A. Electrospraying of polymers with therapeutic molecules: State of the art. *Prog. Polym. Sci.* **2012**, *37*, 1510–1551. [CrossRef]
16. Kamaraj, S.; Palanisamy, U.M.; Mohamed, M.S.B.K.; Gangasalam, A.; Maria, G.A.; Kandasamy, R. Curcumin drug delivery by vanillin-chitosan coated with calcium ferrite hybrid nanoparticles as carrier. *Eur. J. Pharm. Sci.* **2018**, *116*, 48–60. [CrossRef]
17. Mabrouk, M.; Kumar, P.; Choonara, Y.; du Toit, L.; Pillay, V. Artificial, triple-layered, nanomembranous wound patch for potential diabetic foot ulcer intervention. *Materials* **2018**, *11*, 2128. [CrossRef]
18. Mabrouk, M.; Beherei, H.H.; ElShebiney, S.; Tanaka, M. Newly developed controlled release subcutaneous formulation for tramadol hydrochloride. *Saudi Pharm. J.* **2018**, *26*, 585–592. [CrossRef]
19. Vasa, P.; Demuth, B.; Hirsch, E.; Nagy, B.; Andersen, S.K.; Vigh, T.; Verreck, G.; Csontos, I.; Nagy, Z.K.; Marosi, G. Drying technology strategies for colon-targeted oral delivery of biopharmaceuticals. *J. Control. Release* **2019**, *296*, 162. [CrossRef]
20. Liao, Y.; Loh, C.H.; Tian, M.; Wang, R.; Fane, A.G. Progress in electrospun polymeric nanofibrous membranes for water treatment: Fabrication, modification and applications. *Prog. Polym. Sci.* **2018**, *77*, 69. [CrossRef]
21. Wang, S.; Cao, X.; Shen, M.; Guo, R.; Bányai, I.; Shi, X. Fabrication and morphology control of electrospun poly(γ-glutamic acid) nanofibers for biomedical applications. *Colloid Surf. B-Biointerfaces* **2012**, *89*, 254–264. [CrossRef] [PubMed]
22. Tapia-Hernández, J.A.; Del-Toro-Sánchez, C.L.; Cinco-Moroyoqui, F.J.; Ruiz-Cruz, S.; Juárez, J.; Castro-Enrıquez, D.D.; Barreras-Urbina, C.G.; López-Ahumada, G.A.; Rodríguez-Felix, F. Gallic acid-loaded zein nanoparticles by electrospraying process. *J. Food Sci.* **2019**, *84*, 818–831. [CrossRef]
23. Tapia-Hernández, J.A.; Rodríguez-Félix, D.E.; Plascencia-Jatomea, M.; Rascón-Chu, A.; López-Ahumada, G.A.; Ruiz-Cruz, S.; Barreras-Urbina, C.G.; Rodríguez-Félix, F. Porous wheat gluten microparticles obtained by electrospray: Preparation and characterization. *Adv. Polym. Technol.* **2018**, *37*, 2314–2324. [CrossRef]
24. Nhat Nguyen, D.; Clasen, C.; Van den Mooter, G. Pharmaceutical applications of electrospraying. *J. Pharm. Sci.* **2016**, *105*, 2601–2620. [CrossRef]
25. Loscertales, I.G.; Barrero, A.; Guerrero, I.; Cortijo, R.; Marquez, M.; Ganan-Calvo, A.M. Micro/nano encapsulation via electrified coaxial liquid jets. *Science* **2002**, *295*, 1695–1698. [CrossRef]
26. Parhizkar, M.; Reardon, P.J.T.; Knowles, J.C.; Browning, R.J.; Stride, E.; Pedley, R.B.; Grego, I.; Edirisinghe, M. Performance of novel high throughput multi electrospray systems for forming of polymeric micro/nanoparticles. *Mater. Des.* **2017**, *126*, 73–84. [CrossRef]
27. Li, J.J.; Yang, Y.Y.; Yu, D.G.; Du, Q.; Yang, X.L. Fast dissolving drug delivery membrane based on the ultra-thin shell of electrospun core-shell nanofibers. *Eur. J. Pharm. Sci.* **2018**, *122*, 195–204. [CrossRef]

28. Liu, Z.P.; Zhang, L.L.; Yang, Y.Y.; Wu, D.; Jiang, G.; Yu, D.G. Preparing composite nanoparticles for immediate drug release by modifying electrohydrodynamic interfaces during electrospraying. *Powder Technol.* **2018**, *327*, 179–187. [CrossRef]
29. Huang, W.; Yang, Y.; Zhao, B.; Liang, G.; Liu, S.; Liu, X.L.; Yu, D.G. Fast dissolving of ferulic acid via electrospun ternary amorphous composites produced by a coaxial process. *Pharmaceutics* **2018**, *10*, 115. [CrossRef]
30. Kadivar, N.; Tavanai, H.; Allafchian, A. Fabrication of cellulose nanoparticles through electrospraying. *IET Nanobiotechnol.* **2018**, *12*, 807–813. [CrossRef]
31. Liu, X.; Yang, Y.; Yu, D.G.; Zhu, M.J.; Zhao, M.; Williams, G.R. Tunable zero-order drug delivery systems created by modified triaxial electrospinning. *Chem. Eng. J.* **2019**, *356*, 886–894. [CrossRef]
32. Yang, Y.; Li, W.; Yu, D.G.; Wang, G.; Williams, G.R.; Zhang, Z. Tunable drug release from nanofibers coated with blank cellulose acetate layers fabricated using tri-axial electrospinning. *Carbohydr. Polym.* **2019**, *203*, 228–237. [CrossRef]
33. Hai, T.; Wan, X.; Yu, D.G.; Wang, K.; Yang, Y.; Liu, Z.P. Electrospun lipid-coated medicated nanocomposites for an improved drug sustained-release profile. *Mater. Des.* **2019**, *162*, 70–79. [CrossRef]
34. Wu, Y.H.; Yang, C.; Li, X.Y.; Zhu, J.Y.; Yu, D.G. Medicated nanofibers fabricated using NaCl solutions as shell fluids in a modified coaxial electrospinning. *J. Nanomater.* **2016**, *2016*, 8970213. [CrossRef]
35. Wu, Y.H.; Li, H.P.; Shi, X.X.; Wan, J.; Liu, Y.F.; Yu, D.G. Effective utilization of the electrostatic repulsion for improved alignment of electrospun nanofibers. *J. Nanomater.* **2016**, *2016*, 2067383. [CrossRef]
36. Yew, C.; Azari, P.; Choi, J.; Muhamad, F.; Pingguan-Murphy, B. Electrospun Polycaprolactone Nanofibers as a Reaction Membrane for Lateral Flow Assay. *Polymers* **2018**, *10*, 1387. [CrossRef]
37. Yu, D.G.; Li, J.J.; Williams, G.R.; Zhao, M. Electrospun amorphous solid dispersions of poorly water-soluble drugs: A review. *J. Control. Release* **2018**, *292*, 91–110. [CrossRef] [PubMed]
38. Mao, Z.; Li, J.; Huang, W.; Jiang, H.; Zimba, B.L.; Chen, L.; Wan, J.L.; Wu, Q. Preparation of poly (lactic acid)/graphene oxide nanofiber membranes with different structures by electrospinning for drug delivery. *RSC Adv.* **2018**, *8*, 16619. [CrossRef]
39. Wang, S.; Hu, F.; Li, J.; Zhang, S.; Shen, M.; Huang, M.; Shi, X. Design of electrospun nanofibrous mats for osteogenic differentiation of mesenchymal stem cells. *Nanomedicine* **2019**, *147*, 2505–2520. [CrossRef]
40. Wang, S.; Zhu, J.; Shen, M.; Zhu, M.; Shi, X. Poly(amidoamine) dendrimer-enabled simultaneous stabilization and functionalization of electrospun poly(γ-glutamic acid) nanofibers. *ACS Appl. Mater. Interfaces* **2014**, *6*, 2153–2161. [CrossRef]
41. Shams, T.; Parhizkar, M.; Illangakoon, U.E.; Orlu, M.; Edirisinghe, M. Core/shell microencapsulation of indomethacin/paracetamol by co-axial electrohydrodynamic atomization. *Mater. Des.* **2017**, *136*, 204–213. [CrossRef]
42. Tapia-Hernandez, J.A.; Torres-Chavez, P.I.; Ramirez-Wong, B.; Rascon-Chu, A.; Plascencia-Jatomea, M.; Barreras-Urbina, C.G.; Rangel-Vázquez, N.A.; Rodriguez-Felix, F. Micro-and nanoparticles by electrospray: Advances and applications in foods. *J. Agric. Food Chem.* **2015**, *63*, 4699–4707. [CrossRef] [PubMed]
43. Yu, D.G.; Zheng, X.L.; Yang, Y.; Li, X.Y.; Williams, G.R.; Zhao, M. Immediate release of helicid from nanoparticles produced by modified coaxial electrospraying. *Appl. Surf. Sci.* **2019**, *473*, 148–155. [CrossRef]
44. Li, X.Y.; Zheng, Z.B.; Yu, D.G.; Liu, X.K.; Qu, Y.L.; Li, H.L. Electrosprayed sperical ethylcellulose nanoparticles for an improved sustained-release profile of anticancer drug. *Cellulose* **2017**, *24*, 5551–5564. [CrossRef]
45. Wang, K.; Wen, H.F.; Yu, D.G.; Yang, Y.; Zhang, D.F. Electrosprayed hydrophilic nanocomposites coated with shellac for colon-specific delayed drug delivery. *Mater. Des.* **2018**, *143*, 248–255. [CrossRef]
46. Mantas, A.; Mihranyan, A. Immediate-release nifedipine binary dry powder mixtures with nanocellulose featuring enhanced solubility and dissolution rate. *Pharmaceutics* **2019**, *11*, 37. [CrossRef] [PubMed]
47. Marzoli, F.; Marianecci, C.; Rinaldi, F.; Passeri, D.; Rossi, M.; Minosi, P.; Carafa, M.; Pieretti, S. Long-lasting, antinociceptive effects of pH-sensitive niosomes loaded with ibuprofen in acute and chronic models of pain. *Pharmaceutics* **2019**, *11*, 62. [CrossRef]
48. Han, F.; Zhang, W.; Wang, Y.; Xi, Z.; Chen, L.; Li, S.; Xu, L. Applying supercritical fluid technology to prepare ibuprofen solid dispersions with improved oral bioavailability. *Pharmaceutics* **2019**, *11*, 67. [CrossRef]
49. Wang, Q.; Yu, D.G.; Zhang, L.L.; Liu, X.K.; Deng, Y.C.; Zhao, M. Electrospun hypromellose-based hydrophilic composites for rapid dissolution of poorly water-soluble drug. *Carbohydr. Polym.* **2017**, *174*, 617–625. [CrossRef]

50. Adrover, A.; Varani, G.; Paolicelli, P.; Petralito, S.; Di Muzio, L.; Casadei, M.; Tho, I. Experimental and modeling study of drug release from HPMC-based erodible oral thin films. *Pharmaceutics* **2018**, *10*, 222. [CrossRef]
51. Nakayama, S.; Ihara, K.; Senna, M. Structure and properties of ibuprofen–hydroxypropyl methylcellulose nanocomposite gel. *Powder Technol.* **2009**, *190*, 221–224. [CrossRef]
52. Plakkot, S.; De Matas, M.; York, P.; Saunders, M.; Sulaiman, B. Comminution of ibuprofen to produce nano-particles for rapid dissolution. *Int. J. Pharm.* **2011**, *415*, 307–314. [CrossRef] [PubMed]
53. Caparica, R.; Júlio, A.; Baby, A.; Araújo, M.; Fernandes, A.; Costa, J.; Santos de Almeida, T. Choline-amino acid ionic liquids as green functional excipients to enhance drug solubility. *Pharmaceutics* **2018**, *10*, 288. [CrossRef] [PubMed]
54. Zhang, X.; Xing, H.; Zhao, Y.; Ma, Z. Pharmaceutical dispersion techniques for dissolution and bioavailability enhancement of poorly water-soluble drugs. *Pharmaceutics* **2018**, *10*, 74. [CrossRef]
55. Bhakay, A.; Rahman, M.; Dave, R.; Bilgili, E. Bioavailability enhancement of poorly water-soluble drugs via nanocomposites: Formulation–Processing aspects and challenges. *Pharmaceutics* **2018**, *10*, 86. [CrossRef]
56. Gao, Y.; Bai, Y.; Zhao, D.; Chang, M.W.; Ahmad, Z.; Li, J.S. Tuning microparticle porosity during single needle electrospraying synthesis via a non-solvent-based physicochemical approach. *Polymers* **2015**, *7*, 2701–2710. [CrossRef]
57. Yao, Z.C.; Jin, L.J.; Ahmad, Z.; Huang, J.; Chang, M.W.; Li, J.S. Ganoderma lucidum polysaccharide loaded sodium alginate micro-particles prepared via electrospraying in controlled deposition environments. *Int. J. Pharm.* **2017**, *524*, 148–158. [CrossRef] [PubMed]
58. Zhang, C.; Yao, Z.C.; Ding, Q.; Choi, J.J.; Ahmad, Z.; Chang, M.W.; Li, J.S. Tri-needle coaxial electrospray engineering of magnetic polymer yolk–shell particles possessing dual-imaging modality, multiagent compartments, and trigger release potential. *ACS Appl. Mater. Inter.* **2017**, *9*, 21485–21495. [CrossRef] [PubMed]
59. Sanchez-Vazquez, B.; Amaral, A.J.; Yu, D.G.; Pasparakis, G.; Williams, G.R. Electrosprayed Janus particles for combined photo-chemotherapy. *AAPS PharmSciTech* **2017**, *18*, 1460–1468. [CrossRef] [PubMed]
60. Zhang, C.; Chang, M.W.; Li, Y.; Qi, Y.; Wu, J.; Ahmad, Z.; Li, J.S. Janus particle synthesis via aligned non-concentric angular nozzles and electrohydrodynamic co-flow for tunable drug release. *RSC Adv.* **2016**, *6*, 77174–77178. [CrossRef]
61. Zhang, C.; Li, Y.; Hu, Y.; Peng, Y.; Ahmad, Z.; Li, J.S.; Chang, M.W. Porous yolk-shell particle engineering via nonsolvent assisted tri-needle co-axial electrospraying for burn related wound healing. *ACS Appl. Mater. Inter.* **2019**, *11*, 8–7823.
62. Xing, Z.; Zhang, C.; Zhao, C.; Ahmad, Z.; Li, J.S.; Chang, M.W. Targeting oxidative stress using tri-needle electrospray engineered Ganoderma lucidum polysaccharide-loaded porous yolk-shell particles. *Eur. J. Pharm. Sci.* **2018**, *125*, 64–73. [CrossRef] [PubMed]
63. Haj-Ahmad, R.; Rasekh, M.; Nazari, K.; Onaiwu, E.V.; Yousef, B.; Morgan, S.; Evans, D.; Chang, M.W.; Hall, J.; Samwell, C.; et al. Stable increased formulation atomization using a multi-tip nozzle device. *Drug Del. Transl. Res.* **2018**, *8*, 1815–1827. [CrossRef] [PubMed]
64. Zhang, C.; Gao, C.; Chang, M.W.; Ahmad, Z.; Li, J.S. Continuous micron-scaled rope engineering using a rotating multi-nozzle electrospinning emitter. *Appl. Phys. Lett.* **2016**, *109*, 151903. [CrossRef]

Article

Nanoemulsion Based Vehicle for Effective Ocular Delivery of Moxifloxacin Using Experimental Design and Pharmacokinetic Study in Rabbits

Jigar Shah [1], Anroop B. Nair [2,*], Shery Jacob [3], Rakesh K. Patel [4], Hiral Shah [5], Tamer M. Shehata [2,6] and Mohamed Aly Morsy [2,7]

1. Department of Pharmaceutics, Institute of Pharmacy, Nirma University, Ahmedabad, Gujarat 382481, India; jigsh12@gmail.com
2. Department of Pharmaceutical Sciences, College of Clinical Pharmacy, King Faisal University, Al-Ahsa 31982, Saudi Arabia; momorsy@kfu.edu.sa
3. Department of Pharmaceutical Sciences, College of Pharmacy, Gulf Medical University, Ajman 4184, UAE; sheryjacob6876@gmail.com
4. Shree S.K. Patel College of Pharmaceutical Education and Research, Kherva, Ganpat Vidyanagar, Mehsana, Gujarat 384012, India; rakesh.patel@ganpatuniversity.ac.in
5. Arihant School of Pharmacy & BRI, Gandhinagar, Gujarat 382421, India; jigshir123@gmail.com
6. Department of Pharmaceutics and Industrial Pharmacy, Faculty of Pharmacy, University of Zagazig, Zagazig 44519, Egypt; tamershehata@zu.edu.eg
7. Department of Pharmacology, Faculty of Medicine, Minia University, El-Minia 61511, Egypt

* Correspondence: anair@kfu.edu.sa; Tel.: +966-536-219-868

Received: 13 April 2019; Accepted: 8 May 2019; Published: 11 May 2019

Abstract: Nanoemulsion is one of the potential drug delivery strategies used in topical ocular therapy. The purpose of this study was to design and optimize a nanoemulsion-based system to improve therapeutic efficacy of moxifloxacin in ophthalmic delivery. Moxifloxacin nanoemulsions were prepared by testing their solubility in oil, surfactants, and cosurfactants. A pseudoternary phase diagram was constructed by titration technique and nanoemulsions were obtained with four component mixtures of Tween 80, Soluphor® P, ethyl oleate and water. An experiment with simplex lattice design was conducted to assess the influence of formulation parameters in seven nanoemulsion formulations (MM1–MM7) containing moxifloxacin. Physicochemical characteristics and in vitro release of MM1–MM7 were examined and optimized formulation (MM3) was further evaluated for ex vivo permeation, antimicrobial activity, ocular irritation and stability. Drug pharmacokinetics in rabbit aqueous humor was assessed for MM3 and compared with conventional commercial eye drop formulation (control). MM3 exhibited complete drug release in 3 h by Higuchi diffusion controlled mechanism. Corneal steady state flux of MM3 (~32.01 μg/cm^2/h) and control (~31.53 μg/cm^2/h) were comparable. Ocular irritation study indicated good tolerance of MM3 and its safety for ophthalmic use. No significant changes were observed in the physicochemical properties of MM3 when stored in the refrigerator for 3 months. The greater aqueous humor concentration (C_{max}; 555.73 ± 133.34 ng/mL) and delayed T_{max} value (2 h) observed in MM3 suggest a reduced dosing frequency and increased therapeutic efficacy relative to control. The area under the aqueous humor concentration versus time curve ($AUC_{0–8\,h}$) of MM3 (1859.76 ± 424.51 ng·h/mL) was ~2 fold higher ($p < 0.0005$) than the control, suggesting a significant improvement in aqueous humor bioavailability. Our findings suggest that optimized nanoemulsion (MM3) enhanced the therapeutic effect of moxifloxacin and can therefore be used as a safe and effective delivery vehicle for ophthalmic therapy.

Keywords: nanoemulsion; mixture design; aqueous humor; antimicrobial activity

---

1. Introduction

The ocular drug delivery system is one of the most attractive and challenging drug delivery systems for pharmaceutical scientists [1]. Conventional eye drops account for more than 90% of the available ophthalmic formulations but the efficiency of these products is limited by transient residence time, low permeability of corneal epithelium, rapid pre-corneal loss and high tear fluid turnover [2]. However, less than 5% of the drugs contained in eye drops penetrates the corneal membrane and reaches the intraocular tissues; and the remaining dose usually undergoes trans-conjunctival absorption or trans-nasal absorption or drainage via the nasolacrimal duct. Consequently, extensive research has been conducted to improve the effectiveness of topical ocular therapy by developing drug delivery systems which can increase ocular retention, improve trans-corneal drug absorption and reduce systemic adverse effects while retaining the ease of application and benefit of eye drops. Various drug delivery strategies such as bioadhesive hydrogels, temperature or pH-sensitive in situ gel forming systems, collagen shields, colloidal dosage forms like microparticles, microemulsions, nanoparticles, nanosuspensions, nanoemulsions, liposomes, niosomes, nanomicelles, and dendrimers have been formulated and evaluated to partially or fully achieve these objectives [3]. Among various delivery approaches, the colloidal systems received greater attention due to their potential to improve corneal penetration, greater retention at ocular surface, as well as ease of administration similar to eye drop solutions [4]. Despite the fact that the colloidal ocular drug delivery systems like liposomes and niosomes offer certain improvement over conventional liquid dosage forms, the major limitations which have limited their prospects include their inherent tendency to aggregate, the instability and leakage of entrapped drug have limited their future prospect [5,6].

Nanoemulsions are thermodynamically stable and optically transparent fine dispersion (10–200 nm) of multi-component fluids, and frequently consist of an aqueous phase, an oily phase, a primary surfactant as an emulsifying agent, a cosurfactant generally an alkanol of intermediate chain length and occasionally an electrolyte [7]. The major benefits of this colloidal dispersion include enhancement in ocular retention and extended duration, reduced drug protein binding, improved corneal drug permeation, sustained drug release, reduced systemic adverse effects and ease of use for the incorporation of both hydrophilic and lipophilic drugs. In addition, nanoemulsions can interact with the lipid layer of the tear film, can stay in the conjunctival sac for longer times, and subsequently act as a drug depot [8]. The potential of nanoemulsions as a promising alternative for conventional eye formulation in treating various ocular diseases of both the anterior and posterior ocular segments has been described in literature [4]. In this context, moxifloxacin, a fourth-generation fluoroquinolone antibiotic is commercially available as an ophthalmic solution (0.5% *w*/*v*) and is used for the treatment of bacterial conjunctivitis or other bacterial infections of the eyes. Moxifloxacin acts by inhibiting the DNA gyrase and topoisomerase IV required for bacterial DNA replication, repair, and recombination. It was reported that moxifloxacin has enhanced efficacy, safety and tolerance in comparison to older fluoroquinolone derivatives [9]. However, the conventional therapy of this drug in ocular therapy is limited by short residence time. Hence, encapsulating moxifloxacin in droplets that form a nanoemulsion could be an alternative for its ophthalmic use. The objective of this investigation was to optimize the moxifloxacin-loaded nanoemulsion system, characterize and compare the in vivo ocular efficacy with the commercial eye drop. A pseudoternary phase diagram of four component mixtures was constructed by titration technique. Selected nanoemulsions were characterized for pH, droplets size, polydispersity index, zeta potential, conductivity, viscosity, drug content, and dilution potential. The optimized nanoemulsion (MM3) was further evaluated for corneal permeation, antimicrobial effect, ocular irritation and in vivo drug pharmacokinetics in the aqueous humor of rabbits.

2. Materials and Methods

2.1. Materials

Moxifloxacin (molecular weight of 437.9 Da) was received as an in-kind gift from Zydus Cadila Ltd., Ahmedabad, India with a purity of 99.99%. Soluphor® P was generously supplied by BASF, Ludwigshafen, Germany. Tween 80 was purchased from S.D Fine Chem, Mumbai, India and ethyl oleate was purchased from Central Drug House Pvt. Ltd., Mumbai, India. All other chemicals used were of analytical grade.

2.2. Drug Analysis

The amount of moxifloxacin was determined by the high-performance liquid chromatography (HPLC) method described in the literature [10]. A HPLC system (1525, Waters, Milford, MA, USA) consisting of a Discovery C18 column (250 mm × 4.6 mm, i.d, 5 μm) was used. Chromatographic separation was carried out using a mixture of methanol: acetonitrile: water (85:5:10, *v*/*v*/*v*), pH 2.75 adjusted with phosphoric acid was delivered at a flow rate of 1 mL min^{-1} with an injection volume of 25 μL. The isocratic elution was done at 25 °C and monitored using UV detector at 290 nm.

2.3. Development of Pseudoternary Phase Diagram

The pseudoternary phase diagram of four component mixtures of oil (ethyl oleate), surfactant (Tween 80), cosurfactant (Soluphor P) and water was constructed by titration technique to obtain concentration ranges that can result in large existence with the nanoemulsion region at room temperature [11]. A titration technique was employed for the preparation of the pseudoternary phase diagrams [12]. Briefly, Tween 80 was blended with Soluphor P in a fixed weight ratio of 1:1, 1:1.5, 1:2, 1:2.5, 1:3, 1:4 and 1:5, respectively. Aliquots of each Tween 80 and cosurfactant mixture (S_{mix}) were then mixed with oil phase at room temperature (25 °C). The ratio of oil to S_{mix} was varied as 9:1, 8:2, 7:3, 6:4, 5:5, 4:6, 7:3, 2:8 and 1:9. Then, water was added to the above mixture in 5% increments and checked for formation of emulsion or liquid crystal or gel. The phases were identified using visual inspection, microscopic inspection, and measurement of droplets size. The resulting nanoemulsion was tightly sealed and stored at ambient temperature. The points corresponding to mixture components that resulted in nanoemulsion were noted and marked in a phase diagram region. The resulted phase diagrams of different surfactant: Cosurfactant ratios were compared and the ratio that could result in the large existence area of nanoemulsion domain in phase diagram was selected for further optimization.

2.4. Preparation of Moxifloxacin Nanoemulsion

Moxifloxacin nanoemulsions (o/w) were formulated by appropriate quantity of deionized water, ethyl oleate, Tween 80 and Soluphor P. Required amount of moxifloxacin (0.5%, *w*/*v*) was dissolved in ethyl oleate. To S_{mix}, an adequate amount of ethyl oleate was added and mixed. Then the mixture was titrated by gradual addition and mixing of distilled water in order to achieve the equilibrium immediately. Further, it was thoroughly stirred and vortexed to obtain nanoemulsion. To maintain sterility of the prepared formulation, benzalkonium chloride (0.005%, *w*/*w*) was used as a preservative and stored in multi-dose containers [13].

2.5. Experimental Design

The objective of modeling the phase diagrams is to quantify the effect of formulation composition on the droplets size. A successful "mixture design" shows the statistical approach to obtain the relationship between the droplets size distribution and the amounts of various components. In this method, the pseudoternary phase diagrams were plotted and several points were selected within the nanoemulsion region for particle size measurement. The points (black dots) represent the nanoemulsion region in Figure 1A. Within the nanoemulsion region, a triangular area was arbitrarily selected and marked with a red triangle (Figure 1A) which shows different runs of mixture design and the enlarged version is represented in Figure 1B. Hence, Figure 1B shows the sketch diagram of Run (trial) 1 to Run 7 of mixture design. The constraint that the proportions of different components must sum to 100% should be satisfied. According to Figure 1B, the points can be chosen such as three vertexes, three halfway points between vertices and the center point. Each vertex represents a formulation containing the maximum quantity of one component, with the other two components at a minimum amount. The halfway point between the two vertices illustrates a formulation incorporating the average of the minimum and maximum quantity of the two constituents represented by the respective apex. The centre point shows a formulation containing one third of the individual component. A total of seven formulations (MM1–MM7) were opted from the nanoemulsion region for further study and their compositions are summarized in Table 1.

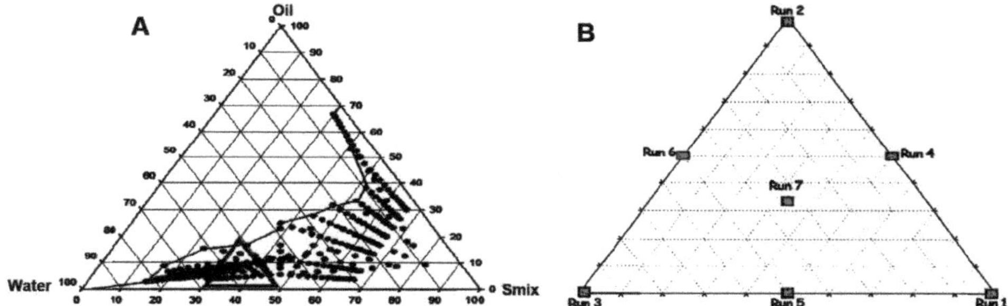

Figure 1. Pseudo ternary phase diagram showing nanoemulsion region (**A**) and distribution for each of run in a mixture design (**B**).

Table 1. Formulation composition used for preparing nanoemulsions after applying simplex lattice design.

Formulations	Run	Formulation Components			Transformed Proportion		
		S_{mix} (%)	Oil (%)	Water (%)	Smix	Oil	Water
MM1	1	52	4	44	1	0	0
MM2	2	36	20	44	0	1	0
MM3	3	36	4	60	0	0	1
MM4	4	44	12	44	0.5	0.5	0
MM5	5	44	4	52	0.5	0	0.5
MM6	6	36	12	52	0	0.5	0.5
MM7	7	41.33	9.33	49.33	0.33	0.33	0.33
MM8 *	8 *	37	18	45	0.063	0.875	0.063

* Check point batch.

2.6. Characterization of Moxifloxacin Nanoemulsions

2.6.1. Drug Content and pH

Drug content in the prepared nanoemulsions (MM1–MM7) was determined using HPLC. The pH of formulations (MM1–MM7) were measured by a calibrated pH meter (Mettler Toledo MP-220, Greifensee, Switzerland).

2.6.2. Transmittance, Conductivity and Dilution Potential

The percentage of the transmittance of nanoemulsions was measured using colorimeter (Photoelectric Colorimeter 113, Systronics, Ahmedabad, India). Percentage transmission was set to zero using filter and 100% using transparent cuvette filled with water. Then, different nanoemulsion samples were kept in the transparent cuvette and percentage transmission was measured. Electrical conductivity of nanoemulsions was studied using a conductometer to determine the type. Briefly, an electrode was totally immersed and fixed in the nanoemulsion (20 mL) and the temperature was raised to 1 °C/min steadily. The nanoemulsion was agitated with a stirrer, and the change in the conductivity was recorded. To determine dilution potential, the nanoemulsion was diluted 10 times with continuous media and the occurrence of phase separation was noted.

2.6.3. Particle Size Characterization and Zeta Potential

The droplets size, size distribution and polydispersity index of nanoemulsions were analyzed employing a dynamic light scattering technique using Malvern Zetasizer (Nano ZS90, Malvern Instruments, Malvern, UK) at 25 °C. In brief, a few drops of respective samples were added directly to a polystyrene disposable cuvette and fixed in the direction of laser light beam. The scattered light signal was measured with a detector placed at a right angle and the droplets size were determined based on the physical properties of the scattered light such as the angular distribution, frequency shift, the polarization and the intensity of the light [14]. For the zeta potential measurement, samples were diluted with deionized water and the electrophoretic mobility values was determined at 25 °C using the software DTS, version 4.1 (Malvern, England, UK, 2009).

2.6.4. Viscosity

Viscosity of nanoemulsions was measured at different angular velocities at a temperature of 25 °C using the Brookfield synchro-Lectric viscometer (LVDVI prime, Middleborough, MA, USA).

2.7. Transmission Electron Microscopy (TEM)

Structural morphology of nanoemulsion droplets was investigated with TEM (Tecnai 20, Philips, Holland, OR, USA) operated at 200 kV and of a 0.15 nm efficient in comprehensive resolution. The bright field imaging technique with high magnification and diffraction modes was used to examine the morphology and structure of the nanoemulsion globules [15]. TEM imaging was carried out by staining a drop of the nanoemulsion with phosphotungstic acid solution (2% *w*/*v*) and directly placing on the copper grids, subsequently dried at room temperature (25 ± 2 °C).

2.8. In Vitro Release

The drug release studies were performed employing a vertical Franz diffusion cell having an effective surface area of 1.13 cm^2 and simulated tear fluid (pH 7.4) as the dissolution medium [16]. Briefly, 1 mL of nanoemulsion (equivalent to 5 mg of moxifloxacin per ml) or drug solution (control) in the stimulated tear fluid was taken on a previously soaked cellophane dialyzing membrane (MWCO 12–14 kDa, Spectra/por® Spectrum Laboratories Inc., Rancho Dominguez, Berkeley, CA, USA) that separates the donor and receptor cell. The entire assembly was kept on a thermostatically controlled water bath set at 37 ± 0.5 °C and receptor medium was stirred at 50 rpm. Aliquots of sample (1 mL) were drawn at regular time intervals (0.5, 1, 2, 3, 4, 5 and 6 h) and replaced with the same volume of fresh media. The samples were subsequently diluted and analyzed for moxifloxacin content by HPLC. The data were analyzed to determine correlation coefficient (r^2) and release kinetics using various mathematical models [17];

a. Zero order model $Q = Q_0 + kt$
b. First order model $Q = Q_0 \times e^{kt}$
c. Higuchi model $Q = k \times t^{0.5}$

d. Hixson-Crowell model $Q^{1/3} = kt + Q_0{}^{1/3}$
e. Korsmeyer–Peppas model $Q = k \times t^n$
f. Weibull model $Q = 1 - \exp[-(t)^{b/a}]$

where Q represents quantity of drug released in time t, Q_0 represents value of Q at zero time, k represents the rate constant, n represents the diffusional exponent, a represents the time constant and b represents the shape parameter. The correlation coefficient (r^2) and the order of release pattern was calculated in each case.

2.9. Ex Vivo Permeation

The permeation studies were carried out using the Franz diffusion cell with a standard setup previously used in our earlier study [18]. Briefly, optimized formulation (MM3) or control (commercial eye drops; Vigamox™) (equivalent to 5 mg of moxifloxacin per mL) was placed in donor compartment and simulated tear fluid (pH 7.4) in the receptor compartment. Between donor and receptor compartment, the isolated rabbit cornea membrane was placed. The temperature of the medium was maintained at 37 ± 0.5 °C by the circulation of warm water through the outer jacket. Samples were withdrawn at predetermined time intervals (up to 6 h) and the same volume of fresh medium was replaced. The withdrawn samples were diluted and analyzed by HPLC.

2.10. Antimicrobial Efficacy

The antimicrobial efficacy of MM3 was determined by the agar diffusion test employing the cup plate technique and compared with control (commercial eye drops of moxifloxacin). The sterile products (MM3 or control) were poured (0.01 mL) into the cups of sterile nutrient agar previously inoculated with susceptible test organisms (*Pseudomonas aeruginosa* and *Staphylococcus aureus*) under horizontal laminar air flow hood. After allowing diffusion of the solutions for 2 h, the agar plates were incubated at 37 °C for 24 h. The zone of inhibition measured around each cup was compared with the control. Both positive (with test organisms) and negative controls (without test organisms) were maintained throughout the study.

2.11. Ocular Irritation

Albino rabbits (2–3 kg) placed in an animal house under observation were given ad libitum access to water and diet for 24 h. The guidelines of the Institutional Animal Ethics Committee were strictly followed while performing experiments (IPS/PCEU/PHD10/002). In vivo ocular irritation studies were performed according to the Draize technique [19]. Single administration of 60 μL was instilled in the left eye of each rabbit while keeping the untreated eye as the control. The sterile MM3 was administered twice daily for a period of 21 days (1, 2, 3, 4, 7, 10, 15, 18 and 21 days). The rabbits were observed periodically for redness, swelling and watering of the eye.

2.12. Pharmacokinetics

In vivo pharmacokinetic studies were carried out in Albino rabbits (2–3 kg). Animals were separated into two groups of each containing six rabbits. All animal procedures were performed at the Nirma University following the guidelines (IPS/PCEU/PHD10/002, dated; 23/01/2010, Scientific Research Ethics Committee, Nirma University). Rabbits were treated with MM3 or control (commercial eye drops). For the first group, drops (50 μL of 0.5% *w*/*v* moxifloxacin) of MM3 was instilled in the lower cul-de-sac of both eyes of each rabbit [20], and the upper and lower eyelids were gently held closed for 2 min to maximize drug cornea contact. In a similar manner, both eyes of each rabbit of the second group were given single topical instillation (50 μL of 0.5% *w*/*v* moxifloxacin) of commercial eye drops. Each rabbit was anaesthetized by intra muscular injection of ketamine (50 mg/kg) and xylazine (10 mg/kg) [21]. Additionally, lidocaine (2% *w*/*v*) was applied at the injection site to provide local anesthesia. Then eyelash and eye liners/lids were wiped with povidone solution (5% *w*/*v*) to maintain standard of care to give intra-ocular injection. A 29-gauge insulin syringe needle was used to collect aqueous humor (0.5, 1, 2, 4 and 8 h) and samples (50 μL) were mixed with acetonitrile, and stored at −80 °C. The mixture was then centrifuged at 3000 rpm for 15 min and the supernatant was analyzed for drug content using HPLC.

2.13. Stability

Sterile nanoemulsion (MM3; 5 mL) was stored in an amber colored container for a period of three months in the refrigerator (2–8 °C). Stability of MM3 was routinely evaluated by visual inspection of the samples initially on a daily and later on a weekly basis for pH, phase separation, flocculation or precipitation [22]. Stability of nanoemulsion was also checked by centrifugation (Remi Centrifuge, Mumbai, India) by 12,000 rpm for 30 min and then the clarity, phase separation and concentration of drug were investigated. Droplet size of the nanoemulsion upon storage was also determined to assess the stability in terms of drastic changes in the mean droplet diameter due to droplet coalescence or aggregation.

2.14. Data Analysis

A plot of the cumulative amount of moxifloxacin permeated across the cornea was plotted against time and the slope was measured as flux [23]. The statistical evaluation of the data was analyzed using one-way analysis of variance (ANOVA) (SPSS 23, SPSS Inc., Chicago, IL, USA). The statistical differences between values showing $p < 0.05$ were considered as significant.

3. Results and Discussion

3.1. Pseudoternary Phase Diagram

Preliminary studies were carried out to select ingredients for preparing moxifloxacin nanoemulsion. Indeed, the selection of the vehicle is extremely important as it provides good solubilizing efficiency of the drug, which is essential for constituting a nanoemulsion [24]. Hence, the solubility of moxifloxacin in various oils, surfactants, and cosurfactants was determined by the standard procedures. Based on the high moxifloxacin solubility, ethyl oleate (oil; solubility ~28.37 mg/mL), Tween 80 (surfactant; solubility ~10.56 mg/mL) and Soluphor P (cosurfactant; solubility ~7.22 mg/mL) were selected for nanoemulsion preparation. The utility of the selected vehicles in preparing nanoemulsions is described in the literature. For instance, ethyl oleate has been used in nanoemulsions due to its lower molecular size (310.51 g/mol) as related to medium chain triglycerides (~800 g/mol). Furthermore, nanoemulsions formulated with ethyl oleate have demonstrated enhancement in corneal permeation of drugs [25]. Similarly, Tween 80 (up to 10% *w*/*w*) has been used as a surfactant in many commercial ophthalmic preparations which usually do not cause ocular irritation [26]. Due to the longer hydrocarbon chain length of Tween 80, it can broaden the area of the nanoemulsion region [27]. In addition, Soluphor P has demonstrated improved drug penetration across the biological membranes [28].

The pseudoternary phase diagram of the systems consisting of surfactant: Cosurfactant (Tween 80: Soluphor P) mixture (S_{mix}), oil phase and water was illustrated in Figure 1A. The binodal curve separating two phase and one phase in the pseudoternary phase diagram was indicated by the visual observation of the sample appearance from turbid to transparent or vice versa. To complete the entire nanoemulsion domain, in addition to the water titration method, oil and water components were fixed and surfactant component varied. Similarly, water and S_{mix} components were fixed and the oil component varied. The nanoemulsion domain obtained by these trials at ratios of 1:1, 1:2 and 2:1 with surfactant (Tween 80) to cosurfactant (Soluphor P) ratio were plotted in the phase diagram (Figure 1A). The phase behavior study revealed that, when the surfactant to cosurfactant ratio was 1:2, the maximum quantity of oil can be included in the nanoemulsion system. From the phase diagram, it was indicated that the change in phase behavior within the nanoemulsion region is mainly due to the hydrophobic carbon chain length of the oil and the ratio between surfactant and cosurfactant mixture used in the formulation. Therefore, it is likely that the hydrocarbon chain length compatibility among surfactant and oil is an important factor that affects the globule size formation and stability of nanoemulsion as suggested by Schneider et al. [29].

3.2. Formulation Optimization

The nanoemulsions selected for the optimization were evaluated for parameters like droplets size, polydispersity index, zeta potential, and conductivity. The droplets size was optimized as a response in simplex design. The droplets size was in the range of 28.78–81.04 nm. Regression summary output, full model confirmed that components A (S_{mix}), B (oil), C (water), AB and BC significantly affect the response (droplets size). Whereas AC has p value >0.05 suggests that it does not significantly affect the response. Therefore, reduced model was developed eliminating AC term and simplex coefficients were obtained from regressions summary output using Microsoft Office Excel® 2007. R Square value of 0.999 suggests the suitability of the model and the lower stand error value of 1.9505 suggests minimum error in the model.

Response was calculated over the simplex space, and a contour diagram is plotted (Figure 2). It is inferred from the contour plot that particle size is minimum, when the water component is at higher level (1) and oil and surfactant components are at lower level (0). The optimized formulation of the nanoemulsion consists of ethyl oleate (4%), Tween 80 (12%), Soluphor P (24%) and water (60%). Moxifloxacin was included at a concentration of 0.5% w/w in all tested formulations.

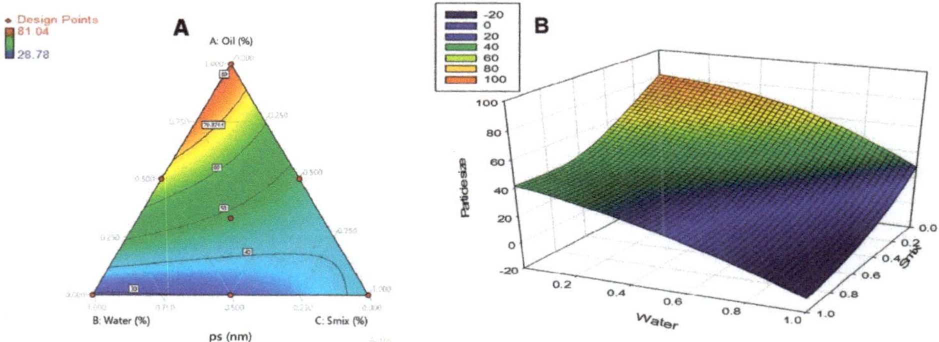

Figure 2. Contour plot (**A**) over the simplex space and response surface graph (**B**) representing nanoemulsion particle size (nm).

3.3. Validation of Applied Design

The simplex equation of response, droplets size is derived as follows:

$$Y = 40.76\,A + 81.16\,B + 27.72\,C - 55.72\,AB + 48.71\,BC$$

The result at the extra-design checkpoint is predicted based on the equation to confirm the validity of the applied design. It was found that the observed value was close to the practically measured value and hence establishes the effectiveness of the tested mathematical model.

3.4. Characterization of Nanoemulsion

Various physicochemical characteristics of prepared nanoemulsions (MM1–MM7) were determined and summarized in Table 2. The drug content of nanoemulsions were in the range of 90–105%. The pH of formulations was ~6–7 and was comparable to commercial ophthalmic drops (pH 6.8), which is likely to be buffered by the lachrymal fluid and may not induce any ocular irritation, reflex tears or rapid blinking of the eye. The developed nanoemulsion shows percentage transmittance value >90% which proves the transparency of the system and droplets are in the nanometer dimensions, confirmed by the average value obtained for droplets size (29–81 nm). The conductivity of nanoemulsions was in the range of 0.08–0.2 mS/cm. No phase separation was observed when diluted 10 times with continuous media in all prepared nanoemulsions. The polydispersity index value of 0–0.5 indicates homogenous, uniformly sized, spherical vesicles. As shown in Table 2, MM1–MM7 have polydispersity indices ranged from 0.24–0.39, which indicate that they were more uniform and homogenous. According to the classical electrical double layer theory, zeta potential value above ±30 mV demonstrates moderate repulsion between similar charged particles, thereby decreasing flocculation or aggregation and potentially stabilizes the dispersion. The observed zeta potential values of MM1–MM7 (0.28–0.38 mV) were comparable, signify the formulation components have not influenced the zeta potential in the current experimental conditions. Viscosity is another important parameter as it can significantly influence the corneal retention time as well as ocular bioavailability from an ophthalmic product. The prepared nanoemulsions have viscosity range from 3.2–6.5 cP, which renders it easily pourable during instillation into the eye and is comparable to normal human tear fluids [30]. Based on the physiochemical properties observed, it seems nanoemulsion MM3 possesses the smallest droplets size and the lowest viscosity as compared to other nanoemulsions prepared. Hence, the nanoemulsion MM3 was selected for further investigation.

3.5. TEM

A representative TEM image of prepared moxifloxacin nanoemulsion (MM3) is shown in Figure 3. It is evident from Figure 3 that the prepared system has droplets in circular shape with uniform size and can be easily distinguished. The droplets appear darker with bright background and are randomly dispersed without any agglomeration throughout the field.

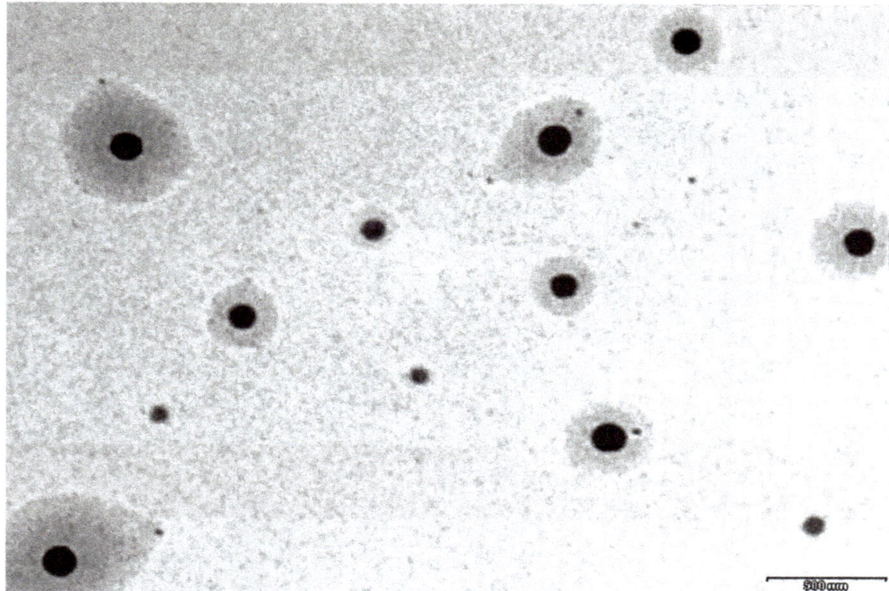

Figure 3. A representative transmission electron microscopy image of moxifloxacin nanoemulsion (MM3).

Table 2. Physicochemical characteristics of prepared nanoemulsions *.

Parameter	MM1	MM2	MM3	MM4	MM5	MM6	MM7	MM8 **
Drug content (%)	96.05 ± 4.01	93.38 ± 2.92	99.90 ± 2.62	94.47 ± 3.48	101.62 ± 2.29	102.06 ± 2.47	98.25 ± 3.78	95.34 ± 4.62
pH	6.33 ± 0.41	6.70 ± 0.52	6.64 ± 0.43	6.85 ± 0.32	6.19 ± 0.28	6.22 ± 0.44	7.04 ± 0.54	6.51 ± 0.36
Transmittance (%)	97.22 ± 5.72	95.38 ± 5.29	97.82 ± 3.83	96.17 ± 3.92	97.71 ± 3.94	95.46 ± 4.42	96.43 ± 4.85	97.62 ± 3.63
Conductivity (mS/cm)	0.11 ± 0.02	0.08 ± 0.01	0.20 ± 0.08	0.14 ± 0.04	0.15 ± 0.07	0.16 ± 0.06	0.14 ± 0.03	0.11 ± 0.03
Dilution potential	>10 times	>10 times	>10 times	>10 times	>10 times	>10 times	>10 times	>10 times
Droplets size (nm)	41.82 ± 13.71	81.04 ± 15.35	28.78 ± 10.34	47.56 ± 16.28	32.41 ± 14.70	67.15 ± 15.84	47.42 ± 14.11	75.99 ± 16.36
Polydispersity index	0.35 ± 0.05	0.26 ± 0.03	0.38 ± 0.06	0.39 ± 0.04	0.34 ± 0.02	0.30 ± 0.05	0.39 ± 0.03	0.24 ± 0.02
Zeta potential (mV)	−0.33 ± 0.01	−0.35 ± 0.02	−0.38 ± 0.012	−0.28 ± 0.02	−0.32 ± 0.03	0.37 ± 0.03	−0.29 ± 0.02	−0.32 ± 0.01
Viscosity (cP)	4.81 ± 1.67	6.50 ± 1.17	3.28 ± 1.42	5.80 ± 1.37	4.57 ± 1.47	5.86 ± 2.44	4.92 ± 1.85	6.39 ± 2.24

* Mean ± SD (n = 3); ** Check point batch.

3.6. In Vitro Release

The release of drug from nanoemulsions is essential for absorption as well as its therapeutic effect. Figure 4 compares the cumulative percentage of moxifloxacin released at periodic time intervals from MM1–MM7 and control. It is evident from Figure 4 that the drug release profile showed a similar trend, increased steadily over time and is more than 90% in 3 h for all formulations. However, the release of moxifloxacin from control was rapid and complete in 45 min. Amongst all designed o/w nanoemulsion formulations, MM3 showed complete drug release in 150 min. Overall, the release profiles indicate the ability of prepared nanoemulsions to provide a steady drug release for 3 h, which in turn can prolong the therapy. The release mechanism for MM3 was studied using various models and the values are summarized in Table 3. It is evident from the Table 3 that the release kinetics of moxifloxacin from MM3 was fitted into the Higuchi model displaying high r^2 value (0.9486), least SSR value (613.14) and F value (87.59). Thus, the release of moxifloxacin from MM3 was due to Higuchi diffusion controlled mechanism. Further, the n value was less than 0.5 which indicates that drug release in MM3 is mainly by Fickian diffusion.

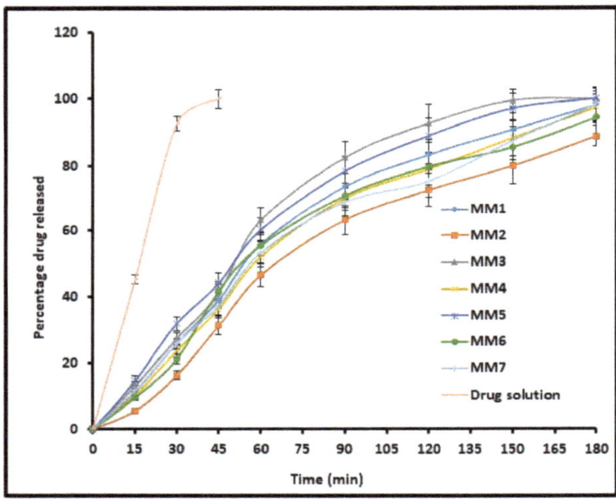

Figure 4. Comparison of percentage moxifloxacin release from prepared nanoemulsions (MM1–MM7) and drug solution (control). The data represents average ± SD of six trials.

Table 3. Model fitting for selected nanoemulsion (MM3).

Model Name	Multiple *R*	*R* Square	*X* Variable	Slope	SSR	Fischer Ratio
Zero order	0.9486	0.8999	0.5862	12.7455	1194.4235	170.6319
First order	0.9558	0.9135	−0.0166	2.3706	20804.4358	2972.0623
Higuchi	0.9740	0.9486	8.7765	−10.4853	613.1492	87.5927
Korsmeyer–Peppas	0.9766	0.9538	0.8446	−1.8055	1086.6082	155.2297
Weibull Model	0.9904	0.9810	1.6232	−2.8693	4932.8215	704.6888
Hixson–Crowell	0.9940	0.9880	0.0271	−0.2801	5019.6483	717.0926

3.7. Ex Vivo Permeation

The diffusion of therapeutic molecules into and across the biological membrane is mainly influenced by the drug's physicochemical properties, physiological characteristics of the membrane and different transport routes available for permeation [18]. Figure 5 compares the amount of moxifloxacin transported across the isolated rabbit cornea membrane from MM3 and control (commercial eye drops). A typical permeation profile was exhibited by both MM3 and control and the steady state flux values were comparable (MM3; 32.01 μg/cm^2/h and control; 31.53 μg/cm^2/h). The flux value observed here signifies that the physicochemical characteristics of MM3 are suitable for cornea permeation. The permeation rate was relatively high in the first hour with control as compared to MM3. This is probably due to the unique structure of moxifloxacin along with biphasic solubility (both lipophilic and aqueous solubility) and high lipophilicity which would have assisted its easy permeation through the corneal membrane [31].

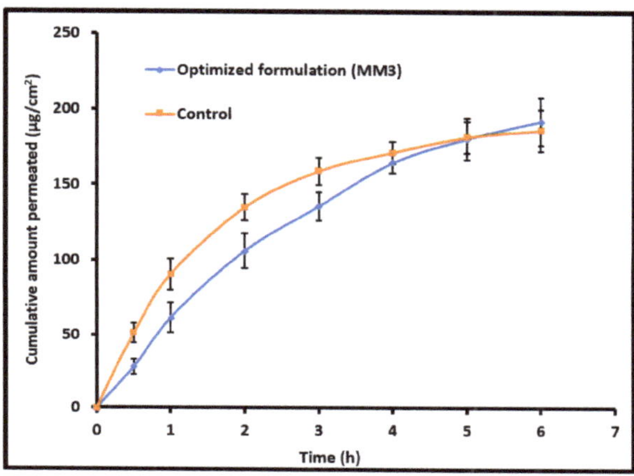

Figure 5. Comparison of moxifloxacin ex vivo permeation across the isolated rabbit cornea membrane from optimized nanoemulsion (MM3) and control (commercial eye drops). The data represents average ± SD of six trials.

3.8. Antimicrobial Efficacy

It is essential to perform microbiological efficacy studies to demonstrate the activity of the drug in the nanoemulsion against the commonly susceptible microorganisms [32]. The optimized formulation (MM3) showed microbicidal activity, when microbiological testing was performed by the cup plate technique. Visible zones of inhibition were noticed in case of MM3 and control formulation. The diameter of the zone of inhibitions generated by MM3 against both test organisms (*Pseudomonas aeruginosa* and *Staphylococcus aureus*) were either similar or more than that produced by commercial ophthalmic drops (Table 4). The broad-spectrum antibacterial activity of moxifloxacin in the nanoemulsion against the susceptible pathogens was similar to the reference formulation in terms of antimicrobial efficacy. This clearly indicates that the prepared formulation did not change the inherent bactericidal effect of the incorporated moxifloxacin.

Table 4. Antimicrobial efficacy of the optimized formulation (MM3).

Concentration (μg/mL)	Zone of Inhibition in (cm)	
	Control *	MM3
	Staphylococcus aureus	
1	1.63 ± 0.22	1.63 ± 0.26
10	2.44 ± 0.31	2.58 ± 0.18
100	3.52 ± 0.25	3.94 ± 0.24
	Pseudomonas aeruginosa	
1	1.82 ± 0.34	1.97 ± 0.16
10	2.77 ± 0.23	3.05 ± 0.21
100	4.05 ± 0.25	4.71 ± 0.29

* Control: Commercial eye drops of moxifloxacin.

3.9. Ocular Irritation

The eye irritation potential of the inducing agent was classified into four grades; practically non-irritating, score 0–3; slightly irritating, score 4–8; moderately irritating, score 9–12; and severely irritating (or corrosive), score 13–16 [20]. The eye irritation score was calculated by dividing the total score for all rabbits by the total number of rabbits tested. The observed eye irritation score in the control was 0.33 and for MM3 was 0.66, which signifies excellent ocular tolerance. Moreover, no ocular damage or abnormal clinical signs pertaining to the cornea, iris, or conjunctivae were visible. In addition, no signs of redness, watering of the eye or swelling were noticed for both MM3 and control. Overall the results of this study revealed that MM3 is safe for ocular application.

3.10. Pharmacokinetics in the Aqueous Humor

The concentration of the drug permeated into the aqueous humor after administration to the rabbit eyes was quantified to evaluate the ocular bioavailability of moxifloxacin from MM3 as well as the control. Different pharmacokinetic properties were analyzed by using a non-compartmental method [33]. The determined pharmacokinetic parameters are summarized in Table 5. Figure 6 compares the mean moxifloxacin concentration in the aqueous humor following topical installation of MM3 and control in rabbits. It is evident from Figure 6 that the kinetic profiles are distinct for MM3 and control. Indeed, moxifloxacin absorption was rapid and available in the aqueous humor after 30 min in MM3 (113.98 ± 51.45 ng/mL) and control (209.44 ± 64.53 ng/mL), however, the amount of drug was different. At 1 h, the drug level in the aqueous humor increased to 305.99 ± 94.95 ng/mL and 454.19 ± 126.91 ng/mL in MM3 and control, respectively. The drug absorption in MM3 was further prolonged and the drug level in the aqueous humor continued to rise to attain the peak drug concentration (C_{max}; 555.73 ± 133.34 ng/mL) and the time corresponding to peak concentration (T_{max}) was 2 h (which is 1 h in control). Comparing the drug absorption with the ex vivo permeation data, one can easily corroborate that the delay in absorption of moxifloxacin into the aqueous humor in MM3 is probably due to its slow permeation rate in the first hour observed in Figure 5. Followed by the rapid absorption, the drug level declined in both groups. At 4 h, the drug level in aqueous humor was considerably higher in MM3 ($p < 0.0001$) as compared to control. At 8 h, the drug level in aqueous humor for MM3 was 35.90 ± 13.01 ng/mL while no moxifloxacin was detected in the control. The mean value of area under the aqueous humor concentration versus time curve ($AUC_{0–8\ h}$) for MM3 (Table 5) was ~2 fold higher ($p < 0.001$), relative to the control, suggesting significant improvement in ocular bioavailability by MM3 in comparison to the control. The possible reason for the greater observed AUC could be due to the longer retention of vesicles (MM3) in the ocular surface which in turn prolongs the contact time of medicament with the eye and thereby improve corneal penetration, when compared to conventional eye drops, as described in the literature [34].

Table 5. Mean pharmacokinetic parameters of moxifloxacin in aqueous humor following topical installation of nanoemulsion (MM3) and control in rabbits.

Parameter	Nanoemulsion (MM3)	Control
T_{max} (h)	2	1
C_{max} (ng/mL)	555.73 ± 133.34	454.19 ± 126.91
$AUC_{0–8}$ (ng.h/mL)	1859.76 ± 424.51 *	958.63 ± 206.84

* Significant difference ($p < 0.001$) observed in moxifloxacin level in nanoemulsion (MM3) group compared to control. Area under the aqueous humor concentration versus time curve ($AUC_{0–8\ h}$).

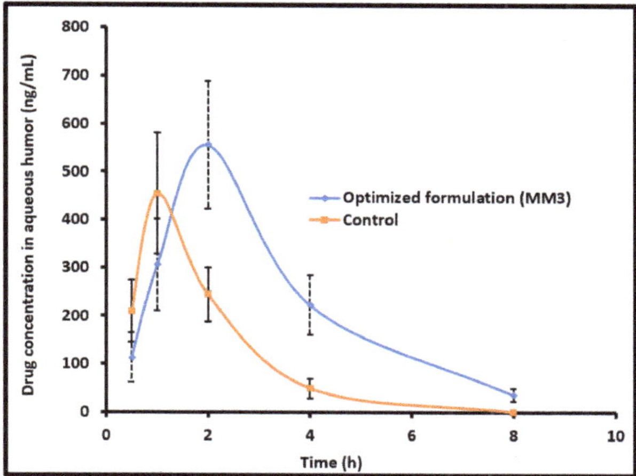

Figure 6. Comparison of mean moxifloxacin concentration in the aqueous humor following topical installation of optimized nanoemulsion (MM3) and control (commercial eye drops) in rabbits. The data represents average ± SD of six trials.

3.11. Stability Assessment

Stability studies were conducted to determine the influence of various excipients on drug in MM3 and also to determine physicochemical characteristics of the finished product during storage. After stability testing, the nanoemulsion was found to be a transparent biphasic solution and exhibited no coagulation, aggregation or precipitation. No significant change in pH of MM3 was observed during the stability period. Drug content of MM3 was more than 95% during storage period. There was no phase separation or creaming on visual observation and it was found stable after cooling centrifugation. The data observed indicate minimum changes in stability indicating parameters of nanoemulsion systems like droplets size of MM3 after three months of storage.

4. Conclusions

The nanoemulsion system of moxifloxacin was successfully formulated by developing pseudoternary phase diagram of different proportions of ethyl oleate, Tween 80, Soluphor P and water. It is inferred from the contour plot that particle size is minimum when the water component is at a higher level (1) and oil and surfactant components are at lower level (0). The developed nanoemulsions have optimum viscosity for instillation in eye. They show transmittance value >95% which proves the transparency of the system and the droplets are in nanometer dimensions. The optimized nanoemulsion (MM3) showed adequate antimicrobial effect of the entrapped moxifloxacin and is safe for ocular application. The moxifloxacin level in the aqueous humor was prolonged by MM3 which is greater than the minimum concentration necessary for the therapeutic efficacy (i.e., 0.2 μg/mL or >>>MIC_{90} for aerobic gram-positive and gram-negative pathogens responsible for causing eye infections), signifying the importance of nanoemulsion for ocular therapy. In a nutshell, the optimized nanoemulsion formulation (MM3) could be a promising and viable drug delivery system for effective delivery of moxifloxacin for treatment of various bacterial eye infections.

Author Contributions: Formulation and in vivo evaluation, J.S., R.K.P. and H.S., Experimental design, writing and data processing, A.B.N. and S.J.; In vitro, ex vivo studies and data processing, T.M.S. and M.A.M.

Funding: This research received no external funding.

Acknowledgments: The authors are highly thankful to Institute of Pharmacy, Nirma University, Ahmedabad for support.

Conflicts of Interest: The authors declare no conflict of interest.

References

1. Gaudana, R.; Ananthula, H.K.; Parenky, A.; Mitra, A.K. Ocular drug delivery. *AAPS J.* **2010**, *12*, 348–360. [CrossRef]
2. Agrahari, V.; Mandal, A.; Agrahari, V.; Trinh, H.M.; Joseph, M.; Ray, A.; Hadji, H.; Mitra, R.; Pal, D.; Mitra, A.K. A comprehensive insight on ocular pharmacokinetics. *Drug Deliv. Transl. Res.* **2016**, *6*, 735–754. [CrossRef]
3. Patel, A.; Cholkar, K.; Agrahari, V.; Mitra, A.K. Ocular drug delivery systems: An overview. *World J. Pharmacol.* **2013**, *2*, 47–64. [CrossRef] [PubMed]
4. Reimondez-Troitiño, S.; Csaba, N.; Alonso, M.J.; De La Fuente, M. Nanotherapies for the treatment of ocular diseases. *Eur. J. Pharm. Biopharm.* **2015**, *95*, 279–293. [CrossRef] [PubMed]
5. Abdelkader, H.; Alani, A.W.; Alany, R.G. Recent advances in non-ionic surfactant vesicles (niosomes): Self-assembly, fabrication, characterization, drug delivery applications and limitations. *Drug Deliv.* **2014**, *21*, 87–100. [CrossRef]
6. Agarwal, R.; Iezhitsa, I.; Agarwal, P.; Abdul Nasir, N.A.; Razali, N.; Alyautdin, R.; Ismail, N.M. Liposomes in topical ophthalmic drug delivery: An update. *Drug Deliv.* **2016**, *23*, 1075–1091. [CrossRef]
7. Ali, A.; Ansari, V.A.; Ahmad, U.; Akhtar, J.; Jahan, A. Nanoemulsion: An advanced vehicle for efficient drug delivery. *Drug Res.* **2017**, *67*, 617–631. [CrossRef] [PubMed]
8. Alany, R.G.; Rades, T.; Nicoll, J.; Tucker, I.G. Davies NM. W/O microemulsions for ocular delivery: Evaluation of ocular irritation and precorneal retention. *J. Control. Release* **2006**, *111*, 145–152. [CrossRef] [PubMed]
9. Miller, D. Review of moxifloxacin hydrochloride ophthalmic solution in the treatment of bacterial eye infections. *Clin. Ophthalmol.* **2008**, *2*, 77–91. [CrossRef]
10. Sultana, N.; Arayne, M.S.; Akhtar, M.; Shamim, S.; Gul, S.; Khan, M.M. High-performance liquid chromatography assay for moxifloxacin in bulk, pharmaceutical formulations and serum: Application to in-vitro metal interactions. *J. Chin. Chem. Soc.* **2010**, *57*, 708–717. [CrossRef]
11. Ma, Y.J.; Yuan, X.Z.; Huang, H.J.; Xiao, Z.H.; Zeng, G.M. The pseudo-ternary phase diagrams and properties of anionic–nonionic mixed surfactant reverse micellar systems. *J. Mol. Liq.* **2015**, *203*, 181–186. [CrossRef]
12. Syed, H.K.; Peh, K.K. Identification of phases of various oil, surfactant/co-surfactants and water system by ternary phase diagram. *Acta Pol. Pharm.* **2014**, *71*, 301–309. [PubMed]
13. Fialho, S.L.; Da Silva-Cunha, A. New vehicle based on a microemulsion for topical ocular administration of dexamethasone. *Clin. Exp. Ophthalmol.* **2004**, *32*, 626–632. [CrossRef] [PubMed]
14. Al-Dhubiab, B.E.; Nair, A.B.; Kumria, R.; Attimarad, M.; Harsha, S. Development and evaluation of buccal films impregnated with selegiline-loaded nanospheres. *Drug Deliv.* **2016**, *23*, 2154–2162. [CrossRef]
15. Morsy, M.A.; Nair, A.B. Prevention of rat liver fibrosis by selective targeting of hepatic stellate cells using hesperidin carriers. *Int. J. Pharm.* **2018**, *552*, 241–250. [CrossRef] [PubMed]
16. Shah, J.N.; Patel, R.K.; Shah, H.J.; Mehta, T.A. Beyond the blink: Using in-situ gelling to optimize ophthalmic drug delivery. *Pharm. Technol.* **2015**, *39*, 1–7.
17. Shah, H.; Nair, A.B.; Shah, J.; Bharadia, P.; Al-Dhubiab, B.E. Proniosomal gel for transdermal delivery of lornoxicam: Optimization using factorial design and in vivo evaluation in rats. *DARU J. Pharm. Sci.* **2019**. [CrossRef] [PubMed]
18. Al-Dhubiab, B.E.; Nair, A.B.; Kumria, R.; Attimarad, M.; Harsha, S. Formulation and evaluation of nano based drug delivery system for the buccal delivery of acyclovir. *Colloids Surf. B Biointerfaces* **2015**, *136*, 878–884. [CrossRef] [PubMed]
19. Gonzalez-Mira, E.; Egea, M.A.; Garcia, M.L.; Souto, E.B. Design and ocular tolerance of flurbiprofen loaded ultrasound-engineered NLC. *Colloids Surf. B Biointerfaces* **2010**, *81*, 412–421. [CrossRef] [PubMed]
20. Gan, L.; Gan, Y.; Zhu, C.; Zhang, X.; Zhu, J. Novel microemulsion in situ electrolyte-triggered gelling system for ophthalmic delivery of lipophilic cyclosporine A: In vitro and in vivo results. *Int. J. Pharm.* **2009**, *365*, 143–149. [CrossRef] [PubMed]
21. Nair, A.; Morsy, M.A.; Jacob, S. Dose translation between laboratory animals and human in preclinical and clinical phases of drug development. *Drug Dev. Res.* **2018**, *79*, 373–382. [CrossRef]
22. Jacob, S.; Nair, A.B.; Al-Dhubiab, B.E. Preparation and evaluation of niosome gel containing acyclovir for enhanced dermal deposition. *J. Liposome Res.* **2017**, *27*, 283–292. [CrossRef]

23. Kumria, R.; Al-Dhubiab, B.E.; Shah, J.; Nair, A.B. Formulation and evaluation of chitosan-based buccal bioadhesive films of zolmitriptan. *J. Pharm. Innov.* **2018**, *13*, 133–143. [CrossRef]
24. Wadhwa, J.; Nair, A.; Kumria, R. Self-emulsifying therapeutic system: A potential approach for delivery of lipophilic drugs. *Braz. J. Pharm. Sci.* **2011**, *47*, 447–465. [CrossRef]
25. Hegde, R.R.; Verma, A.; Ghosh, A. Microemulsion: New insights into the ocular drug delivery. *ISRN Pharm.* **2013**, *2013*, 826798. [CrossRef] [PubMed]
26. Jiao, J. Polyoxyethylated nonionic surfactants and their applications in topical ocular drug delivery. *Adv. Drug Deliv. Rev.* **2008**, *60*, 1663–1673. [CrossRef] [PubMed]
27. Lv, F.F.; Zheng, L.Q.; Tung, C.H. Phase behavior of the microemulsions and the stability of the chloramphenicol in the microemulsion-based ocular drug delivery system. *Int. J. Pharm.* **2005**, *301*, 237–246. [CrossRef]
28. Shah, S.M.; Jain, A.S.; Kaushik, R.; Nagarsenker, M.S.; Nerurkar, M.J. Preclinical formulations: Insight, strategies, and practical considerations. *AAPS PharmSciTech.* **2014**, *15*, 1307–1323. [CrossRef]
29. Schneider, K.; Ott, T.M.; Schweins, R.; Frielinghaus, H.; Lade, O.; Sottmann, T. Phase behavior and microstructure of symmetric nonionic microemulsions with long-chain *n*-alkanes and waxes. *Ind. Eng. Chem. Res.* **2019**, *58*, 2583–2595. [CrossRef]
30. Tiffany, J.M. The viscosity of human tears. *Int. Ophthalmol.* **1991**, *15*, 371–376. [CrossRef]
31. Robertson, S.M.; Curtis, M.A.; Schlech, B.A.; Rusinko, A.; Owen, G.R.; Dembinska, O.; Liao, J.; Dahlin, D.C. Ocular pharmacokinetics of moxifloxacin after topical treatment of animals and humans. *Surv. Ophthalmol.* **2005**, *50*, S32–S45. [CrossRef] [PubMed]
32. Bassetti, M.; Dembry, L.M.; Farrel, P.A.; Callan, D.A.; Andriole, V.T. Comparative antimicrobial activity of gatifloxacin with ciprofloxacin and beta-lactams against gram-positive bacteria. *Diagn. Microbiol. Infect. Dis.* **2001**, *41*, 143–148. [CrossRef]
33. Nair, A.B.; Kaushik, A.; Attimarad, M.; Al-Dhubiab, B.E. Enhanced oral bioavailability of calcium using bovine serum albumin microspheres. *Drug Deliv.* **2012**, *19*, 277–285. [CrossRef] [PubMed]
34. Liu, Q.; Wang, Y. Development of an ex vivo method for evaluation of precorneal residence of topical ophthalmic formulations. *AAPS PharmSciTech.* **2009**, *10*, 796–805. [CrossRef]

Article

Development and In Vitro-In Vivo Evaluation of a Novel Sustained-Release Loxoprofen Pellet with Double Coating Layer

Dongwei Wan [1], Min Zhao [1], Jingjing Zhang [1] and Libiao Luan [2,*]

1 College of Pharmacy, China Pharmaceutical University, No. 639 Longmian Road, Nanjing 211100, China; 710176323@163.com (D.W.); 19850856528@163.com (M.Z.); 15651915012@163.com (J.Z.)

2 College of Pharmacy, China Pharmaceutical University, Xuanwumen Campus, No. 24 Tongjiaxiang, Nanjing 210009, China

* Correspondence: luanlibiao@126.com; Tel.: +86-18851106518

Received: 7 May 2019; Accepted: 3 June 2019; Published: 5 June 2019

Abstract: This study aimed to develop a novel sustained release pellet of loxoprofen sodium (LXP) by coating a dissolution-rate controlling sub-layer containing hydroxypropyl methyl cellulose (HPMC) and citric acid, and a second diffusion-rate controlling layer containing aqueous dispersion of ethyl cellulose (ADEC) on the surface of a LXP conventional pellet, and to compare its performance in vivo with an immediate release tablet (Loxinon®). A three-level, three-factor Box-Behnken design and the response surface model (RSM) were used to investigate and optimize the effects of the citric acid content in the sub-layer, the sub-layer coating level, and the outer ADEC coating level on the in vitro release profiles of LXP sustained release pellets. The pharmacokinetic studies of the optimal sustained release pellets were performed in fasted beagle dogs using an immediate release tablet as a reference. The results illustrated that both the citric acid (CA) and ADEC as the dissolution- and diffusion-rate controlling materials significantly decreased the drug release rate. The optimal formulation showed a pH-independent drug release in media at pH above 4.5 and a slightly slow release in acid medium. The pharmacokinetic studies revealed that a more stable and prolonged plasma drug concentration profile of the optimal pellets was achieved, with a relative bioavaibility of 87.16% compared with the conventional tablets. This article provided a novel concept of two-step control of the release rate of LXP, which showed a sustained release both in vitro and in vivo.

Keywords: sustained release pellets; double coating layer; loxoprofen; citric acid; pharmacokinetic studies

---

1. Introduction

Pellets, as multiple unit preparations, offer a lot of clinical benefits compared with single unit dosage forms, such as reduced intra- and inter-subject variability on drug plasma, decreased local irritations, less dose dumping risk, and stable plasma concentrations [1,2]. To prepare sustained release pellets, film coating is an ideal method. With the development of aqueous-based dispersion systems, film coating technologies have shifted from organic-based polymeric solutions to aqueous-based polymeric dispersion systems [3]. As one of the aqueous ethyl cellulose dispersions, Surelease® (ADEC) could be used alone or combined with other polymers to obtain satisfactory release profiles [4–7]. Additionally, most of these release profiles showed a diffusion-controlled release mechanism, which meant a predictable release pattern could be achieved by altering the ADEC coating weight gain [8,9].

Loxoprofen sodium (LXP), as a 2-phenylpropinate non-steroidal anti-inflammatory drug (NSAD), was first introduced by Sankyo Company in Japan. It has been widely used for the treatment of osteoarthritis, scapulohumeral periarthritis, rheumatoid arthritis, arthritis, toothache, and post-operation pain [10]. As a pro-drug, LXP is converted to its active metabolite (trans-OH LXP) in vivo to inhibit the activity of cycloosygenase (COX), which mediates the production of inflammatory prostaglandins [11]. Due to the short elimination half-life of approximately 65 min [12], the commercial tablet of LXP has to be administrated three times a day to maintain the therapeutic concentration in plasma, which might cause high risks of gastrointestinal (GI) lesions and systemic side effects [13,14]. Several studies have been reported on the preparation of LXP sustained release dosage forms [15–17]. However, due to its high solubility, most of the preparations, especially for the matrix-based formulations, showed a burst release (drug release >30%) during the first 2 h [18], which could cause unexpected GI mucosal injury for patients. Therefore, a sustained release dosage form with decreased initial release would be necessary.

As a weakly acidic drug, loxoprofen shows good solubility at high pH, while poor solubility at low pH. Several strategies have been developed to prepare sustained release formulations of the pH-sensitive drugs [19–22]. Among them, incorporation of pH-modifiers into the preparation was a common approach in matrix or coating systems. These pH modifiers could significantly modify the micro-environmental pH (pH_M) inside the systems, and result in a decrease or increase of the drug solubility, leading to a modified drug dissolution rate [22–24]. In addition, their extent and duration played an important role on the drug release rate [25]. Approaches like using coated pH-modifier as the starting core [23], blending pH-modifier with drugs into the core with a subsequent coating [26], or incorporating pH-modifiers into the matrix formulations [20], have been proposed and studied to achieve the sustained release of pH-sensitive drugs. However, for the maintenance of an appropriate pH_M inside the dosage forms, more than 20% pH-modifiers in the preparations were often needed [20,23,26], which might cause undesired GI irritations, especially for patients with GI ulcers. In order to reduce the usage of pH-modifiers and maintain an appropriate pH_M in the dosage form, citric acid (CA) as the pH-modifier was first proposed to be incorporated into the dissolution-rate controlling layer to decrease the dissolution rate of LXP.

Drug delivery systems (DDS), based on their system design or rate-controlling mechanism, can be divided into models such as dissolution, diffusion, erosion, osmosis, and swelling [27]. As for the film coating systems, the diffusion or osmosis mechanisms were often applied to elucidate the drug release profiles [5,28–30], while the influence of drug dissolution rate was often omitted or just attributed to the drug diffusion rate [27]. In the most common cases, only one of these mechanisms was applied to control the drug release rate in DDS, except for the bio-erodible or hydrogel matrix systems, where the drug release rate was controlled by two or three of these release mechanisms [31,32]. Although theoretical approaches regarding the dissolution-diffusion mechanism have been extensively reported [33,34], a combination of the dissolution and diffusion release mechanisms as a rate-controlling strategy was seldom reported.

In this study, a novel concept of two-step control of the drug release rate is proposed. A schematic diagram of this hypothesis is illustrated in Figure 1. In this system, the first-step control was to reduce the dissolution rate of LXP by creating a sub-coating layer containing pH-modifier CA, while the second-step control was to decrease the diffusion rate of LXP by creating a non-soluble polymeric film. Furthermore, a three-level, three-factor Box-Behnken experiments design was conducted to optimize and evaluate the effects of different parameters on the drug release. Additionally, the pharmacokinetic studies of the optimal formulation were performed in fasted beagles to compare its in vivo performance with the conventional tablet.

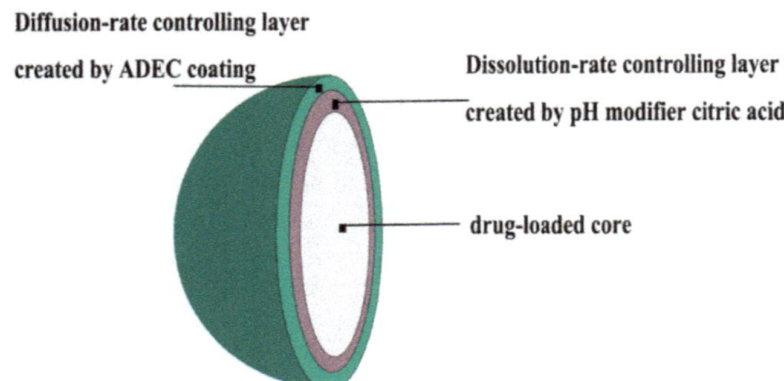

Figure 1. Schematic diagram of the sustained release pellets.

2. Materials and Methods

2.1. Materials

LXP dihydrate was purchased from Fujian Hui Tian biological Pharmaceutical Co., Ltd. (99.1% purity, China). Loxonin® tablets were purchased from Daiichi Sankyo Co., Ltd. (Tokyo, Japan). Ketoprofen was purchased from Sigma (St. Louis, MO, USA). Microcrystalline cellulose SH-101 and corn starch were supplied by Sunhere Pharmaceutical Co., Ltd. (Huainan, China). Hydroxypropylmethyl cellulose (Methocel E5LV) and Surelease® E-7-7050 (aqueous ethyl cellulose dispersion) were supplied by Colorcon Ltd. (Dartford, Kent, UK). Hard gelatin capsules were from Suzhou Capsugel Ltd. (Suzhou, China). Other chemicals were of reagent grade or higher grades.

2.2. HPLC-Assay for LXP and CA Contents

HPLC methods were used for the determination of LXP and CA. Chromatograph was carried out on Shimadzu LC-2030 (Shimadzu, Japan), equipped with an autosampler and an SPD-20A UV detector. The mobile phase used for the determination of LXP consisted of a mixture of methanol-water-acetic acid-triethylamine (600:400:1:1, *v/v*). Separation was achieved by applying an Inertsil C18 column (5 μm, 4.6 × 150 mm, Shimadzu, Japan), and the chromatogram was recorded at 223 nm. The limit of determination (LOD) and limit of quantitation (LOQ) for LXP were 0.03 and 0.1 μg/mL respectively. The mobile phase performed for the determination of CA was carried out with one part isopropanol and 999 parts 0.018 M phosphate buffer, adjusted to pH 2.5 with phosphoric acid. An Inertsil C18 column (5 μm, 4.6 × 150 mm, Shimadzu, Japan) was employed to separate the CA content. The flow rate was 1 mL/min, and the column temperature was controlled at 40 °C with the UV detection at 210 nm. The LOD and LOQ for CA were 0.2 and 0.6 μg/mL respectively. In addition, the calibration curve over the concentration range of 0.6–60 μg/mL had a regression coefficient of 0.9996.

2.3. Solubility/pH Profiles of LXP

Solubilities of LXP in different pH media were determined by adding excess of LXP to different buffer solutions: hydrochloric acid solutions (pH 1.2, 2.2), phosphate buffers (pH 3.0, 3.5, 4.0, 4.5 and 6.8). After vibrated at 25 °C for 24 h in a constant-temperature shaker (SHZ-82, Guohua Co., Ltd., Changzhou, China), 2 mL of the saturated solution was filtered through a membrane filter, and was diluted to avoid crystallization. Sample was determined by HPLC method.

2.4. Preparation of Drug-Loaded Pellets

The core consisted of loxoprofen sodium (37.1%, *w/w*), microcrystalline cellulose (41.9%, *w/w*), corn starch (19.9%, *w/w*), and talc (1.1%, *w/w*). Briefly, the mixture was blended in an ERWEKA mixer (Type AR YB5, Heusenstamm, Germany) at a speed of 40 rpm for 30 min. Then it was kneaded with the ethanol water solution (30%, v/v) in a laboratory kneader (Type LK5, Heusenstamm, Germany) for 10 min. The obtained moist mass was extruded at a speed of 500 rpm through a stainless steel barrel with a die of 0.8 mm diameter. Then, 300 g of the extrudates were processed in a 23 cm radial cut plate of spheronizer (JBZ-300, Liaoning New Drug Research Institute, China) at 1000 rpm for 10 min. The obtained pellets were collected on a Teflon tray and were dried in a hot oven at 60 °C for 24 h to remove the residual water and ethanol. Pellets with fraction size between 800 and 1250 μm were used for the following procedure.

2.5. Preparation of Sustained-Release Pellets

2.5.1. Preparation of the Dissolution-Rate Controlling Layer

Methocel E5 LV was added to purified water to achieve a hydroxypropylmethyl cellulose concentration of 4% (w/w). Then the solution was mixed with critic acid, with a concentration range from 0% to 6% (w/w), and was plasticized with PEG 6000 (0.5%, w/w) for 40 min [9]. Then, talc was added to the polymer mixture at a concentration of 2.5% (w/w). The aqueous suspension was stirred during the coating process. Using bottom spray with a Wurster insert, 50 g pellets were coated in a laboratory-scale fluid bed coater (Hanse, Changzhou, China). The process parameters were as follows: inlet temperature 50 °C, material temperature 40 °C, atomization pressure 0.15 MPa, spray rate 0.8 mL/min, and air flow rate 35 m^3/h.

2.5.2. Preparation of the Diffusion-Rate Controlling Layer

ADEC was separately used as the diffusion-rate controlling layer. Mainly, aqueous dispersion of Surelease® was diluted in purified water to achieve a solid content of 15% (w/w), and was stirred for 45 min before coating. Then the aqueous dispersion was sprayed on pellets sub-coated with the dissolution-rate controlling layer in the same equipment. The process parameters for ADEC coating were as follows: inlet temperature 54 °C, material temperature 42 °C, atomization pressure 0.18 MPa, spray rate 1.0 mL/min and air flow rate 40 m^3/h. After the coating process, pellets were cured in a hot oven at 60 °C for 16 h.

2.6. Experimental Design

A three-factor, three-level Box-Behnken experiment design was applied to evaluate the effects of different parameters on drug release rate. Briefly, the design is equal to the three replicated centre points and the set of points lying at the midpoint of each surface of the three-dimensional cube that defines the region of interest of each parameter. The three independent variables (X_1, X_2 or X_3) were the concentration of CA in the sub-coating aqueous dispersion (X_1), the sub-coating weight gain based on the dry uncoated pellet mass (X_2), and the ADEC coating weight gain based on the sub coated pellets mass (X_3). Each variable was coded to be in the range of −1, 0, 1, which represented different variable levels. Levels of the factors and constraints for the in vitro drug release based on preliminary pharmacokinetic study are listed in Table 1. The design required 15 experimental formulations. The independent variables of each formulation and their responses are listed in Table 2. The response surface model generated by the design is given as Equation (1):

$$Y = a_0 + a_{1\times 1} + a_2X_2 + a_3X_3 + a_4X_1X_2 + a_5X_2X_3 + a_6X_1X_3 + a_7X_1{}^2 + a_8X_2{}^2 + a_9X_3{}^2 \quad (1)$$

where Y is the response parameter, X_1, X_2, and X_3 are the independent parameters, a_0 is the intercept, $a_1 - a_3$ are the main effect coefficients, $a_4 - a_9$ are coefficients of parameters with interaction or quadratic effects. Statistical analysis of the model was performed in Design-Expert software (V.8.0.6, Stat-Ease Inc., Minneapolis, MN, USA). The regression models of Y_1, Y_2, and Y_3 were evaluated in terms of statistically significant coefficients using analysis of variance (ANOVA) and r^2 values. Only coefficients with p values less than 0.05 were constructed in the models. In addition, response surface plots were performed to visualize the effect of parameters and their interactions on the responses. Design space, which was determined from the common region of successful operating ranges for the responses, was established following the obtained response surface to clarify the optimal formulation.

Table 1. The factors and responses of the Box-Behnken design.

Independent Variables	Levels Used		
	−1	0	1
X_1 = citric acid concentration (%)	1	2	3
X_2 = subcoating weight (%)	6	8	10
X_3 = ADEC coating weight (%)	10	13	16
Responses		Constraints	
Y_1 = the drug release within 2 h		<30%	
Y_2 = the drug release within 6 h		50–70%	
Y_3 = the drug release within 12 h		>90%	

Table 2. Independent variables and observed responses of the Box-Behnken design.

Formulations	Factors (%)			Responses (%)		
	X_1	X_2	X_3	Y_1	Y_2	Y_3
1	2.0	10.0	10.0	21.5	71.2	93.8
2	3.0	10.0	13.0	8.4	46.0	78.5
3	2.0	6.0	16.0	7.4	40.9	68.9
4	2.0	8.0	13.0	17.1	54.8	87.0
5	3.0	6.0	13.0	13.1	53.8	81.0
6	1.0	6.0	13.0	31.4	75.0	91.0
7	1.0	8.0	16.0	21.0	58.0	86.0
8	2.0	8.0	13.0	17.7	59.7	89.0
9	1.0	10.0	13.0	27.8	75.5	92.6
10	3.0	8.0	16.0	13.6	33.1	57.7
11	1.0	8.0	10.0	56.0	88.0	99.0
12	2.0	10.0	16.0	8.6	40.2	68.3
13	3.0	8.0	10.0	20.5	60.2	91.0
14	2.0	6.0	10.0	38.2	81.6	90.0
15	2.0	8.0	13.0	13.9	60.7	89.0

2.7. In Vitro Release of LXP and CA

A dissolution test was carried out at 37 °C in 900 mL water, using a dissolution apparatus (78X-6A, Huanghai medicine inspecting institute, China) with the basket rotation speed of 100 rpm, which is specified in China Pharmacopoeia. The prepared sustained-release pellets containing 90 mg of anhydrous LXP were added to the dissolution apparatus. At pre-determined intervals, 5 mL of the sample was withdrawn and replaced with fresh medium. Then the samples were analyzed by HPLC.

In order to better understand the impact of CA on the drug release rate, simultaneous release profiles of CA and LXP in formulations with different CA concentrations were conducted. Additionally, the impact of dissolution media on the release of CA and LXP were investigated by performing the dissolution tests in the following media: pH 1.0 HCl, pH 4.5 and 6.8 phosphate buffers, and water. The contents of CA and LXP in the formulations were determined by HPLC.

2.8. Release Mechanism Studies

The in vitro release mechanisms of LXP were analyzed by seven kinetics models. As shown in Table 3, Q_t is the release amount of LXP at time t, Q_0 is the initial amount of LXP in the pellets, k_0 is the zero order release constant and k_1 is the first order release constant, k_H is the Higuchi dissolution constant, n is exponent constant characterizing different release mechanisms, a is a time scale parameter and b is a shape parameter that characterizes the curves of the release profiles. The dissolution data of LXP were fitted to these models by linear or non-linear least-squares fitting methods. The correlation coefficients calculated by regression analysis were used to evaluate the goodness of fit for each model.

Table 3. Models for drug release.

Model Name	Equation
Zero-order model	$Q_t = k_0 t$
First-order model	$ln(Q_0 - Q_t) = -k_1 t + Q_0$
Higuchi diffusion model	$Q_t = k_H t^{1/2}$
Ritger–Peppas model	$lnQ_t = n\, lnt + k$
Weibull distribution model	$log[-ln(1 - Q_t)] = b\, logt - loga$
Hixson–Crowell model	$(1 - Q_t)^{1/3} = 1 - kt$
Baker–Lonsdale model	$3/2\, [1 - (1 - Q_t)^{2/3}] - Q_t = kt$

2.9. Morphology Study

Scanning electron microscopy (S-8000; Hitachi High-Technologies Europe, Krefeld, Germany) was used to evaluate the morphology of the surface and cross-section of coating pellets. Samples were fitted on the copper sample holder with a double sided adhesive tape, sputter coated with a 10-nm thick gold layer under argon atmosphere.

2.10. The Pharmacokinetic Studies

2.10.1. Administration Programme

All animal treatments were performed in accordance with the Regulations of the Administration of Affairs Concerning Experimental Animals and the study protocol was admitted by the Ethics Committee of China Pharmaceutical University (Approval No. 2018-0315). An open label, randomized, two-period crossover experiment design with one week wash-out period was used in this study. Six male beagle dogs (weight 8.7 ± 1.1 kg), fasted but free access to water for 12 h prior to the experiment, were used in the study. Pellets of the optimal formulation were filled into hard gelatin capsules. The immediate release tablet (Loxonin®, 60 mg anhydrous LXP) and the capsule of the optimal formulation (90 mg anhydrous LXP) were administered to beagles in the morning with 100 mL water. Then, 6 h after dosing, dogs were provided with standard food.

A total of 2 mL of the blood samples were withdrawn before and then 0.5, 1.0, 2.0, 3.0, 4.0, 5.0, 6.0, 8.0, 10.0, and 12.0 h after dosing via cannulated needle from front legs. Plasma was obtained by centrifuging the blood at 4000 rpm for 15 min, and was kept frozen at −20 °C before analysis.

2.10.2. Determination of LXP in Plasma

A stable and selective HPLC method, modified by previous papers [12,35], was applied for the analysis of LXP in dog plasma. Pretreatment was carried out by adding 50 µL of internal standard (100 µg/mL ketoprofen in acetonitrile), 50 µL of zinc sulfate solution (10%, *w/w*) and 750 µL acetonitrile into 500 µL plasma sample. After vortex-mixing for 1 min, sample was centrifuged at 8000 rmp for 15 min. Then 10 µL of the supernatant was injected into the HPLC system. The separation was performed on an Inertsil C18-ODS column (5 µm, 4.6 × 150 mm, Shimadzu, Japan) with a guard column (4.6 × 10 mm, 5 µm particle size, ANPEL Laboratory Technologies Inc, Shanghai, China) at a flow rate of 1 mL/min. Additionally, the mobile phase was a mixture of acetonitrile and 0.05 M monopotassium phosphate (35:65, *v/v*), adjusted to pH 3.0 with phosphoric acid. Chromatograms were recorded at 223 nm with a Shimadzu-SPD detector. The linear range of this method was 0.1–20.0 µg/mL with an r^2 value of not less than 0.999. The lower limit of quantification (LLOQ) was 100 ng/mL and the extraction recoveries of high, middle, and low concentrations of LXP were 102.5 ± 2.0%, 97.1 ± 2.5%

and 97.1 ± 5.9%, respectively. The R.S.D.s reflecting the intra-day and inter-day precision of LXP were less than 11.77%.

2.10.3. Bioavailability Study

Non-compartmental pharmacokinetic analysis was applied to calculate parameters such as T_{max}, C_{max}, AUC_{0-t}, and $AUC_{0-\infty}$ from the plasma concentration-time curve data using WinNonlin software (version 1.5, Pharsight Corp. Mountain View, CA, USA). The relative bioavailability of the optimal formulation to the commercial tablet (reference) was calculated using the following Equation (2):

$$\text{Relative.bioavaibility} = \frac{AUC_{T0-\infty} \times X_R}{AUC_{R0-\infty} \times X_T} \times 100\% \quad (2)$$

where X_R and X_T were the administered dose of the reference and test respectively. Results were presented as means ± standard deviation. A one-way ANOVA (SPSS, version 19) with $p < 0.05$ as a level of significance was applied to examine the differences of C_{max} and $AUC_{0-\infty}$ between the test and reference.

3. Results and Discussions

3.1. Impact of CA on Drug Release

Formulations with different concentrations of CA in the sub-layer were developed to evaluate the effect of pH-modifier on the drug release rate, while the dissolution-controlling layer and ADEC coating levels were kept at 8% and 11% respectively. The results in Figure 2 illustrated that formulation without CA showed a fast release of LXP (>80% within 2 h), while the drug release within 2 h was decreased to 40% at a CA concentration of 1%. Additionally, the release rate continued to decrease with the increase of CA concentration, which showed 16.37%, 11.34%, and 7.77% of LXP release within the first 2 h. At the CA concentration of 1%, a completed drug release was finished within 6 h. While at higher CA concentrations, there were still 21.80% (2.5% CA) and 32.43% (4.0% CA) of the initial drug amount released after 6 h.

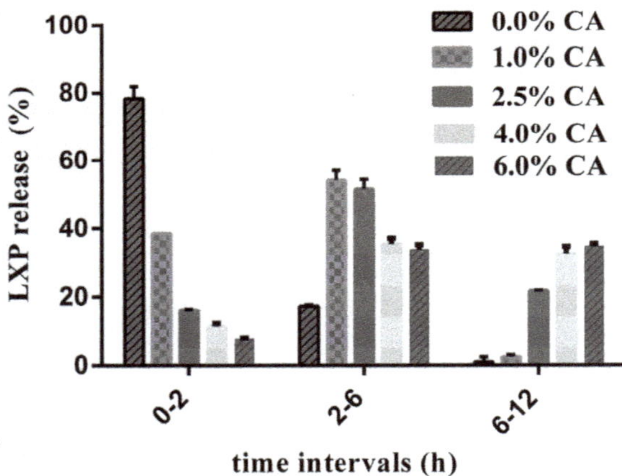

Figure 2. Effect of citric acid concentration on the drug release within different intervals.

As a pH modifier, CA was aimed to modulate the pH_M inside the systems. For pH-sensitive compound, its solubility is more appropriate to be described as the solubility in the diffusion layer at the surface of the dissolving particles [25]. Therefore, according to the Noyes–Whitney theory, the dissolution rate of LXP was much more dependent on the solubility in the low pH_M beneath the diffusion-controlling layer, other than the dissolution media. Theoretically, drug release rate from a coherent film coating system is controlled by both the coating level and the drug concentration gradient across the coating film, which obeyed the Fick's diffusion law. As the film coating level was kept constant, drug release rate was predominantly controlled by the drug concentration gradient, which was determined by the dissolution rate of LXP inside the pellets. Therefore, as the drug release was significantly decreased with the increase of CA concentrations (Figure 2), the first step of developing a dissolution-rate controlling layer proved to work.

Furthermore, simultaneous release profiles in Figure 3 were constructed to investigate the impact of dynamic release process of CA on the drug release rate. In formulations with lower CA concentrations, with the release of CA during the dissolution period, pH_M could be changed from 0.4 (the saturated solution pH of CA) to the approximate equilibrium pH of the dissolution medium [36]. As the solubility of LXP changed approximately 300 times within this pH range (Supplementary Materials Figure S1), the drug dissolution rate was significantly dependent on the amount of CA left inside the pellets. Therefore, matching release profiles of LXP and CA were observed at low concentrations of CA.

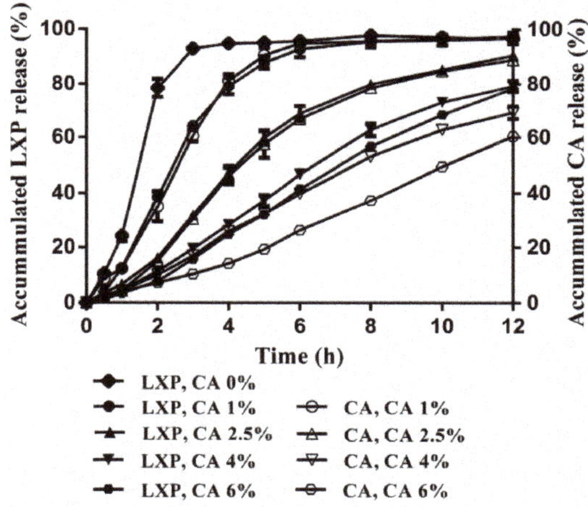

Figure 3. The simultaneous release profiles of citric acid and loxoprofen from sustained release pellets at different concentrations of citric acid ($n = 3$).

While in formulations with higher CA concentrations, as a sufficient amount of CA remained inside the pellets at the end of the dissolution period, e.g., 30% or 40% of the initial CA amount left at the concentration of 4% or 6% respectively, a constant and effective pH_M was achieved inside the pellets, which resulted in no further decrease of the drug release rate (Figure 3). A similar phenomenon was observed in a matrix tablet containing dipyridamole, due to a constant and effective pH_M maintained by fumaric acid inside the dosage form, no further enhancement of the dipyridamole release was observed after increasing the fumaric acid concentration to 40% [37]. Therefore, a discrepancy of the release profiles between CA and LXP was observed in formulations with higher CA concentrations.

3.2. Release Experiments and Statistical Evaluation

3.2.1. Testing of Drug Release

Experimental variables and observed responses of all the 15 formulations were listed in Table 2. And their drug dissolution profiles were displayed in Figure 4. At a low level of CA concentration, most of the formulations (Formulation Nos. 6, 9, 11) showed a fast drug release except Formulation No. 7, which had a high coating level of ADEC. The fast release in Formulation Nos. 6, 9, 11 was attributed to an inefficient pH_M inside the pellets and the short diffusion pathway of ADEC coating, which could be identified as the failure of the first- and second-step control. As a prolonged diffusion pathway was developed in Formulation No. 7, the release rate of LXP was significantly decreased. While at a low coating level of ADEC, most of the formulations (Formulation Nos. 1, 11, 14) showed a fast drug release rate, expect Formulation No. 13, which was incorporated with a high concentration of CA. The fast release in Formulation Nos. 1, 11, 14 could be explained by a failure of the second-step control, as a short diffusion pathway created by the low coating level of ADEC was unable to retard the release rate of LXP. However, when 3% of the CA concentration was applied in Formulation No. 13, the effective control of the drug dissolution-rate could compensate for the failure of the diffusion-rate control to achieve a sustained release of LXP.

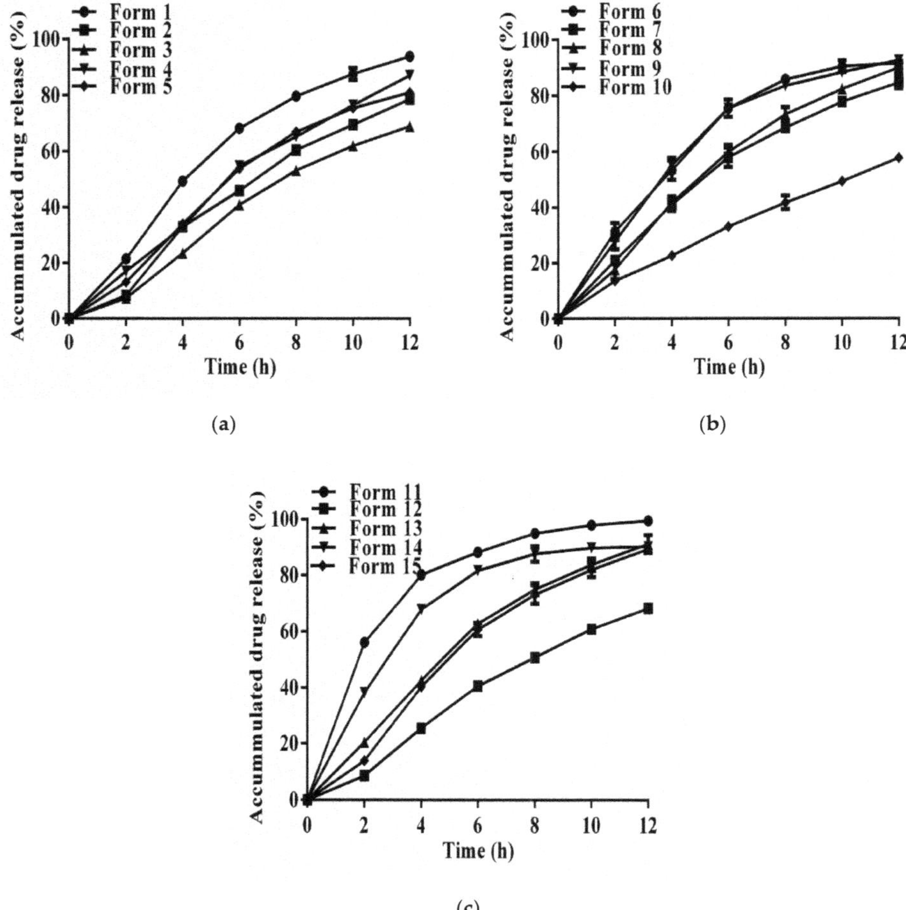

Figure 4. Dissolution profiles of loxoprofen in formulations prepared by the Box-Behnken design experiments (**a**) Formulation Nos. 1–5, (**b**) Formulation Nos. 6–10, (**c**) Formulation Nos. 11–15.

3.2.2. Regression Equations

Based on the experiment data, the coefficients and their *p*-values of the fitted full quadratic equations calculated by Expert-Design 8.0.6 (Stat-Ease Inc., Minneapolis, MN, USA) are listed in Table 4.

The final equations consisted of only statistically significant coefficients. It is clear that the citric acid concentration (X_1) and ADEC coating weight gain (X_3) showed significant effects on the drug release rate throughout the dissolution period, while the weight gain of the dissolution-rate controlling layer (X_2) showed only a weak effect during the dissolution period. The impact of X_1 on the drug release rate verified the effectiveness of the first-step control on the drug release rate, which could be explained by its impact on the pH_M, as extensively reported in the literature [36,37]. The effect of X_3 on the drug release rate could be attributed to its control on the drug diffusion rate, which was considered as the second-step control [38]. Besides, an interaction effect of X_1 and X_3 was observed on the response of Y_1 and Y_3, which might be explained by a speculation that X_3 also showed an effect on the release rate of citric acid. Additionally, it seemed that X_3 played a more dominant role on the final drug release, as high coefficients of the main, interaction, and quadratic effects of X_3 were observed in Y_3. From the statistic results in Table 4, we could conclude that both the dissolution-rate and diffusion-rate controlling steps have significant effects on the drug release rate.

Table 4. Regression coefficients and associated *p*-values of the fitted models.

Term	Drug Release Within 2 h (Y_1)		Drug Release Within 6 h (Y_2)		Drug Release Within 12 h (Y_3)	
	Cofficient	***p*-Value**	**Cofficient**	***p*-Value**	**Cofficient**	***p*-Value**
Constant	16.23	0.000	58.40	0.000	88.33	0.000 *
X_1	−10.08	0.001 *	−12.91	0.001 *	−7.38	0.000 *
X_2	−2.98	0.007 *	−2.30	0.091	0.29	0.697
X_3	−10.70	0.001 *	−16.11	0.001 *	−11.79	0.000 *
X_1*X_2	−0.27	0.787	−2.07	0.240	−1.02	0.347
X_2*X_3	4.47	0.006 *	2.42	0.180	−1.10	0.316
X_1*X_3	7.03	0.000 *	0.70	0.672	−4.72	0.005 *
X_1*X_1	6.40	0.001 *	2.78	0.148	0.13	0.902
X_2*X_2	−2.45	0.058	1.40	0.427	2.69	0.047 *
X_3*X_3	5.15	0.004 *	−1.33	0.451	5.39	0.003 *
Regression equation	$Y_1 = 16.23 - 10.08X_1 - 2.98X_2 - 10.7X_3 - 4.47X_2X_3 + 7.03X_1X_3 + 6.4X_1^2 + 5.15X_3^2$		$Y_2 = 58.40 - 12.91X_1 - 16.11X_3$		$Y_3 = 88.33 - 7.38X_1 - 11.79X_3 - 4.72X_1X_3 - 2.69X_2^2 - 5.39X_3^2$	
R-Squared	0.9921		0.9865		0.9891	

* *p*-value < 0.05.

3.2.3. Response Surface Plots

The relationship between the dependent and independent variables was further elucidated using a 3D response surface plot, which is useful to see the effect of two factors on the response at one time while the third factor is kept at a constant level. The effects and interactions between concentration of citric acid (X_1), the sub coating weight gain (X_2), and ADEC coating weight gain (X_3) on the finial drug release (Y_3) are given in Figure 5. The similar impacts of the three factors on the other responses (Y_1 and Y_2) can be seen in Supplementary Materials Figures S2 and S3. As illustrated in Figure 5a,b, it was clear to see that X_2 showed little effect on Y_3 irrespective of the levels of other two factors. This was attributed to the fact that the dissolution-rate controlling layer is made up of aqueous polymer, which was dissolved before 12 h.

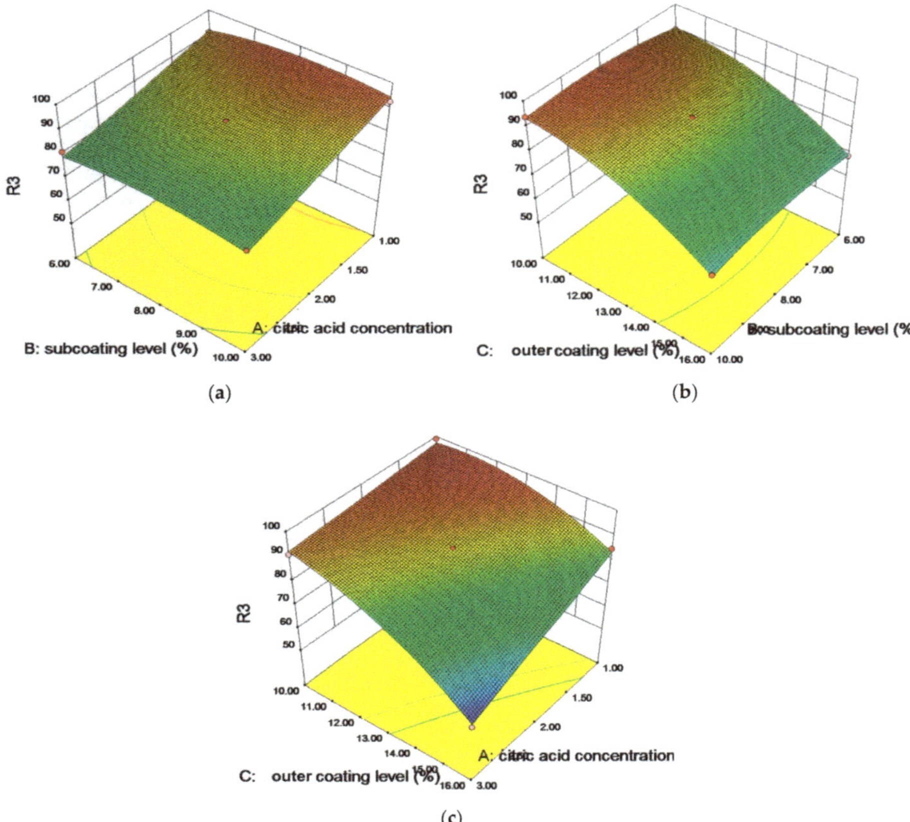

Figure 5. Contour plots showing the effects of (**a**) X_1 and X_2, (**b**) X_2 and X_3, and (**c**) X_1 and X_3 on the response Y_3.

The effects of citric acid concentration (X_1) and the ADEC coating weight gain (X_3) on Y_3 are depicted in Figure 5c. While X_3 was kept at low level, the increase of X_1 from 1% to 3% showed no effect on Y_3, which kept nearly constant at above 90%. While at a high level of X_3, the increase of X_1 from 1% to 3% resulted in a significant decrease of Y_3 from 85% to 60%. The result indicated that an interaction effect of the two factors existed on the drug release rate, as mentioned in Section 3.2.2. At a high level of X_3, an effective diffusion barrier was formed on the surface of the pellets [38], which significantly reduced the diffusion rate of loxoprofen. However, the release rate of loxoprofen was not solely controlled by the diffusion-controlling layer. For example, a nearly complete release of loxoprofen (>85%) was observed at a high level of X_3 (Figure 5c), when X_1 was kept at a low level of 1%. Therefore, the release rate of loxoprofen was a combined result of the two controlling steps. At a low level of X_3, the citric acid was soon released regardless of its concentrations, which resulted in a quick increase of the pH_M and a fast drug release rate. When the concentration of citric acid and ADEC coating weight gain were kept at high levels, both the drug dissolution and diffusion rate were reduced, which resulted in a prominent decrease of the drug release rate.

3.2.4. Design Space and Formulation Parameters Optimization

Design space was defined by the ICH Q8 as the relationship between the process inputs (material attributes and process parameters) and the critical quality attributes that have been demonstrated to provide assurance of quality [39]. The wider the design space is, the more robust and flexible the process is to resist variations [40]. As the response surface models of the output parameters as a function of selected variables were given, design space of X_1, X_2, and X_3 was determined by applying constraints on Y_1 (<30%), Y_2 (50–70%), and Y_3 (>90%). The yellow overlap region of ranges for the three responses in Figure 6a–c show the proposed design space of the citric acid concentration X_1 and the ADEC coating weight gain X_3 at three different levels of the sub-coating weight gain X_2. As shown in Figure 6a, there was no design space of X_1 and X_3 at the low level of X_2. Additionally, Figure 6c depicted a narrow design space of X_1 and X_3 at high level of X_2, which would increase the difficulty of the operation process since an accurate coating load of ADEC must be achieved during the manufacturing process. While at the medium level of X_2 (Figure 6b), the design space was expanded, which showed a less strict field of ADEC coating level. As the design space depicted the ranges of the formulation parameters for achieving the desired quality of product, the levels of the three factors for the optimal formulation must be set within the design space. Considering the robustness and flexibility, parameters of the optimal formulation were set at the medium level of sub coating weight gain with the CA concentration and coating level of ADEC at 2.5% and 11.0% respectively. The model predicted a release profile of 19.87% at 2 h, 64.48% at 6 h, and 91.71% at 12 h. To verify these values, a new batch of the optimal formulation was prepared. The obtained release data of the optimal formulation were in close agreement with the predicted values with a maximum percentage error of 11.73% at the initial release (data not showed).

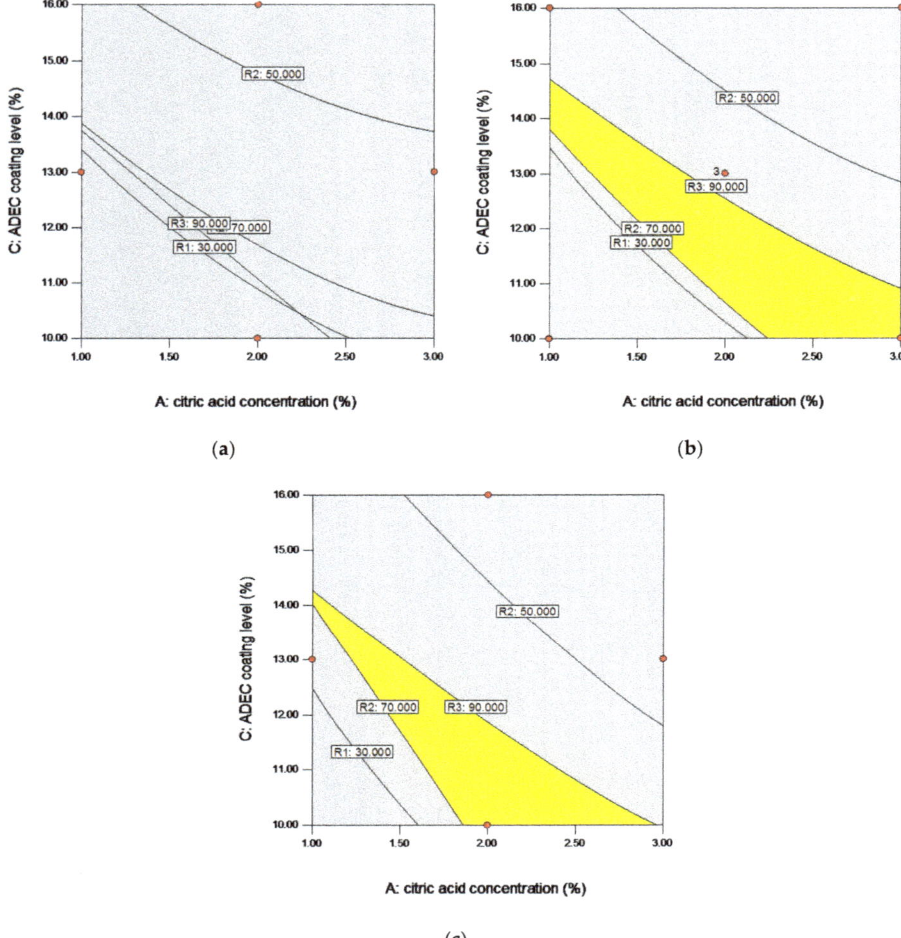

Figure 6. Design space of operating variables of the citric acid (CA) concentration and aqueous dispersion of ethyl cellulose (ADEC) coating level (**a**) at the low level of the sub-layer coating weight gain, (**b**) at the medium level of the sub-layer coating weight gain, and (**c**) at the high level of the sub-layer coating weight gain (yellow zone: design space; grey zone: failure space).

3.3. Simultaneous Release of CA and LXP from the Optimal Formulation in Different Dissolution Media

In order to evaluate the effect of pH on drug release, various media simulating different physiology pH values were applied. As shown in Figure 7, dissolution tests were performed in pH 1.0 HCl, pH 4.5 and pH 6.8 phosphate buffer solutions and water. Furthermore, the release profiles of CA were also investigated in these media. As illustrated in Figure 7, drug release profiles were pH-independent at pH above 4.5, and showed similar release profiles to that of CA.

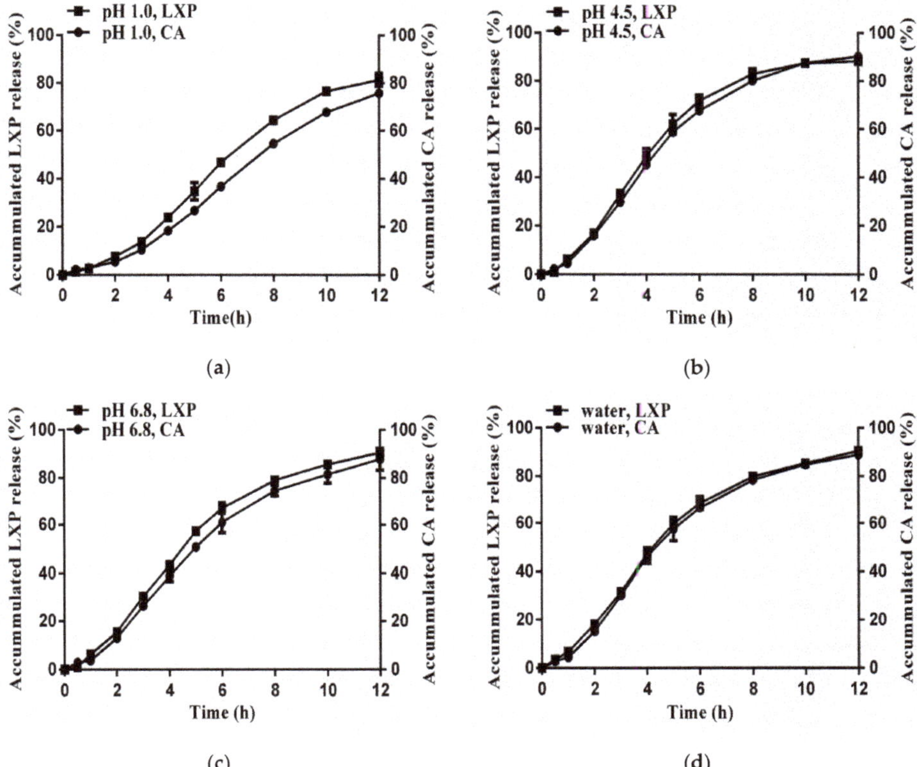

Figure 7. Loxoprofen and citric acid released from sustained release pellets in different dissolution media. (**a**) pH 1.0 HCl (**b**) pH 4.5 phosphate buffer (**c**) pH 6.8 phosphate buffer (**d**) water (means ± SD, $n = 3$).

Although the solubility of LXP was pH-independent in media with pH above 4.5 (Supplementary Materials Figure S1), it seemed that the dissolution media were not the reason for this pH-independent release behavior. As the drug showed a completed release within 3 h without the incorporation of CA inside the pellets (Figure 3, formulation with the CA concentration of 0%), it should exhibit a similar release for the optimal formulation as the dissolution media were also above 4.5. In the contrast, the optimal formulation showed sustained release for almost 12 h. It was the pH_M created by CA, which showed similar release profiles at pH above 4.5, that accounted for the pH-independent release profiles of LXP (Figure 7). As mentioned before, the saturated solution pH of CA was 0.4 [36], which was much lower than that of the dissolution media except the pH 1.0 HCl. However, with the release of CA during the dissolution period, the amount of CA left inside the optimal pellets was insufficient to maintain a constant pH_M inside the pellets. The pH_M was gradually increased, which resulted in a consistent enhancement of the drug solubility. In addition, matching sustained release profiles of LXP and CA were achieved in Figure 7.

While at pH 1.0, it was the dissolution media that showed a major effect on the drug release rate. With the decrease of CA, the pH_M would soon be increased up to above 1.0. However, the dissolution media that penetrated into the pellets provided a stable pH_M inside the pellets, which in turn resulted in a different release profile to the other three. In conclusion, the dissolution media and the incorporated CA played a combined effect on the drug release rate. In media with high pH values, the CA showed a greater effect on the drug release rate. While in media with low pH values, it was the dissolution media that dominated the drug release rate.

3.4. Release Mechanism Studies

In order to elucidate the transport mechanism of LXP in the optimal formulation, different mathematical models were applied to analyze the kinetics of the release data. As shown in Table 5, the incorporation of CA into the sub-layer resulted in abnormal release kinetics of this ADEC coating system, as the n value for Ritger–Peppas was 0.7422, which is between 0.45 and 0.89, indicating a non-Fick diffusion [41]. Additionally, a general empirical equation of Weibull distribution model with r^2 of 0.9944 was more appropriate to describe the release process of the optimal formulation. In the model, the derived estimate of b value was calculated to be 1.38, which represented a sigmoid shape curve ($b > 1$) for the release profile [41]. The initial slow release representing the starting part of the sigmoid curve was a result of several factors. As reported previously, the permeability of water through the EC coating is much faster than the permeability of the compound [38], which might contribute to the initial slow release of LXP. Besides, the decreased dissolution rate created by the initial low pH_M inside the pellets also played an important role on the initial slow release. Furthermore, the hydration of the hydroxypropyl methyl cellulose (HPMC) inside the pellets during the initial dissolution period could also inhibit the initial drug release rate. Thereafter, due to the saturation of water inside the pellets and the disruption of the dissolution-rate controlling layer, the drug release rate was dominated by the dissolution- and diffusion-rate control, which resulted in a sustained release profile.

Table 5. Models simulated for the drug release profiles of the optimal formulation.

Content	Model	Equation	r^2
loxoprofen	Zero-order model	$Q_t = 0.0833t + 0.0238$	0.9288
	First-order model	$ln(Q_0 - Q_t) - lnQ_0 = -0.1787t + 0.0814$	0.9794
	Higuchi diffusion model	$Q_t = 0.3126t^{1/2} - 0.1496$	0.9454
	Ritger–Peppas model	$lnQ_t = 0.7422\ lnt + 2.7562$	0.9597
	Weibull distribution model	$log[-ln(1 - Q_t)] = 1.3840\ logt - 1.0449$	0.9944
	Hixson–Crowell model	$(1 - Q_t)^{1/3} = -0.0514t + 1.0167$	0.9874

3.5. Scanning Electron Photomicrographs

Figure 8 shows the scanning electron photomicrographs of the optimal pellets. The surface of pellets were smooth (Figure 8a,b), and no crack could be seen. Besides, layers of the dissolution-rate controlling layer and the diffusion-rate controlling layer were clearly seen in the cross-section of the coating pellets (Figure 8c,d). These results indicated that a successful procedure had been developed for manufacturing the sustained release pellets.

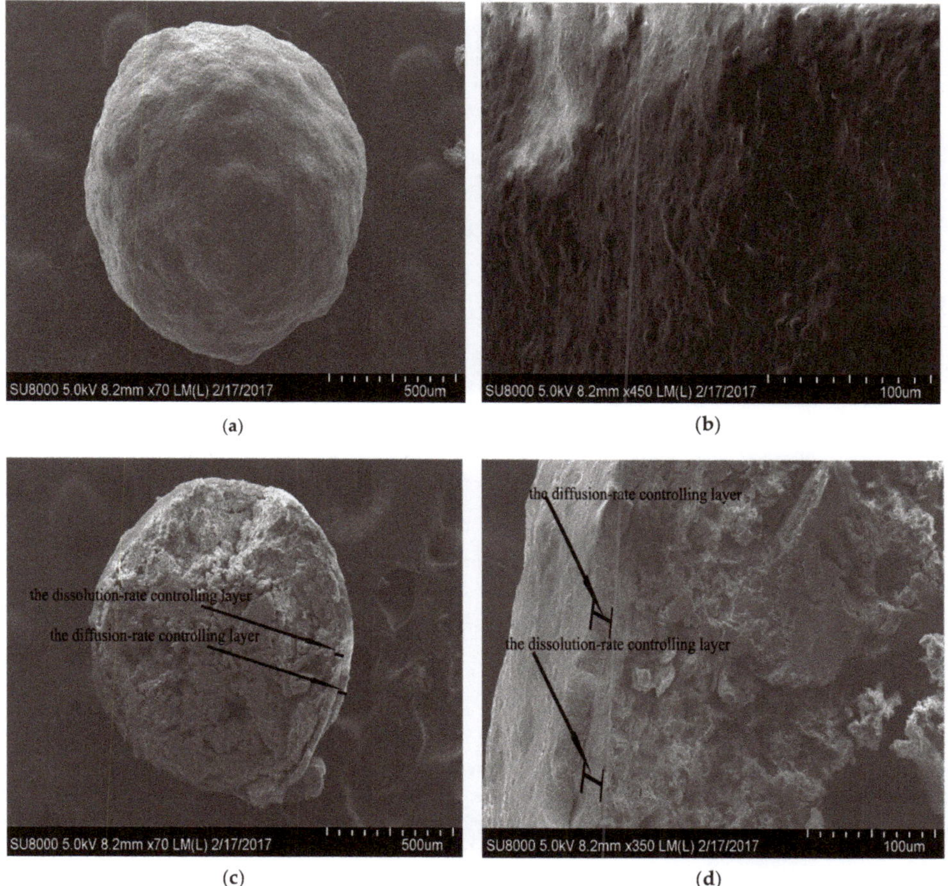

(a) (b) (c) (d)

Figure 8. SEM photographs of pellets with double coating layers: (**a**) Surface of sustained release pellets with 70 magnifications, (**b**) surface of sustained release pellets with 450 magnifications, (**c**) cross-section of sustained release pellets with 70 magnifications, (**d**) cross-section of sustained release pellets with 350 magnifications.

3.6. Pharmacokinetic Studies

The pharmacokinetic studies of the optimal pellets and the commercial tablets were investigated on fasted beagles. The profile of mean plasma concentrations of LXP versus time is shown in Figure 9. The main pharmacokinetic parameters are summarized in Table 6. As shown in Figure 9, the plasma concentration of the commercial tablet quickly increased and reached the peak concentration of 5.16 μg/mL at 0.5 h after administration. Then it dropped down and was only 0.2 μg/mL at 6 h. This was attributed to the short half-life ($t_{1/2}$ = 64.46 min) of LXP [12], which resulted in a quick elimination of the drug in vivo. The optimal formulation reached the maximum plasma concentration of 2.40 μg/mL at 5 h after administration, and the drug concentration fell slowly even at 12 h, when the drug concentration was 0.15 μg/mL, in contrast with the undetectable drug concentration in plasma for the commercial tablet 8 h after administration.

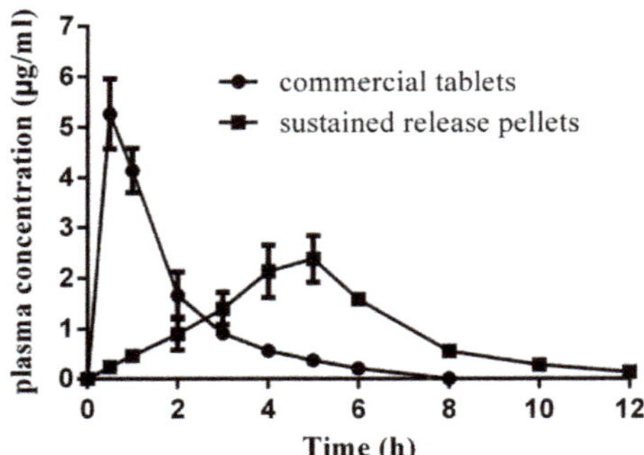

Figure 9. Plasma drug concentrations vs. time after oral administration of conventional tablets (60 mg) and sustained release pellets (90 mg).

Table 6. The pharmacokinetic parameters of loxoprofen after oral administration of the optimal sustained release pellets and commercial tablets in beagle dogs ($n = 6$).

Pharmacokinetic Parameters	C_{max} (μg/mL)	T_{max} (h)	AUC_{0-12} (μg h/mL)	$AUC_{0-\infty}$ (μg h/mL)	Relative Bioavailability (%)
Optimal pellets (90 mg)	2.60 ± 0.23	4.80 ± 0.57	12.77 ± 0.88	13.48 ± 0.94	87.16 ± 0.07
Commercial tablets (60 mg)	5.16 ± 0.60	0.60 ± 0.22	9.73 ± 0.61	10.31 ± 0.45	-

As a pro-drug, LXP inhibits the activities of cyclooxygenase-1 and -2 (COX-1 and COX-2) by its active metabolite trans-OH LXP, of which the IC_{50} values for COX-1 and COX-2 were 0.38 and 0.12 μM, respectively [42]. It has been reported that the concentration of trans-OH LXP, which was the major metabolite of LXP, was equal to more than half of the LXP concentration detected in plasma [12]. As the LXP concentration range in plasma of the optimal pellets was 0.6–10.0 μM, we can deduce that the concentration of trans-alcohol LXP after administration of the optimal pellet would be higher than the IC_{50} of the trans-OH LXP. Therefore, a therapy concentration of trans-OH LXP in plasma after administration of the optimal pellets would be maintained for almost 12 h. As the frequency of dosage of the optimal formulation was reduced to two times a day, the patient's compliance would be better improved. It has been reported that the incidence of gastric lesions after administration of LXP in rats showed a dose-dependent manner. Additionally, the amount of PGE_2, which has a strong protective effect on the GI mucosa, also decreased in a concentration-dependent manner after treatment of LXP within the concentration range of 1.0 μM to 1.0 M [14,43]. Therefore, the risk of GI lesions would be significantly decreased, since the initial burst release disappeared in vivo and the C_{max} of LXP was significantly decreased from 20 to 10 μM after administration of the optimal formulation. Besides, a less fluctuant drug concentration in plasma was achieved for the optimal pellets in Figure 9. The significant difference ($p < 0.05$) of $AUC_{0-\infty}$/dose between the test formulation and conventional tablet might be caused by the limited GI absorption site of LXP, which would need further investigation. The relative bioavailability of the test formulation was 87.16% compared with the reference, and it would be improved in patients as a prolonged GI transit time has been reported in humans [44].

4. Conclusions

In conclusion, this article provided a novel concept of two-step control of the release rate of pH-sensitive drugs. Additionally, the results of the drug in vitro release profiles proved that both the dissolution-rate controlling step created by the sub-layer containing CA and the diffusion-rate controlling step developed by the ADEC coating showed significant effects on the release rate of LXP. In addition, the amount of the acid modifier in the optimal formulation, which accounted for only approximately 3% of the total preparation weight, was dramatically decreased compared with other formulations containing acidic modifiers. The in vivo studies revealed that this novel two-step control system could achieve a more stable and sustained release plasma concentration of LXP compared with the immediate release tablet.

Supplementary Materials: The following are available online at http://www.mdpi.com/1999-4923/11/6/260/s1, Figure S1: pH solubility profiles of loxoprofen at 25 °C. Figure S2: Contour plots showing the effects of (**A**) X_1 and X_2, (**B**) X_2 and X_3, and (**C**) X_1 and X_3 on the response Y_1. Figure S3: Contour plots showing the effects of (**A**) X_1 and X_2, (**B**) X_2 and X_3, and (**C**) X_1 and X_3 on the response Y_2.

Author Contributions: Conceptualization, D.W. and L.L.; methodology, D.W., M.Z., J.Z.; and L.L.; formal analysis, L.L.; investigation, D.W.; writing—original draft preparation, D.W. and M.Z, writing—review and editing, D.W. and M.Z.

Funding: This research no external funding.

Acknowledgments: We are thankful to Colorcon Coating School® for technical and excipient supports.

Conflicts of Interest: The authors declare no conflicts of interest.

References

1. Peter, J.C. Tramadol SR formulations: Pharmacokinetic comparison of a multiple-units dose (capsule) versus a single-unit dose (tablet). *Clin. Drug Investig.* **2005**, *7*, 435–443.
2. Chen, T.; Li, J.; Chen, T.; Sun, C.C.; Zheng, Y. Tablets of multi-unit pellet system for controlled drug delivery. *J. Control. Release* **2017**, *262*, 222–231. [CrossRef] [PubMed]
3. Felton, L.A. Mechanisms of polymeric film formation. *Int. J. Pharm.* **2013**, *457*, 423–427. [CrossRef] [PubMed]
4. Howick, K.; Alam, R.; Chruscicka, B.; Kandil, D.; Fitzpatrick, D.; Ryan, A.M.; Cryan, J.F.; Schellekens, H.; Griffin, B.T. Sustained-release multiparticulates for oral delivery of a novel peptidic ghrelin agonist: Formulation design and in vitro characterization. *Int. J. Pharm.* **2018**, *536*, 63–72. [CrossRef] [PubMed]
5. Thapa, P.; Thapa, R.; Choi, D.H.; Jeong, S.H. Effects of pharmaceutical processes on the quality of ethylcellulose coated pellets: Quality by design approach. *Powder Technol.* **2018**, *339*, 25–38. [CrossRef]
6. López, E.V.; Luzardo Álvarez, A.; Blanco Méndez, J.; Otero Espinar, F.J. Cellulose-polysaccharide film-coating of cyclodextrin based pellets for controlled drug release. *J. Drug Deliv. Sci. Technol.* **2017**, *42*, 273–283. [CrossRef]
7. Dekyndt, B.; Verin, J.; Neut, C.; Siepmann, F.; Siepmann, J. How to easily provide zero order release of freely soluble drugs from coated pellets. *Int. J. Pharm.* **2015**, *478*, 31–38. [CrossRef] [PubMed]
8. Muschert, S.; Siepmann, F.; Leclercq, B.; Carlin, B.; Siepmann, J. Prediction of drug release from ethylcellulose coated pellets. *J. Control. Release* **2009**, *135*, 71–79. [CrossRef] [PubMed]
9. Sadeghi, F.; Ford, J.L.; Rajabi-Siahboomi, A. The influence of drug type on the release profiles from Surelease-coated pellets. *Int. J. Pharm.* **2003**, *254*, 123–135. [CrossRef]
10. Terada, A.; Naruto, S.; Wachi, K.; Tanaka, S.; Iizuka, Y.; Misaka, E. Synthesis and antiinflammatory activity of [(cycloalkylmethyl)phenyl]acetic acids and related compounds. *J. Med. Chem.* **1984**, *27*, 212–216. [CrossRef] [PubMed]
11. Mu, R.; Bao, C.D.; Chen, Z.W.; Zheng, Y.; Wang, G.C.; Zhao, D.B.; Hu, S.X.; Li, Y.J.; Shao, Z.W.; Zhang, Z.Y. Efficacy and safety of loxoprofen hydrogel patch versus loxoprofen tablet in patients with knee osteoarthritis: A randomized controlled non-inferiority trial. *Clin. Rheumatol.* **2016**, *35*, 165–173. [CrossRef] [PubMed]
12. Cho, H.Y.; Park, C.H.; Lee, Y.B. Direct and simultaneous analysis of loxoprofen and its diastereometric alcohol metabolites in human serum by on-line column switching liquid chromatography and its application to a pharmacokinetic study. *J. Chromatogr. B Analyt. Technol. Biomed. Life Sci.* **2006**, *835*, 27–34. [CrossRef] [PubMed]
13. Mizukami, K.; Murakami, K.; Yamauchi, M.; Matsunari, O.; Ogawa, R.; Nakagawa, Y.; Okimoto, T.; Kodama, M.; Fujioka, T. Evaluation of selective cyclooxygenase-2 inhibitor-induced small bowel injury: Randomized cross-over study compared with loxoprofen in healthy subjects. *Dig. Endosc.* **2013**, *25*, 288–294. [CrossRef] [PubMed]

14. Yamakawa, N.; Suemasu, S.; Matoyama, M.; Kimoto, A.; Takeda, M.; Tanaka, K.; Ishihara, T.; Katsu, T.; Okamoto, Y.; Otsuka, M.; et al. Properties and synthesis of 2-{2-fluoro (or bromo)-4-[(2-oxocyclopentyl)methyl]phenyl}propanoic acid: Nonsteroidal anti-inflammatory drugs with low membrane permeabilizing and gastric lesion-producing activities. *J. Med. Chem.* **2010**, *53*, 7879–7882. [CrossRef]
15. Tak, J.W.; Gupta, B.; Thapa, R.K.; Woo, K.B.; Kim, S.Y.; Go, T.G.; Choi, Y.; Choi, J.Y.; Jeong, J.H.; Choi, H.G.; et al. Preparation and optimization of immediate release/sustained release bilayered tablets of loxoprofen using box-behnken design. *AAPS PharmSciTech* **2017**, *18*, 1125–1134. [CrossRef]
16. Venkatesan, P.; Manavalan, R.; Valliappan, K. Preparation and evaluation of sustained release loxoprofen loaded microspheres. *J. Basic. Clin. Pharm.* **2011**, *2*, 159–162.
17. Zaman, M.; Rasool, S.; Ali, M.Y.; Qureshi, J.; Adnan, S.; Hanif, M.; Sarfraz, R.M.; Ijaz, H.; Mahmood, A. Fabrication and analysis of hydroxypropylmethyl cellulose and pectin-based controlled release matrix tablets loaded with loxoprofen sodium. *Adv. Polym. Technol.* **2015**, *34*, 1624–1631. [CrossRef]
18. Khanum, H.; Ullah, K.; Murtaza, G.; Khan, S.A. Fabrication and in vitro characterization of HPMC-g-poly(AMPS) hydrogels loaded with loxoprofen sodium. *Int. J. Biol. Macromol.* **2018**, *120*, 1624–1631. [CrossRef]
19. Sonawane, R.O.; Patil, S.D. Fabrication and statistical optimization of starch-kappa-carrageenan cross-linked hydrogel composite for extended release pellets of zaltoprofen. *Int. J. Biol. Macromol.* **2018**, *120*, 2324–2334. [CrossRef]
20. Patil, H.; Tiwari, R.V.; Upadhye, S.B.; Vladyka, R.S.; Repka, M.A. Formulation and development of pH-independent/dependent sustained release matrix tablets of ondansetron HCl by a continuous twin-screw melt granulation process. *Int. J. Pharm.* **2015**, *496*, 33–41. [CrossRef]
21. Kasashima, Y.; Uchida, S.; Yoshihara, K.; Yasuji, T.; Sako, K.; Namiki, N. Oral sustained-release suspension based on a lauryl sulfate salt/complex. *Int. J. Pharm.* **2016**, *515*, 677–683. [CrossRef] [PubMed]
22. Kang, W.H.; Nguyen, H.V.; Park, C.; Choi, Y.W.; Lee, B.J. Modulation of microenvironmental pH for dual release and reduced in vivo gastrointestinal bleeding of aceclofenac using hydroxypropyl methylcellulose-based bilayered matrix tablet. *Eur. J. Pharm. Sci.* **2017**, *102*, 85–93. [CrossRef] [PubMed]
23. Ploen, J.; Andersch, J.; Heschel, M.; Leopold, C.S. Citric acid as a pH-modifying additive in an extended release pellet formulation containing a weakly basic drug. *Drug Dev. Ind. Pharm.* **2009**, *35*, 1210–1218. [CrossRef] [PubMed]
24. Healy, A.M.; Corrigan, O.I. Predicting the dissolution rate of ibuprofen-acidic excipient compressed mixtures in reactive media. *Int. J. Pharm.* **1992**, *84*, 167–173. [CrossRef]
25. Badawy, S.I.; Hussain, M.A. Microenvironmental pH modulation in solid dosage forms. *J. Pharm. Sci.* **2007**, *96*, 948–959. [CrossRef] [PubMed]
26. Xu, L.; Luo, Y.; Feng, J.; Xu, M.; Tao, X.; He, H.; Tang, X. Preparation and in vitro–in vivo evaluation of none gastric resident dipyridamole (DIP) sustained-release pellets with enhanced bioavailability. *Int. J. Pharm.* **2012**, *422*, 9–16. [CrossRef]
27. Frenning, G. Modelling drug release from inert matrix systems: From moving-boundary to continuous-field descriptions. *Int. J. Pharm.* **2011**, *418*, 88–99. [CrossRef] [PubMed]
28. Cuppok, Y.; Muschert, S.; Marucci, M.; Hjaertstam, J.; Siepmann, F.; Axelsson, A.; Siepmann, J. Drug release mechanisms from Kollicoat SR: Eudragit NE coated pellets. *Int. J. Pharm.* **2011**, *409*, 30–37. [CrossRef]
29. Marucci, M.; Ragnarsson, G.; Nilsson, B.; Axelsson, A. Osmotic pumping release from ethyl-hydroxypropyl-cellulose-coated pellets: A new mechanistic model. *J. Control. Release* **2010**, *142*, 53–60. [CrossRef]
30. Marucci, M.; Ragnarsson, G.; Axelsson, A. Evaluation of osmotic effects on coated pellets using a mechanistic model. *Int. J. Pharm.* **2007**, *336*, 67–74. [CrossRef]
31. Ferrero, C.; Massuelle, D.; Doelker, E. Towards elucidation of the drug release mechanism from compressed hydrophilic matrices made of cellulose ethers. II. Evaluation of a possible swelling-controlled drug release mechanism using dimensionless analysis. *J. Control. Release* **2010**, *141*, 223–233. [CrossRef] [PubMed]
32. Siepmann, J.; Kranz, H.; Bodmeier, R.; Peppas, N.A. HPMC-matrices for controlled drug delivery: A new model combining diffusion, swelling, and dissolution mechanisms and predicting the release kinetics. *Pharm. Res.* **1999**, *16*, 1748–1756. [CrossRef] [PubMed]

33. Xiang, A.; Mchugh, A.J. A generalized diffusion–dissolution model for drug release from rigid polymer membrane matrices. *J. Membr. Sci.* **2011**, *366*, 104–115. [CrossRef]
34. Cabrera, M.I.; Luna, J.A.; Grau, R.J.A. Modeling of dissolution-diffusion controlled drug release from planar polymeric systems with finite dissolution rate and arbitrary drug loading. *J. Membr. Sci.* **2006**, *280*, 693–704. [CrossRef]
35. Choo, K.; Kim, I.; Jung, J.; Suh, Y.; Chung, S.; Lee, M.; Shim, C. Simultaneous determination of loxoprofen and its diastereomeric alcohol metabolites in human plasma and urine by a simple HPLC-UV detection method. *J. Pharm. Biomed. Anal.* **2001**, *25*, 639–650. [CrossRef]
36. Badawy, S.I.; Williams, R.C.; Gilbert, D.L. Effect of different acids on solid-state stability of an ester prodrug of a IIb/IIIa glycoprotein receptor antagonist. *Pharm. Dev. Technol.* **1999**, *4*, 325–331. [CrossRef] [PubMed]
37. Siepe, S.; Lueckel, B.; Kramer, A.; Ries, A.; Gurny, R. Strategies for the design of hydrophilic matrix tablets with controlled microenvironmental pH. *Int. J. Pharm.* **2006**, *316*, 14–20. [CrossRef] [PubMed]
38. Kazlauske, J.; Cafaro, M.M.; Caccavo, D.; Marucci, M.; Lamberti, G.; Barba, A.A.; Larsson, A. Determination of the release mechanism of Theophylline from pellets coated with Surelease(R)-A water dispersion of ethyl cellulose. *Int. J. Pharm.* **2017**, *528*, 345–353. [CrossRef] [PubMed]
39. Food and Drug Administration CDER. Guidance for Industry Q8 (R2) Pharmaceutical Development International Conference on Harmonisation of Techniccal Requirements for Registration of Pharmaceuticals for Human Use. 2009; pp. 1–19. Available online: https://www.ich.org/fileadmin/Public_Web_Site/ICH_Products/Guidelines/Quality/Q8_R1/Step4/Q8_R2_Guideline.pdf (accessed on 6 May 2019).
40. Charoo, N.A.; Shamsher, A.A.; Zidan, A.S.; Rahman, Z. Quality by design approach for formulation development: A case study of dispersible tablets. *Int. J. Pharm.* **2012**, *423*, 167–178. [CrossRef]
41. Costa, P.; Lobo, J.M.S. Modeling and comparison of dissolution profiles. *Eur. J. Pharm. Sci.* **2001**, *13*, 123–133. [CrossRef]
42. Kawai, S.; Nishida, S.; Kato, M.; Furumaya, Y.; Okamoto, R.; Koshino, T.; Mizushima, Y. Comparison of cyclooxygenase-1 and -2 inhibitory activities of various nonsteroidal anti-inflammatory drugs using human platelets and synovial cells. *Eur. J. Pharmacol.* **1998**, *347*, 87–94. [CrossRef]
43. Yamakawa, N.; Suemasu, S.; Kimoto, A.; Ishihara, T.; Yokomizo, K.; Okamoto, Y.; Otsuka, M. Low Direct Cytotoxicity of Loxoprofen on Gastric Mucosal Cells. *Biol. Pharm. Bull.* **2010**, *33*, 398. [CrossRef] [PubMed]
44. Hatton, G.B.; Yadav, V.; Basit, A.W.; Merchant, H.A. Animal farm: Considerations in animal gastrointestinal physiology and relevance to drug delivery in humans. *J. Pharm. Sci.* **2015**, *104*, 2747–2776. [CrossRef] [PubMed]

Communication

Enhancement of Magnetic Hyperthermia by Mixing Synthetic Inorganic and Biomimetic Magnetic Nanoparticles

Guillermo R. Iglesias [1,*], Ylenia Jabalera [2], Ana Peigneux [2], Blanca Luna Checa Fernández [1], Ángel V. Delgado [1] and Concepcion Jimenez-Lopez [2,*]

1 Department of Applied Physics, Faculty of Sciences, University of Granada, 18071 Granada, Spain; lunachecaf@gmail.com (B.L.C.F.); adelgado@ugr.es (Á.V.D.)

2 Department of Microbiology, Faculty of Sciences, University of Granada, 18071 Granada, Spain; yjabalera@ugr.com (Y.J.); apn@ugr.es (A.P.)

* Correspondence: iglesias@ugr.es (G.R.I.); cjl@ugr.es (C.J.-L.); Tel.: +34-958-242-734 (G.R.I.); +34-958-249-833 (C.J.-L.)

Received: 30 March 2019; Accepted: 27 May 2019; Published: 11 June 2019

Abstract: In this work we report on the synthesis and characterization of magnetic nanoparticles of two distinct origins, one inorganic (MNPs) and the other biomimetic (BMNPs), the latter based on a process of bacterial synthesis. Each of these two kinds of particles has its own advantages when used separately with biomedical purposes. Thus, BMNPs present an isoelectric point below neutrality (around pH 4.4), while MNPs show a zero-zeta potential at pH 7, and appear to be excellent agents for magnetic hyperthermia. This means that the biomimetic particles are better suited to be loaded with drug molecules positively charged at neutral pH (notably, doxorubicin, for instance) and releasing it at the acidic tumor environment. In turn, MNPs may provide their transport capabilities under a magnetic field. In this study it is proposed to use a mixture of both kinds of particles at two different concentrations, trying to get the best from each of them. We study which mixture performs better from different points of view, like stability and magnetic hyperthermia response, while keeping suitable drug transport capabilities. This composite system is proposed as a close to ideal drug vehicle with added enhanced hyperthermia response.

Keywords: biomimetic magnetite; drug delivery; magnetic hyperthermia; magnetite; MamC; nanoparticles stability

---

1. Introduction

In spite of the certainly wide variety of magnetic nanoparticles (MNPs) with different geometries, compositions and functionalizations [1–8] that have become available in recent years, and of the number of applications that have been devised for them [1,9–12], when the goal is to provide a nanocarrier suitable for targeted chemotherapy, there is still room for progress. On one hand, the methods for obtaining MNPs need to be more cost- and time-effective, eco-friendly, and scalable. On the other, the nanoparticles themselves should be improved in terms of maximizing the magnetic moment per particle and providing novel surface properties while exploring their potential as hyperthermia agents that would allow them to combine therapies in the near future.

In this context, cancer is one of the fields of application where magnetic nanoparticles certainly appear as most promising [12–18]. Aside of drug delivery, whereby functionalized magnetic nanoparticles are loaded with the chosen drug and some targeting molecule, such an antibody, driven to the site of action, maintained there by continuous magnetic fields and eventually set for delivery by some external action, magnetic hyperthermia appears as a realistic application of

MNPs [19–27]. For that purpose, MNPs are dispersed at a suitable concentration in an aqueous solution and located in the target place for action. There, they are subjected to an alternating magnetic field (with induction of tens mT and frequency of several hundred kHz). Subsequently, the magnetic nanoparticles increase the temperature of the microenvironment in which they are immersed inducing apoptosis of the tumor cells, usually more sensitive to temperature increase compared to healthy cells [28–30].

MNPs are generally produced either by the co-precipitation of iron salts in basic aqueous media possibly stabilized by biocompatible surfactants/polymers or by the thermal decomposition of organometallic precursors in high-boiling nonpolar organic solvents at elevated temperatures (200–360 °C), allowing a great control of the size of the MNPs, their monodispersity and uniformity [31]. However, these methods have some drawbacks mainly derived from the high temperatures used, organic solvents or poor solubility of the nanoparticles in water.

Many of these drawbacks are overcome in biomimetic MNPs (BMNPs). These are produced by the mediation of magnetosome membrane-associated proteins (MAPs) from magnetotactic bacteria, and *in vitro* experiments have been demonstrated to control the size (and thus the magnetic moment per particle), shape, and surface properties of the nanoparticles [32,33]. BMNPs production can be scaled up *in vitro* in eco-friendly, cost-effective magnetite precipitation experiments run at room temperature and 1 atm total pressure by the simple addition of the recombinant protein. Promising BMNPs have been obtained by the mediation of MamC protein from *Magnetococcus marinus* MC-1, since these BMNPs are (i) superparamagnetic at room and body temperature while they present a saturation magnetization of 61 emu/g (at 500 Oe and 25 °C); (ii) are larger than most commercial MNPs and/or other biomimetic magnetites, although still single magnetic domain, showing higher blocking temperature and slower magnetization increase, and thus, larger magnetic moment per particle; (iii) contain up to 4.5 wt% of MamC, which provides functional groups allowing for functionalization; and (iv) adopt the isoelectric point of MamC (pH_{iep} 4.4), and are strongly negatively charged at physiological pH (pH 7.4); this property allows the coupling and release of molecules to be pH-dependent. Thus, at physiological pHs, they bind to positively charged molecules (such as doxorubicin, DOXO hereafter) through electrostatic interactions, which are weaker at acidic pHs (such as those found in tumor microenvironments), allowing the release of the adsorbed molecules, and (v); they are fully cytocompatible and hemocompatible, but when they are coupled with DOXO they display dose-dependent cytotoxicity [33].

Recall that single-domain magnetic nanoparticles are characterized by a spin configuration such that in the absence of an external magnetic field, all spins are oriented parallel to each other and parallel or antiparallel to a crystallographic direction, called the easy (or anisotropy) axis [34–36], so that each particle is characterized by a large magnetic moment. Let us imagine for the moment that the particles are immobile, fixed in a non-magnetic matrix, with their easy axes oriented randomly. At very low temperature (or at very high frequencies of the external magnetic field, if this is non-stationary), the transition between the two orientations along the easy axes can only be achieved by the application of sufficiently large external magnetic field, so that the magnetization-H curve will take the form of a square hysteresis cycle, according to the Stoner-Wohlfarth model [35,37]. As temperature rises above 0 K, the transition between the two orientations can be thermally activated, and the coercivity tends to be reduced, eventually making hysteresis negligible. The same effect will be observed when the field frequency or the particle size is very low, since the coercive field H_C is related to the anisotropy field H_k (in turn related to K_{eff}, the effective magnetic anisotropy constant of the particles, and their saturation magnetization M_S as $H_k = 2\ K_{eff}/\mu_0 M_S$) by [35]:

$$H_C = 0.48 H_K (1 - \kappa^{0.8}) \tag{1}$$

with

$$\kappa = \frac{k_B T}{K_{eff} V} \ln\left(\frac{k_B T}{4\mu_0 H_0 M_S f V \tau_0}\right) \tag{2}$$

where k_BT is the thermal energy, V is the particle volume (or, in the case of coated particles, the volume of the magnetizable core), μ_0 is the vacuum magnetic permeability, H_0 is the field amplitude, f is the field frequency, and the characteristic time τ_0 depends on such quantities as temperature, saturation magnetization or anisotropy constant. It will be assumed to be a constant in the range of 10^{-10}–10^{-9} s. If it is admitted that it suffices with H_C being 0.01–0.1H_k to have reversible cycles and absence of hysteresis, then the combination of H_0, f, V, and M_S must be such that κ = 0.97–0.75.

An approximate approach to the general solution of the problem, not using models based on the Stoner-Wohlfarth one is the so-called linear-response theory, where, keeping the assumption of immobile particles, it is found that the hysteresis cycle has its origin in the Néel-Brown relaxation, characterized by a relaxation time τ_N, which is a measure of the time taken by the system to return to equilibrium after application of a step magnetic field, or half the time needed for spontaneous inversion of magnetic moment orientation [38–40]. It is given by:

$$\tau_N = \tau_0 \exp\left(\frac{K_{eff}V}{k_B T}\right) \tag{3}$$

This brings about a delay between magnetization and field, or an imaginary component of the magnetic susceptibility, and manifests again in a finite area hysteresis cycle, as long as the frequency of the field remains in the vicinity of $1/\tau_N$. This approach is only strictly valid for low applied magnetic field strength or highly anisotropic particles.

For hyperthermia applications, the particles are typically dispersed in an aqueous solution, and hence they can rotate under field inversions so that viscous friction is an additional source of phase delay between magnetization and external field, and hence, an additional relaxation contribution to hysteresis. It is called Brownian relaxation, and it is characterized by a time τ_B:

$$\tau_B = \frac{3\eta V_H}{k_B T} \tag{4}$$

where η is the viscosity of the medium, and V_H its hydrodynamic volume (including, if any, that of the coating layer) [19,41,42]. Taking the value of K_{eff} equal to 25 kJ/m^3 for magnetite, the Brownian relaxation times for biomimetic (~35 nm in diameter, reported below) and purely inorganic particles (~28 nm) are, respectively, 14.4 μs, and 7.4 μs, and orders of magnitude higher in the case of magnetization reversal. This means that for the particle sizes and frequencies involved, the magnetic moment is frozen in the particle and the only heating source is friction, according to the linear response model. Considerations on the validity of the relaxation approach just described can be found in References [43–46]. In the context of the linear response theory, if the alternating magnetic field has a frequency in the vicinity of the reciprocal of the mentioned times (f in the order of 70 kHz and 135 kHz, respectively), the imaginary component, χ'', of the complex magnetic susceptibility is maximum, and this is important, as the dissipated power dW/dt is proportional to this quantity [28]:

$$\frac{dW}{dt} = \mu_0 \pi \chi'' H_0^2 \tag{5}$$

Another important aspect, not always considered, regards the stability of the particles in the suspension, compromised not only by the colloidal interactions but also by the magnetic dipolar ones. Hence the need for properly controlling the stability, as the hyperthermia response degrades for aggregated systems [22,47] or when the monodomain range is surpassed [36,48].

Experimentally, the quantity of interest is the so-called Specific Absorption Rate (SAR), or heat released per unit mass of magnetic material (m) and per second:

$$SAR = \left(\frac{CV_s}{m}\right)\frac{dT}{dt} \tag{6}$$

C being the volume heat capacity of the suspension, V_s its volume and dT/dt the rate of temperature increase [28,39]. Typical *SAR* values are in the order of tens to hundreds of W/g. In order to perform a comparison between different materials without the interference of the details of the specific device used (very often lab-made), a quantity is defined as a measure of the magnetothermal performance of a given suspension, namely, the Intrinsic Loss Power (ILP), given as:

$$ILP = \frac{SAR}{fH_0^2} \tag{7}$$

with typical values in the order of 10^{-9} Hm^{2}kg^{-1}.

In this work, we explore the possibility of maximizing the hyperthermia effect (the rate of temperature rise, in fact) by combining the two types of magnetic particles (BMNPs and MNPs), differing in size and other properties. The goal of the present study is to provide a composition that could be used in the future as a platform for combining drug delivery and hyperthermia. The former particles, BMNPs, mediated by MamC have been chosen because of their demonstrated ability to function as nano-transporters of drugs, being the nanoassembly stable at physiological pH while it destabilizes releasing the drug under acidic pH values, which naturally occur in the tumor environment. The second ones, MNPs, have been chosen because of their potential as hyperthermia agents [33]. By mixing the two systems, we expect to deliver a system that, in the future may prove useful to combine both treatments, i.e targeted chemotherapy plus targeted hyperthermia by using the same platform. To the best of our knowledge, the hyperthermia of mixed systems has never been investigated, but advantages are foreseen regarding the possibility of optimizing the response.

2. Experimental

2.1. BMNPs and MNPs Production

Details of the production process are given in Reference [32], and just a short account will be provided here. The MamC gene (ABK44766.1, NCBI Database) was first amplified by the polymerase chain reaction and cloned into pTrcHis-TOPO vector, and expressed in *Escherichia coli* TOP10 competent cells, both from Life Technologies, Invitrogen, Grand Island, NY, USA. The cells were grown at 37 °C in Luria-Bertani (LB) broth supplemented with 100 mg/mL of ampiciline. After 5-h contact with isopropyl-β-D-thiogalactopyranoside (IPTG, Fisher BioReagents, Pittsburgh, PA, USA) the expression of the recombinant MamC was induced. The cell pellet was resuspended in guanidinium lysis buffer overnight. The lysate was centrifuged, and the supernatant loaded onto a HiTrap chelating HP column (GE Healthcare, Chicago, IL, USA) equilibrated with Denaturing Binding Buffer (A). The protein eluate was obtained with Denaturing Elution Buffer (B), and the final step was isolating the MamC protein by successive dialyzations with buffers A and B. The isolated protein was stored at 4 °C until used in the biomineralization process (all buffers were purchased from Sigma Aldrich, St. Louis, MO, USA).

The following step was the production of the MNPs and BMNPs (all reagents needed in this process were purchased from Sigma-Aldrich, Madrid, Spain). Milli-Q water (Millipore, Barcelona, Spain) was first deoxygenated by boiling it for 1 h and then cooling in an ice bath while bubbling nitrogen. The water was immediately stored inside an anaerobic chamber (Coy Laboratory Products, Grass Lake, MI, USA) with a 4% H_2 in N_2 atmosphere. The following solutions were prepared inside the chamber: $NaHCO_3$/Na_2CO_3 (0.15 M/0.15 M), $FeCl_3$ (1 M), $Fe(ClO_4)_2$ (0.5 M), and NaOH (5 M). The precipitation of inorganic magnetite (MNPs) was carried out as described in Reference [49], and consisted in mixing the prepared solutions to the final concentrations of 3.5 mM/3.5 mM, 5.56 mM, and 2.78 mM, respectively, at 25 °C and 1 atm. NaOH was added to reach a pH of 9. The biomimetic nanoparticles were obtained by adding the purified MamC protein (at a final concentration of 10 μg/mL) to the solution used for MNPs. In both cases, the solids were allowed to grow for 30 days, and magnetically decanted and washed three times with deoxygenated water. The magnetic particles were kept in water inside the Coy chamber until further use.

2.2. Nanoparticle Characterization

The morphology and particle size of the synthesized nanocrystals were analyzed by Transmission Electron Microscopy (TEM Philips Model CM20, Eindhoven, The Netherlands) equipped with an energy dispersive X-ray spectrometer (EDAX). The size of the particles was measured by using ImageJ 1.47 software, and size distribution curves and ANOVA statistical analyses were determined from measurements performed on 1000 particles. Averages were considered significantly different if $p < 0.05$. Powder x-ray diffraction (XRD) analysis was carried out on lyophilized samples with an Xpert Pro X-ray diffractometer (PANalytical; Almelo, The Netherlands) using Cu Kα radiation, with the scan range set from 20–60° in 2θ (0.01°/step; 3 s per step). Electrophoretic mobility measurements were carried out in a Zetameter Nano-ZS (Malvern Instruments, Malvern, UK) at 25 °C, in suspensions with 0.01% *w*/*v* solids concentration and constant ionic strength of 5 mM KNO_3. For each suspension, 5 measurement runs were taken. Magnetization cycles and zero-field cooled field-cooled (ZFC-FC) curves were obtained in an MPMS-XL SQUID magnetometer (Quantum Design, San Diego, CA, USA). The stability of the samples was evaluated optically by measuring the time evolution of the phase separation line between particles and medium: The samples were photographed at certain intervals. Afterward, the height and volume of each phase were determined through image processing and analyzed.

Magnetic hyperthermia experiments were carried out using an AC current generator with a double four-turn coil made of water-cooled copper tube, 4 mm in inner diameter, with 800 mL/min flow rate, comparable to other experimental hyperthermia devices [50]. Three frequencies, namely, 197 kHz, 236 kHz, and 280 kHz were selected, with a fixed magnetic field intensity of 18 kA/m, measured at the center of the coil with a NanoScience Laboratories Ltd. Probe (Staffordshire, UK), with 10 μT resolution. These were the combinations accessible with our measurement system, but are close to those used by other authors [45,46,51]. The samples to be evaluated were placed in plastic Eppendorf tubes (1.5 mL sample volume). Four kinds of dispersed systems were evaluated, namely those based on pure MNPs or BMNPs, and mixtures containing 25% BMNPs + 75% MNPs (here referred to as 25 B + 75 M), and 60% BMNPs + 40% MNPs (60 B + 40 M). This selection was based on using a combination with a predominance of inorganic MNPs and another one with a higher fraction of biomimetic particles in order to estimate the relative contribution of each type of particles.

The *SAR* and *ILP* of the dispersions were obtained by measuring the rate of temperature increase as a function of time [19,52], with an optical fiber thermometer (Optocon AG, Dresden, Germany), and using Equations (6) and (7). Considering the rather low concentration of MNPs, the corrections proposed by Gas and Miaskowski [51] in the calculation of the heat capacity of the suspension did not appear necessary. All samples for hyperthermia were prepared with a solids concentration of 25 mg/mL. Note that this concentration is higher than usual in hyperthermia applications (more in the range of 1–10 mg/mL), but it was chosen with the aim of magnifying differences. At such concentrations, magnetic or colloidal interactions might likely affect hyperthermia, as increased stability seems to favor the temperature elevation, if the frequency of the magnetic field is selected in accordance with the size of individual, non-aggregated particles. In fact, in a previous work [22], we found that the hyperthermia response was improved if the suspensions were more stable, although the *SAR* of 20 nm magnetite suspensions was constant up to 2% (*v*/*v*) (or about 100 mg/mL) concentration.

3. Results and Discussion

3.1. Particle Characterization

Figure 1B,D show representative TEM pictures of the two kinds of particles. Size histograms are represented in Figure 1A,C, respectively. The mean (± S.D.) diameters obtained from the histograms were 35 ± 11 nm for BMNPs and 18 ± 6 nm for the purely inorganic nanoparticles. Insets in the pictures also show that the shape of both kinds of particles is rather polyhedral, with better homogeneity in the biomimetic case.

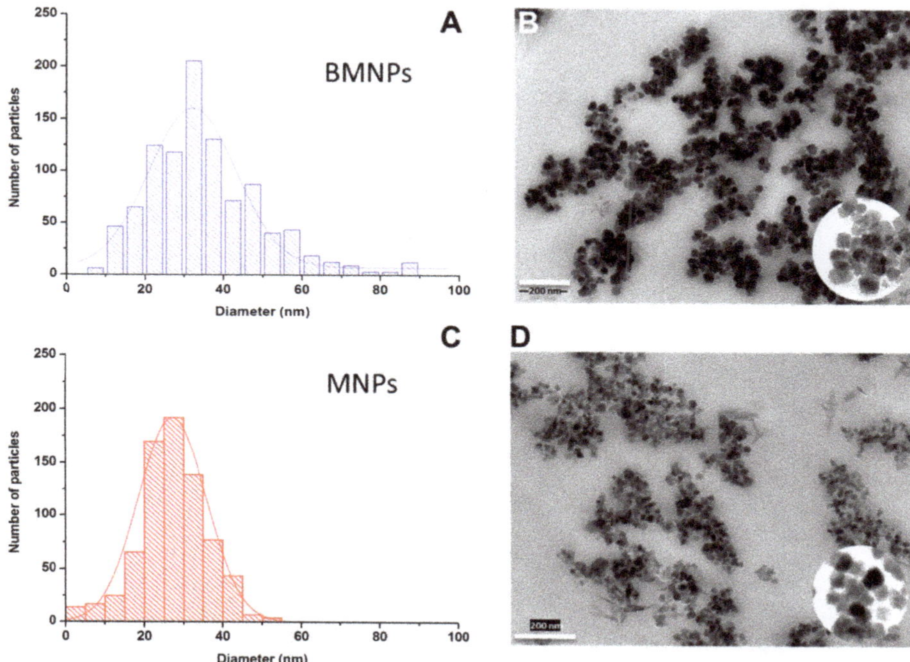

Figure 1. Diameter histograms (**A**,**C**) and TEM pictures (**B**,**D**) of biomimetic (BMNPs) and purely inorganic (MNPs) magnetic nanoparticles. The scale bar in B, D is 200 nm.

Figure 2 shows the XRD patterns of both samples. Note the good crystallinity in the two cases, and the excellent coincidence with the magnetite reference pattern (JCPDS card No 19-0629), being the main reflection for magnetite the 311 (d-spacing = 2.530 Å, for Cu Kα radiation, 2θ is 35.44 degrees). The goethite diffraction lines are also superimposed in the figure, and only a low-angle peak seems to correspond to this oxide. Using Scherrer's formula [53], the crystallite size was obtained for both samples from the half-intensity width of all the lines in the pattern, using specialized Rietveld software (TOPAS 5.0, Bruker, Hamburg, Germany). Calculated crystallite sizes for MNPs vary between 13.0 and 18.3 nm, while those for BMNPs vary between 13.1 and 22.5 nm. Thus the calculated size from XRD data for MNPs matches that measured from TEM images. However, this is not the case for BMNPs for which we get smaller values than those measured with TEM. Recall that the crystallite size calculated from XRD data is a measure of the size of coherent diffraction domains, which can be smaller than the particle size if small discontinuities will make the domain lose such a coherency. It is true that BMNPs could be polycrystals, but this is not what was found by High-Resolution TEM observations of BMNPs in Reference [54]. No discontinuities in lattice fringes were observed, and therefore, the polycrystallinity of BMNPs seems to be ruled out, at least for the majority of the BMNPs analyzed. However, Garcia-Rubia et al. [33] suggested the incorporation of MamC in the outer layers of the BMNPs crystals, that prevented the removal of the protein, and measured, in fact, that 5% of the total mass of the BMNPs is MamC. The presence of the protein would, most probably, induce some defects in the crystal structure, resulting in the loss of coherency in the diffraction. This is why, in BMNPs, the crystal sizes calculated from XRD are lower than those measured in TEM images.

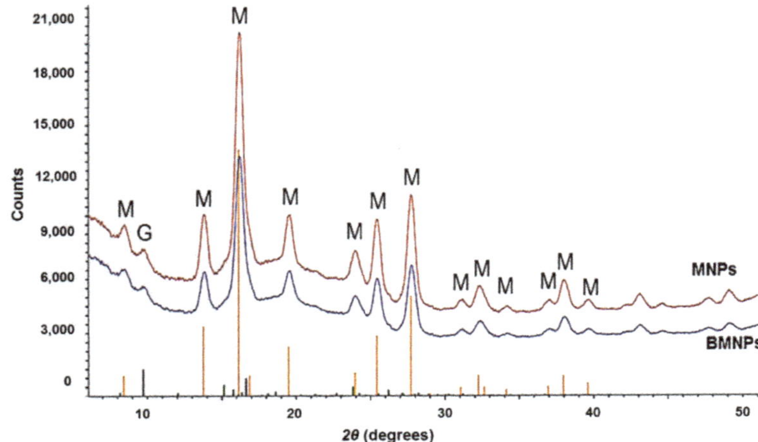

Figure 2. The XRD diffraction patterns of purely inorganic (MNPs) and biomimetic magnetic nanoparticles (BMNPs). The positions and intensities of crystallographically-pure magnetite are labeled as M, and those of goethite as G.

The magnetization of the two kinds of particles and an example of one of the mixtures (25 B + 75 M, in fact the most stable mixture, as will be discussed below) is plotted in Figure 3 for two temperatures, 5 K and 300 K. The detail in Figure 3 shows that some really low remnant magnetization (about 20 emu/g at most), and coercivity can be measured at 5 K, with the interesting feature that this is maximum for the mixed system. This can be ascribed to some degree of aggregation between the two kinds of particles, as at the pH of the aqueous suspensions in which the mixtures were prepared they are oppositely charged (detailed below), and electrostatic attraction cannot be ruled out. For room temperature measurements, the magnetization shows no hysteresis, and the particles behave as paramagnetic. This is characteristic of superparamagnetism, as mentioned. At room temperature, the saturation (mass) magnetization reaches 66 emu/g in the case of MNPs and 25 B + 75 M, and 55 emu/g in the case of BMNPs. Considering the dilution effect of the coating caused by the incorporation of MamC [33], the corrected value of saturation magnetization is 61 emu/g for BMNPs, and 70 emu/g for 25 B + 75 M.

Zero-field cooling-field cooling (ZFC-FC) curves at 500 Oe (39.8 kA/m) (see Figure 4) show that the blocking temperature (maximum in the ZFC cooling curve, the lower branch in each case, corresponding to the rounded appearance of the curve [55]) is 103 K for MNPs, 145 K for BMNPs and 180 K for 25 B + 75 M, and that at temperatures higher than blocking temperature, (including 300 K and above in all cases), these particles will behave as non-magnetic in the absence of an external magnetic field, confirming the superparamagnetic nature of the two kinds of particles. This prevents magnetic aggregation, a very favorable feature of the systems investigated.

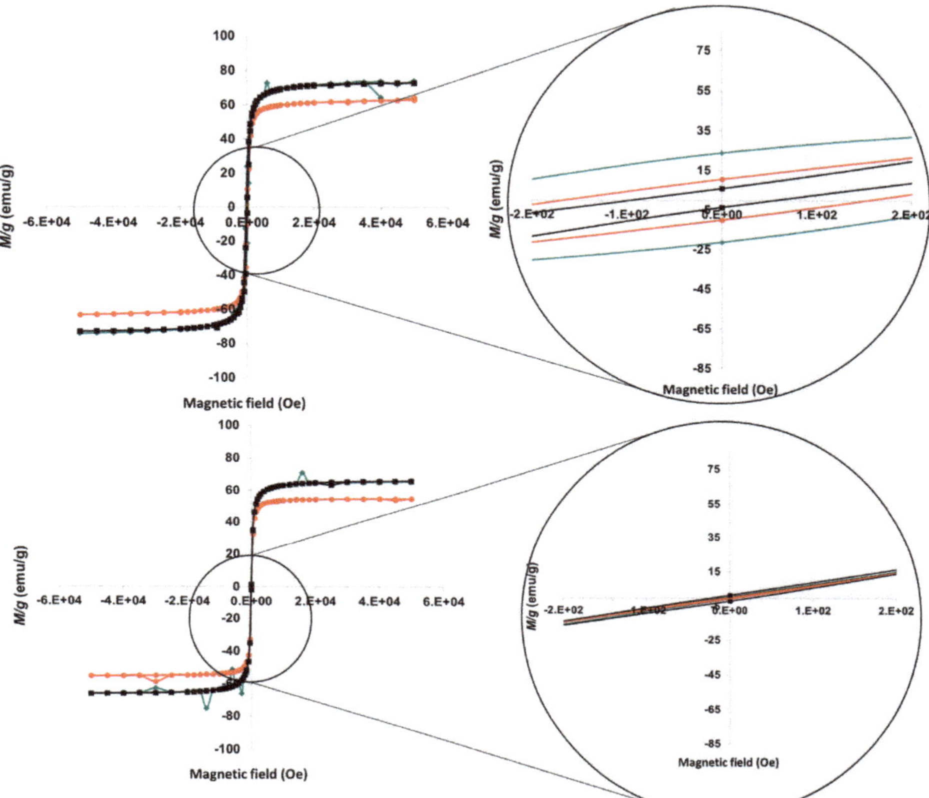

Figure 3. Magnetization cycles of MNPs (■), BMNPs (●), and 25 B + 75 M (♦) at 5 K (**top**), and 300 K (**bottom**). Magnifications of the low-field region are also plotted.

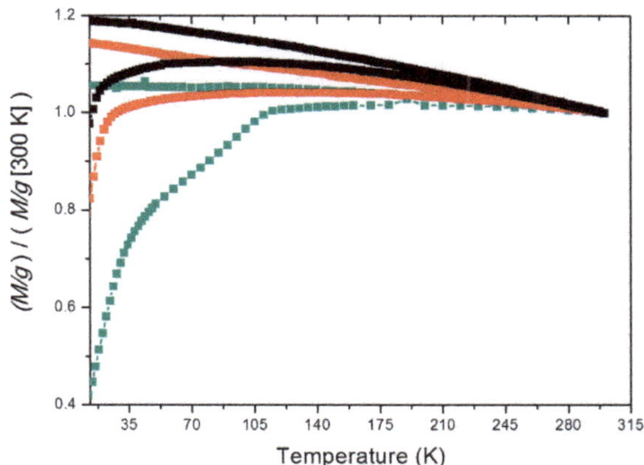

Figure 4. Zero field cooling-field cooling (ZFC-FC) curves at 40 kA/m of MNPs (■), BMNPs (●) and 25 B + 75 M (♦).

However, once an external magnetic field is applied, the nanoparticles will respond efficiently. From these results, it can be inferred that either the bare MNPs and BMNPs, or their mixtures, appear as ideal candidates for the purpose of hyperthermia, drug delivery or combinations thereof.

3.2. Zeta Potential

For the purpose of drug loading and delivery, the ability of the nanoparticles to be coupled to molecules forming stable nanoassemblies at physiological pH and to release the drug at the precise time, as well as the stability, drug-particle or cell membrane-particle interactions, among other properties, are strongly determined by the surface charge, or, alternatively, by the zeta potential ζ, of the particles when dispersed. Calculations of the zeta potential were done from the measurements of the electrophoretic mobility of the particles BMNPs and MNPs by means of O'Brien and White's general theory [56]. As one can see in Figure 5, the isoelectric point (pH_{iep} or pH of zero ζ) of MNPs was obtained at pH 7.0, while for BMNPs it was 4.4. This zeta potential plot reveals significant differences between both types of nanoparticles. They are positively charged at low pH values and negatively charged at high pH, but BMNPs change from positive to negative at quite a different pH: This suggests that the MamC protein is affecting the surface of the particles, and it is probably located mainly at the interface. In fact, the investigations by Nudelman et al. [57], and other authors [54,58] on the biomineralization mechanism induced by this protein demonstrate that the particularity of MamC is the template effect for magnetite nucleation and growth that this protein exerts. The two negatively charged amino acids Asp70 and Glu66, located in MamC loop, are separated by about 8 Å, and this distance can match the disposition of iron cations on (100), (110), and (111) faces. Therefore, the crystal could grow in an orderly manner, first from the Fe cations available in solution and then, since the amount of Fe in solution is limited, at the expense of the release of the Fe cations adsorbed in other negatively charged moieties of MamC. Since the process of nucleation is kinetically favored by this template effect, it is precisely those exposed amino acids that act as nucleation sites. Therefore, the restricted number of nucleation sites in the presence of MamC compared to the process of homogeneous nucleation from the bulk solution allows the formation of larger crystals in the former. Moreover, since (100), (110), and (111) faces show up in the final morphology of the BMNPs [57], it can be concluded that, while exerting this template effect for the nucleation, the protein prevents the growth of the crystal on these specific directions.

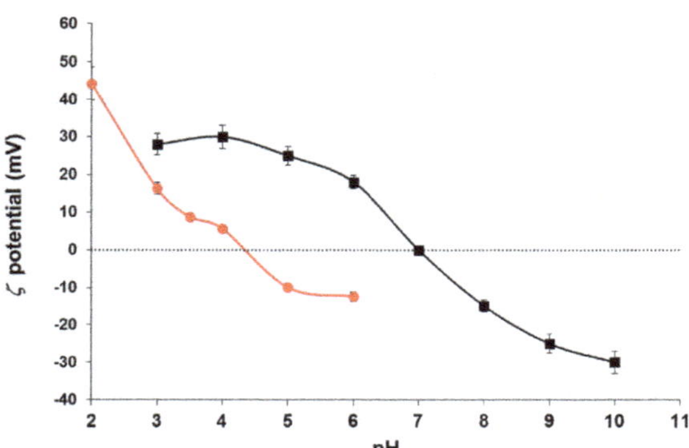

Figure 5. Zeta potential of purely inorganic MNPs (■) and biomimetic magnetic nanoparticles BMNPs (●).

The shift of the isoelectric point of the BMNPs relative to that of MNPs make the former adequate nanocarriers, a fact that is important for the sought application. Since most common drugs for cancer therapy are positively charged at physiological pH, they could be electrostatically attached to BMNPs at such pH values, resulting in a stable nanoassembly. On the contrary, when they eventually reach the tumor (acidic pH values), the drug releases from the BMNPs, as the charge of the latter at acidic pH values is nearly zero [33]. That makes the BMNPs better drug nanocarriers compared to MNPs, as in the former the release of the drug can be controlled by an external factor, as it is a change in the environmental pH value. This feature is absent in MNPs, as their charge is positive anywhere below pH 7.

3.3. *Stability*

Like in many other applications, the use of magnetic nanoparticles in health applications requires to ensure sufficient stability. Previous results from our laboratory demonstrated that, specifically when it comes to hyperthermia applications, the stability, either electrostatically or polymerically achieved, is an essential factor for optimizing the performance of the system [22]. Hence, in this part of the study, we focus on the possibility of improving hyperthermia by favoring stability. It must be recalled that superparamagnetic nanoparticles interact attractively through van der Waals interactions and dipolar magnetic ones (if remanence is not negligible, which is not our case at room temperature, see Figure 3). Furthermore, the repulsive interactions due to the surface charge will be very low at physiological pH values in the case of MNPs, and larger in the case of BMNPs (Figure 5). In addition, steric hindrance could contribute to the stability of the biomimetic nanoparticles through the presence of surface MamC molecules. In order to confirm this possibility, we evaluated the sedimentation behavior vs. time of the samples (pure end members and mixtures of BMNPs and MNPs). The time evolution of the boundary between sedimented volume (height h) and clear supernatant, relative to the initial height h_0, will be used as a simple test of stability. Data are plotted in Figure 6. Surprisingly, it can be observed that, although the stability of bare MNPs is lower than that of BMNPs, as expected from the steric and electrostatic repulsions due to the MamC coating, adding BMNPs to the former at a relative concentration of 25/75 (25 B + 75 M) brings about a measurable increase of stability. This may be the result of a compromise between the smaller size of MNPs favoring stability and the addition of the protective coating represented by the BMNPs. As a result, an ideal system with a mixed composition emerges.

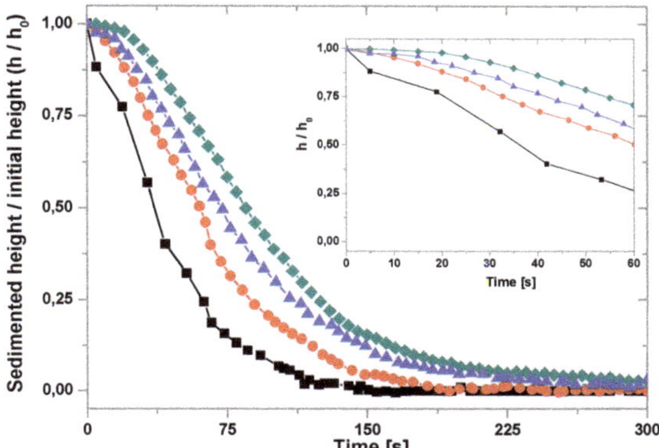

Figure 6. Height of the sedimented volume relative to its initial value as a function of time for MNPs (■), BMNPs (●), 60 B + 40 M (▲), and 25 B + 75 M (♦). Inset: Short-time detail.

3.4. *Performance in Hyperthermia*

As a novel field of application of the two kinds of particles and their mixtures, we consider to what extent they are efficient magnetic hyperthermia agents. Figure 7 shows the time evolution of the temperature of the suspensions of MNPs, BMNPs and the two mixtures 25 B + 75 M and 60 B + 40 M under the influence of an alternating magnetic field at the indicated frequencies and a fixed magnetic field strength of 18 kA/m. All types of nanoparticles are able to raise the temperature, the fastest rise occurring for the highest frequency. This effect is particularly visible at longer times. The rate of temperature increase is maximum (~34 °C/min) in mixture 25 B + 75 M and minimum in BMNPs (~17 °C/min) at the highest frequency. It appears that inorganic MNPs are better hyperthermia agents than BMNPs. This justifies their presence in the mixtures under study. They make it possible to design a composition of nanoparticles that (i) can be guided to the target by the application of an external magnetic field; (ii) behave as suitable drug nanocarriers, stable at physiological pH values and from which the drug release is dependent of an external stimuli, e.g., acidic tumor environment (these are the BMNPs); (iii) produce a fast increase of the temperature of the system upon the application of an external magnetic field (these are the MNPs).

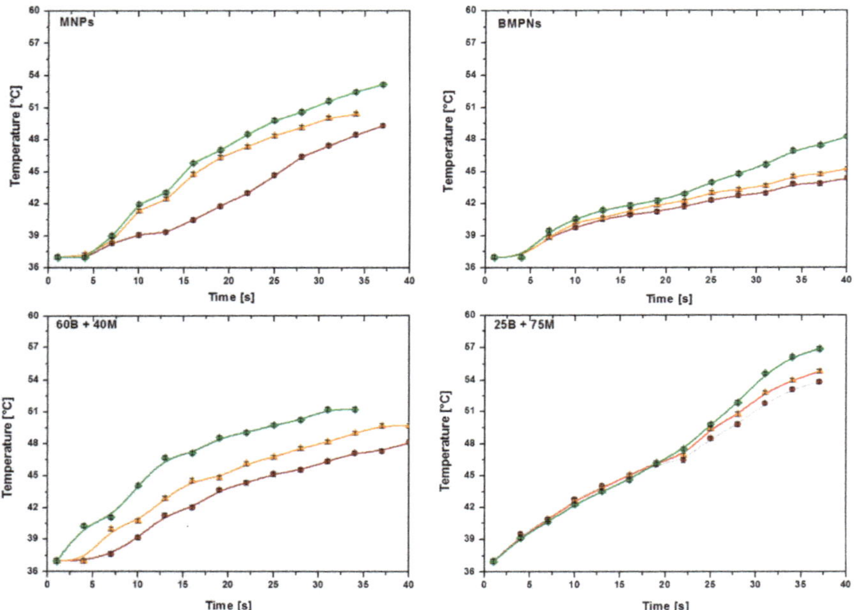

Figure 7. Time evolution of the temperature of the MNP suspensions, for different frequencies: 197 kHz (●), 236 kHz (▲), and 280 kHz (♦). Field strength: H_0 = 18 kA/m. Sample volume 0.5 mL; particle concentration: 25 mg/mL.

In fact, maximum *SAR* (up to 96.2 W/g) and *ILP* (1.26 nHm^{2}kg^{-1}) values were obtained for the 25 B + 75 M sample, followed by MNPs, 60 B + 40 M and, finally, BMNPs (Figure 8, Table 1). The lower heating induced by BMNPs is probably related to the larger size of the particles and to the presence of MamC attached to their surface, which probably hinders the rotation of the nanoparticles. The increased *SAR* and *ILP* values obtained for the 25 B + 75 M samples is promoted by stability, as single particles with close-to-spherical symmetry rotate under the action of the field less impeded than in the case of aggregated particles. In addition, it is likely that the field frequencies used are closer to those of maximum phase lag between magnetization and field, for the MNPs than for the BMNPs, so the latter would play here the role of favoring stability.

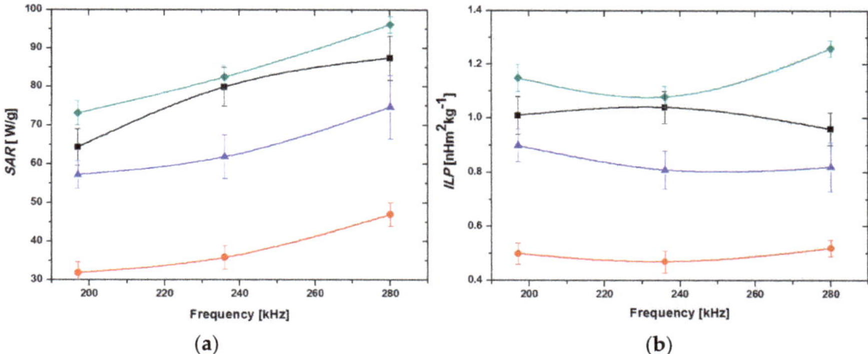

Figure 8. Frequency dependence of (**a**) *SAR* and (**b**) *ILP*, for the investigated systems. MNPs (■), BMNPs (●), 60 B + 40 M (▲), and 25 B + 75 M (♦). Magnetic field strength: 18 kA/m; particle concentration: 25 mg/mL.

Table 1. Summary of Specific Absorption Rate (SAR), Intrinsic Loss Power (ILP) calculations, and temperature increase after 60 s exposition time, ΔT, for the different samples tested [inorganic Magnetic Nanoparticles (MNPs) and MamC-medianted Biomimetic Magnetic Nanoparticles (BMNPs)], and the field frequencies indicated.

System	Frequency *f* [kHz]	*SAR* [W/g]	*ILP* [nHm^2kg^{-1}]	ΔT [°C]
MNPs	197	64 ± 5	1.01 ± 0.07	20.2 ± 0.2
	230	80 ± 5	1.04 ± 0.06	25.2 ± 0.2
	280	87 ± 6	0.96 ± 0.06	28.3 ± 0.2
BMNPs	197	32 ± 3	0.50 ± 0.04	11.6 ± 0.2
	230	36 ± 3	0.47 ± 0.04	12.9 ± 0.2
	280	47 ± 3	0.52 ± 0.03	16.7 ± 0.2
25 B + 75 M	197	73 ± 3	1.15 ± 0.05	28.5 ± 0.2
	230	83 ± 3	1.08 ± 0.04	30.5 ± 0.2
	280	96 ± 2	1.26 ± 0.03	34.1 ± 0.2
60 B + 40 M	197	57 ± 4	0.90 ± 0.06	18.1 ± 0.2
	230	62 ± 6	0.81 ± 0.07	21.6 ± 0.2
	280	75 ± 8	0.82 ± 0.09	27.5 ± 0.2

4. Conclusions

The present study intended to find an appropriate combination of magnetic nanoparticles that provide suitable targeted hyperthermia, and potentially proper conditions for being loaded with a positively-charged drug at a neutral pH while releasing it at the acid tumor environment. Both inorganic (MNPs) and biomimetic (MamC-mediated, BMNPs) nanoparticles have been proven to be superparamagnetic, having a blocking temperature lower than 300 K, therefore behaving as paramagnetic at this temperature (thus preventing magnetic agglomeration) but exhibiting a relatively high saturation magnetization in the presence of an external magnetic field, being thus able to be magnetically guided to a selected target site. BMNPs have been found to bind different molecules based on electrostatic interactions, forming stable nano-assemblies at physiological pH, based on the change in the isoelectric point (iep) of the nanoparticles induced by MamC (iep = 4.4). In contrast, MNPs are not as good candidates for that purpose, since their pH_{iep} is around 7. The adsorption of a positively charged drug as doxorubicin on BMNPs would be favored at neutral pH, and at the same time its release would also be enhanced as the system acidifies (like in tumor microenvironments) approaching the iep of the BMNPs. However, while potentially good nanotransporters, BMNPs are not as good agents for hyperthermia as MNPs. Therefore, the present study offers a composition of nanoparticles, namely,

25% BMNPs + 75% MNPs that, while having the maximum hyperthermia response, likely related to improved stability of the sample, is also suitable as a drug nanocarrier designed to deliver the drug in response to changes in the environmental pH. Therefore, the combination of inorganic and biomimetic nanoparticles potentially allows combined targeted chemotherapy and targeted hyperthermia.

Author Contributions: Conceptualization, G.R.I., C.J.-L. and Á.V.D.; methodology, G.R.I, C.J.-L., Y.J., A.P., B.L.C.F and Á.V.D.; validation, G.R.I, Y.J., A.P. and B.L.C.F.; formal analysis, G.R.I, Y.J. and A.P.; investigation, G.R.I, Y.J., A.P. and B.L.C.F.; resources, G.R.I, C.J.-L. and Á.V.D.; writing—original draft preparation, G.R.I and C.J.-L.; writing—review and editing, G.R.I, C.J.-L. and Á.V.D.; visualization, G.R.I, Y.J. and C.J.-L.; supervision, G.R.I, C.J.-L. and Á.V.D.; project administration, G.R.I, C.J.-L. and Á.V.D.; funding acquisition, G.R.I, C.J.-L. and Á.V.D.

Funding: We wish to thank FPU2016 grant (Ref. FPU16-04580), RYC-2014-6901 (MINECO, Spain), CGL2016-76723 (MINECO, Spain and FEDER, EU), Unidad Científica de Excelencia UCE-PP2016-05 (UGR) and Plan Propio Beca de iniciación a la investigación para estudiantes de master (UGR).

Acknowledgments: C.J.L. wishes to thank Alejandro Rodriguez-Navarro for assistance in the interpretation of DRX data. All the authors want to thank the editor and two anonymous reviewers whose comments have greately improved the manuscript.

Conflicts of Interest: The authors declare no conflict of interest.

References

1. Rezayan, A.H.; Mousavi, M.; Kheirjou, S.; Amoabediny, G.; Ardestani, M.S.; Mohammadnejad, J. Monodisperse magnetite (Fe3O4) nanoparticles modified with water soluble polymers for the diagnosis of breast cancer by MRI method. *J. Magn. Magn. Mater.* **2016**, *420*, 210–217. [CrossRef]
2. Vargas-Osorio, Z.; Argibay, B.; Pineiro, Y.; Vazquez-Vazquez, C.; Lopez-Quintela, M.A.; Alvarez-Perez, M.A.; Sobrino, T.; Campos, F.; Castillo, J.; Rivas, J. Multicore magnetic Fe3O4@C beads with enhanced magnetic response for MRI in brain biomedical applications. *IEEE Trans. Magn.* **2016**, *52*, 1–4. [CrossRef]
3. Portillo, M.A.; Iglesias, G.R. Magnetic nanoparticles as a redispersing additive in magnetorheological fluid. *J. Nanomater.* **2017**, *2017*, 8. [CrossRef]
4. Cardoso, V.F.; Francesko, A.; Ribeiro, C.; Banobre-Lopez, M.; Martins, P.; Lanceros-Mendez, S. Advances in magnetic nanoparticles for biomedical applications. *Adv. Healthc. Mater.* **2018**, *7*, 1700845. [CrossRef]
5. Ghazi, N.; Chenari, H.M.; Ghodsi, F.E. Rietveld refinement, morphology analysis, optical and magnetic properties of magnesium-zinc ferrite nanofibers. *J. Magn. Magn. Mater.* **2018**, *468*, 132–140. [CrossRef]
6. Lisjak, D.; Mertelj, A. Anisotropic magnetic nanoparticles: A review of their properties, syntheses and potential applications. *Prog. Mater. Sci.* **2018**, *95*, 286–328. [CrossRef]
7. Muhlberger, M.; Janko, C.; Unterweger, H.; Schreiber, E.; Band, J.; Lehmann, C.; Dudziak, D.; Lee, G.; Alexiou, C.; Tietze, R. Functionalization of T lymphocytes for magnetically controlled immune therapy: Selection of suitable superparamagnetic iron oxide nanoparticles. *J. Magn. Magn. Mater.* **2019**, *473*, 61–67. [CrossRef]
8. Danilushkina, A.; Rozhina, E.; Kamalieva, R.; Fakhrullin, R. Influence of magnetic nanoparticles stabilized with polyelectrolytes on 2D and 3D cell cultures formation. *Hum. Gene Ther.* **2018**, *29*, A71.
9. Chen, K.L.; Yeh, Y.W.; Chen, J.M.; Hong, Y.J.; Huang, T.L.; Deng, Z.Y.; Wu, C.H.; Liao, S.H.; Wang, L.M. Influence of magnetoplasmonic gamma-Fe2O3/Au core/shell nanoparticles on low-field nuclear magnetic resonance. *Sci. Rep.* **2016**, *6*, 35477. [CrossRef]
10. Wang, W.; Ma, P.X.; Dong, H.; Krause, H.J.; Zhang, Y.; Willbold, D.; Offenhaeusser, A.; Gu, Z.W. A magnetic nanoparticles relaxation sensor for protein-protein interaction detection at ultra-low magnetic field. *Biosens. Bioelectron.* **2016**, *80*, 661–665. [CrossRef]
11. Angelakeris, M. Magnetic nanoparticles: A multifunctional vehicle for modern theranostics. *Biochim. Biophys. Acta Gen. Subj.* **2018**, *1861*, 1642–1651. [CrossRef] [PubMed]
12. Winter, A.; Engels, S.; Reinhardt, L.; Wasylow, C.; Gerullis, H.; Wawroschek, F. Magnetic marking and intraoperative detection of primary draining lymph nodes in high-risk prostate cancer using superparamagnetic iron oxide nanoparticles: Additional diagnostic value. *Molecules* **2017**, *22*, 2192. [CrossRef] [PubMed]

13. Duran, J.D.G.; Arias, J.L.; Gallardo, V.; Delgado, A.V. Magnetic colloids as drug vehicles. *J. Pharm. Sci.* **2008**, *97*, 2948–2983. [CrossRef] [PubMed]
14. Patitsa, M.; Karathanou, K.; Kanaki, Z.; Tzioga, L.; Pippa, N.; Demetzos, C.; Verganelakis, D.A.; Cournia, Z.; Klinakis, A. Magnetic nanoparticles coated with polyarabic acid demonstrate enhanced drug delivery and imaging properties for cancer theranostic applications. *Sci. Rep.* **2017**, *7*, 775. [CrossRef] [PubMed]
15. Gao, Q.; Xie, W.S.; Wang, Y.; Wang, D.; Guo, Z.H.; Gao, F.; Zhao, L.Y.; Cai, Q. A theranostic nanocomposite system based on radial mesoporous silica hybridized with Fe3O4 nanoparticles for targeted magnetic field responsive chemotherapy of breast cancer. *Rsc Adv.* **2018**, *8*, 4321–4328. [CrossRef]
16. Moskvin, M.; Babic, M.; Reis, S.; Cruz, M.M.; Ferreira, L.P.; Carvalho, M.D.; Lima, S.A.C.; Horák, D. Biological evaluation of surface-modified magnetic nanoparticles as a platform for colon cancer cell theranostics. *Colloids Surf. B Biointerfaces* **2018**, *161*, 35–41. [CrossRef]
17. Semkina, A.S.; Abakumov, M.A.; Skorikov, A.S.; Abakumova, T.O.; Melnikov, P.A.; Grinenko, N.F.; Cherepanov, S.A.; Vishnevskiy, D.A.; Naumenko, V.A.; Ionova, K.P.; et al. Multimodal doxorubicin loaded magnetic nanoparticles for VEGF targeted theranostics of breast cancer. *Nanomed. Nanoechnol.* **2018**, *14*, 1733–1742. [CrossRef]
18. Vakilinezhad, M.A.; Alipour, S.; Montaseri, H. Fabrication and in vitro evaluation of magnetic PLGA nanoparticles as a potential Methotrexate delivery system for breast cancer. *J. Drug Deliv. Sci. Technol.* **2018**, *44*, 467–474. [CrossRef]
19. Obaidat, L.M.; Issa, B.; Haik, Y. Magnetic properties of magnetic nanoparticles for efficient hyperthermia. *Nanomaterials (Basel)* **2015**, *5*, 63–89. [CrossRef]
20. Balidemaj, E.; Kok, H.P.; Schooneveldt, G.; van Lier, A.; Remis, R.F.; Stalpers, L.J.A.; Westerveld, H.; Nederveen, A.J.; van den Berg, C.A.T.; Crezee, J. Hyperthermia treatment planning for cervical cancer patients based on electrical conductivity tissue properties acquired in vivo with EPT at 3 T MRI. *Int. J. Hyperth.* **2016**, *32*, 558–568. [CrossRef]
21. Blanco-Andujar, C.; Walter, A.; Cotin, G.; Bordeianu, C.; Mertz, D.; Felder-Flesch, D.; Begin-Colin, S. Design of iron oxide-based nanoparticles for MRI and magnetic hyperthermia. *Nanomedicine (Lond)* **2016**, *11*, 1889–1910. [CrossRef] [PubMed]
22. Iglesias, G.; Delgado, A.V.; Kujda, M.; Ramos-Tejada, M.M. Magnetic hyperthermia with magnetite nanoparticles: Electrostatic and polymeric stabilization. *Colloid Polym. Sci.* **2016**, *294*, 1541–1550. [CrossRef]
23. Sato, L.; Umemura, M.; Mitsudo, K.; Fukumura, H.; Kim, J.H.; Hoshino, Y.; Nakashima, H.; Kioi, M.; Nakakaji, R.; Sato, M.; et al. Simultaneous hyperthermia-chemotherapy with controlled drug delivery using single-drug nanoparticles. *Sci. Rep.* **2016**, *6*, 24629. [CrossRef] [PubMed]
24. Hedayatnasab, Z.; Abnisa, F.; Daud, W.M.A.W. Review on magnetic nanoparticles for magnetic nanofluid hyperthermia application. *Mater. Des.* **2017**, *123*, 174–196. [CrossRef]
25. Nemati, Z.; Salili, S.M.; Alonso, J.; Ataie, A.; Das, R.; Phan, M.H.; Srikanth, H. Superparamagnetic iron oxide nanodiscs for hyperthermia therapy: Does size matter? *J. Alloy. Compd.* **2017**, *714*, 709–714. [CrossRef]
26. Reyes-Ortega, F.; Delgado, A.V.; Schneider, E.K.; Fernandez, B.C.L.; Iglesias, G.R. Magnetic nanoparticles coated with a thermosensitive polymer with hyperthermia properties. *Polymers (Basel)* **2018**, *10*, 10. [CrossRef] [PubMed]
27. Vamvakidis, K.; Mourdikoudis, S.; Makridis, A.; Paulidou, E.; Angelakeris, M.; Dendrinou-Samara, C. Magnetic hyperthermia efficiency and MRI contrast sensitivity of colloidal soft/hard ferrite nanoclusters. *J. Colloid Interface Sci.* **2018**, *511*, 101–109. [CrossRef] [PubMed]
28. Pankhurst, Q.A.; Connolly, J.; Jones, S.K.; Dobson, J. Applications of magnetic nanoparticles in biomedicine. *J. Phys. D* **2003**, *36*, R167–R181. [CrossRef]
29. Soares, P.I.P.; Ferreira, L.M.M.; Igreja, R.; Novo, C.M.M.; Borges, J. Application of hyperthermia for cancer treatment: Recent patents review. *Recent Pat. Anticancer Drug Discov.* **2012**, *7*, 64–73. [CrossRef]
30. Alphandery, E.; Faure, S.; Seksek, O.; Guyot, F.; Chebbi, I. Chains of magnetosomes extracted from AMB-1 magnetotactic bacteria for application in alternative magnetic field cancer therapy. *Acs Nano* **2011**, *5*, 6279–6296. [CrossRef]
31. El-Boubbou, K. Magnetic iron oxide nanoparticles as drug carriers: Preparation, conjugation and delivery. *Nanomedicine (Lond)* **2018**, *13*, 929–952. [CrossRef] [PubMed]
32. Valverde-Tercedor, C.; Montalban-Lopez, M.; Perez-Gonzalez, T.; Sanchez-Quesada, M.S.; Prozorov, T.; Pineda-Molina, E.; Fernandez-Vivas, M.A.; Rodriguez-Navarro, A.B.; Trubitsyn, D.; Bazylinski, D.A.; et al.

Size control of in vitro synthesized magnetite crystals by the MamC protein of *Magnetococcus marinus* strain MC-1. *Appl. Microbiol. Biotechnol.* **2015**, *99*, 5109–5121. [CrossRef] [PubMed]

33. García Rubia, G.; Peigneux, A.; Jabalera, Y.; Puerma, J.; Oltolina, F.; Elert, K.; Colangelo, D.; Morales, J.G.; Prat, M.; Jimenez-Lopez, C. pH-dependent adsorption release of doxorubicin on MamC-biomimetic magnetite nanoparticles. *Langmuir* **2018**, *34*, 13713–13724. [CrossRef] [PubMed]
34. Lacroix, L.M.; Malaki, R.B.; Carrey, J.; Lachaize, S.; Respaud, M.; Goya, G.F.; Chaudret, B. Magnetic hyperthermia in single-domain monodisperse FeCo nanoparticles: Evidences for Stoner-Wohlfarth behavior and large losses. *J. Appl. Phys.* **2009**, *105*, 023911. [CrossRef]
35. Carrey, J.; Mehdaoui, B.; Respaud, M. Simple models for dynamic hysteresis loop calculations of magnetic single-domain nanoparticles: Application to magnetic hyperthermia optimization. *J. Appl. Phys.* **2011**, *110*, 083921. [CrossRef]
36. Mehdaoui, B.; Meffre, A.; Carrey, J.; Lachaize, S.; Lacroix, L.M.; Gougeon, M.; Chaudret, B.; Respaud, M. Optimal size of nanoparticles for magnetic hyperthermia: A combined theoretical and experimental study. *Adv. Funct. Mater.* **2011**, *21*, 4573–4581. [CrossRef]
37. Stoner, E.C.; Wohlfarth, E.P. A mechanism of magnetic hysteresis in heterogeneous alloys. *IEEE Trans. Magn.* **1991**, *27*, 3475–3518. [CrossRef]
38. Pankhurst, Q.A.; Thanh, N.T.K.; Jones, S.K.; Dobson, J. Progress in applications of magnetic nanoparticles in biomedicine. *J. Phys. D Appl. Phys.* **2009**, *42*, 224001. [CrossRef]
39. Ortega, D.; Pankhurst, Q.A. Magnetic hyperthermia. In *Nanoscience: Nanostructures through Chemistry*; O'Brien, P., Ed.; Royal Society of Chemistry: Cambridge, UK, 2013; Volume 1, pp. 60–88.
40. Surowiec, Z.; Miaskowski, A.; Budzynski, M. Investigation of magnetite Fe3O4 nanoparticles for magnetic hyperthermia. *Nukleonika* **2017**, *62*, 183–186. [CrossRef]
41. Laurent, S.; Dutz, S.; Hafeli, U.O.; Mahmoudi, M. Magnetic fluid hyperthermia: Focus on superparamagnetic iron oxide nanoparticles. *Adv. Colloid Interface Sci.* **2011**, *166*, 8–23. [CrossRef]
42. Coffey, W.T.; Crothers, D.S.F.; Kalmykov, Y.P.; Waldron, J.T. Constant-magnetic-field effect in Néel relaxation of single-domain ferromagnetic particles. *Phys. Rev. B Condens. Matter.* **1995**, *51*, 15947–15956. [CrossRef] [PubMed]
43. Hergt, R.; Dutz, S.; Zeisberger, M. Validity limits of the Neel relaxation model of magnetic nanoparticles for hyperthermia. *Nanotechnology* **2010**, *21*, 015706. [CrossRef] [PubMed]
44. Bakoglidis, K.D.; Simeonidis, K.; Sakellari, D.; Stefanou, G.; Angelakeris, M. Size-Dependent mechanisms in AC magnetic hyperthermia response of iron-oxide nanoparticles. *IEEE Trans. Magn.* **2012**, *48*, 1320–1323. [CrossRef]
45. Dutz, S.; Hergt, R. Magnetic nanoparticle heating and heat transfer on a microscale: Basic principles, realities and physical limitations of hyperthermia for tumour therapy. *Int. J. Hyperth.* **2013**, *29*, 790–800. [CrossRef] [PubMed]
46. Dutz, S.; Hergt, R. Magnetic particle hyperthermia-a promising tumour therapy? *Nanotechnology* **2014**, *25*, 452001. [CrossRef] [PubMed]
47. Guibert, C.; Dupuis, V.; Peyre, V.; Fresnais, J. Hyperthermia of magnetic nanoparticles: Experimental study of the role of aggregation. *J. Phys. Chem. C* **2015**, *119*, 28148–28154. [CrossRef]
48. Fortin, J.P.; Wilhelm, C.; Servais, J.; Menager, C.; Bacri, J.C.; Gazeau, F. Size-sorted anionic iron oxide nanomagnets as colloidal mediators for magnetic hyperthermia. *J. Am. Chem. Soc.* **2007**, *129*, 2628–2635. [CrossRef]
49. Perez-Gonzalez, T.; Rodriguez-Navarro, A.; Jimenez-Lopez, C. Inorganic magnetite precipitation at 25 °C: A low-cost inorganic coprecipitation method. *J. Supercond. Nov. Magn.* **2011**, *24*, 549–557. [CrossRef]
50. Gas, P.; Kurgan, E. Cooling effects inside water-cooled inductors for magnetic fluid hyperthermia. In Proceedings of the 2017 Progress in Applied Electrical Engineering (PAEE), Koscielisko, Poland, 25–30 June 2017; pp. 1–4. [CrossRef]
51. Gas, P.; Miaskowski, A. Specifying the ferrofluid parameters important from the viewpoint of Magnetic Fluid Hyperthermia. In Proceedings of the 2015 Selected Problems of Electrical Engineering and Electronics (WZEE), Kielce, Poland, 17–19 September 2015; pp. 1–6. [CrossRef]
52. Wildeboer, R.R.; Southern, P.; Pankhurst, Q.A. On the reliable measurement of specific absorption rates and intrinsic loss parameters in magnetic hyperthermia materials. *J. Phys. D Appl. Phys.* **2014**, *47*, 495003. [CrossRef]

53. Muniz, F.T.L.; Miranda, M.A.R.; dos Santos, C.M.; Sasaki, J.M. The Scherrer equation and the dynamical theory of X-ray diffraction. *Acta Crystallogr. A Found. Adv.* **2016**, *A72*, 385–390. [CrossRef]
54. Lopez-Moreno, R.; Fernandez-Vivas, A.; Valverde-Tercedor, C.; Azuaga Fortes, A.I.; Casares Atienza, S.; Rodriguez-Navarro, A.B.; Zarivach, R.; Jimenez-Lopez, C. Magnetite nanoparticles biomineralization in the presence of the magnetosome membrane protein MamC: Effect of protein aggregation and protein atructure on magnetite formation. *Cryst. Growth Des.* **2017**, *17*, 1620–1629. [CrossRef]
55. Fonseca, F.C.; Goya, G.F.; Jardim, R.F.; Muccillo, R.; Carreno, N.L.V.; Longo, E.; Leite, E.R. Superparamagnetism and magnetic properties of Ni nanoparticles embedded in SiO2. *Phys. Rev. B* **2002**, *66*, 104406. [CrossRef]
56. Obrien, R.W.; White, L.R. Electrophoretic mobility of a spherical colloidal particle. *J. Chem. Soc. Faraday Trans. 2* **1978**, *74*, 1607–1626. [CrossRef]
57. Nudelman, H.; Valverde-Tercedor, C.; Kolusheva, S.; Perez-Gonzalez, T.; Widdrat, M.; Grimberg, N.; Levi, H.; Nelkenbaum, O.; Davidov, G.; Faivre, D.; et al. Structure-function studies of the magnetite-biomineralizing magnetosome-associated protein MamC. *J. Struct. Biol.* **2016**, *194*, 244–252. [CrossRef] [PubMed]
58. Jabalera, Y.; Casares Atienza, S.; Fernandez-Vivas, A.; Peigneux, A.; Azuaga Fortes, A.I.; Jimenez-Lopez, C. Protein conservation method affects MamC-mediated biomineralization of magnetic nanoparticles. *Cryst. Growth Des.* **2019**, *19*, 1064–1071. [CrossRef]

Article

Entrapment of *N*-Hydroxyphthalimide Carbon Dots in Different Topical Gel Formulations: New Composites with Anticancer Activity

Corina-Lenuta Savin [1], Crina Tiron [2,*], Eugen Carasevici [2], Corneliu S. Stan [1], Sorin Alexandru Ibanescu [3], Bogdan C. Simionescu [1,3] and Catalina A. Peptu [1,*]

1 Department of Natural and Synthetic Polymers, Faculty of Chemical Engineering and Environmental Protection, "Gheorghe Asachi" Technical University of Iasi, 71, Prof., Dr. Docent Dimitrie Mangeron Street, 700050 Iasi, Romania
2 Regional Institute of Oncology, 2-4, General Henri Mathias Berthelot Street, 700483 Iasi, Romania
3 "Petru Poni" Institute of Macromolecular Chemistry, 41A Aleea Grigore Ghica Vodă street, 700487 Iasi, Romania
* Correspondence: transcendctiron@iroiasi.ro (C.T.); catipeptu@ch.tuiasi.ro (C.A.P.); Tel.: +40-765-677-356 (C.T.); +40-765-253-915 (C.A.P.)

Received: 7 May 2019; Accepted: 23 June 2019; Published: 1 July 2019

Abstract: In the present study, the antitumoral potential of three gel formulations loaded with carbon dots prepared from *N*-hydroxyphthalimide (CD-NHF) was examined and the influence of the gels on two types of skin melanoma cell lines and two types of breast cancer cell lines in 2D (cultured cells in normal plastic plates) and 3D (Matrigel) models was investigated. Antitumoral gels based on sodium alginate (AS), carboxymethyl cellulose (CMC), and the carbomer Ultrez 10 (CARB) loaded with CD-NHF were developed according to an adapted method reported by Hellerbach. Viscoelastic properties of CD-NHF-loaded gels were analyzed by rheological analysis. Also, for both CD-NHF and CD-NHF-loaded gels, the fluorescence properties were analyzed. Cell proliferation, apoptosis, and mitochondrial activity were analyzed according to basic methods used to evaluate modulatory activities of putative anticancer agents, which include reference cancer cell line culture assays in both classic 2D and 3D cultures. Using the rheological measurements, the mechanical properties of gel formulations were analyzed; all samples presented gel-like rheological characteristics. The presence of CD-NHF within the gels induces a slight decrease of the dynamic moduli, indicating a flexible gel structure. The fluorescence investigations showed that for the gel-loaded CD-NHF, the most intense emission peak was located at 370 nm (upon excitation at 330 nm). 3D cell cultures displayed visibly larger structure of tumor cells with less active phenotype appearance. The in vitro results for tested CD-NHF-loaded gel formulations revealed that the new composites are able to affect the number, size, and cellular organization of spheroids and impact individual tumor cell ability to proliferate and aggregate in spheroids.

Keywords: composite; *N*-hydroxyphthalimide; carbon dots; polymer gels; antitumoral activity

1. Introduction

Statistically, it has been reported that approximately 8 million people die from different types of cancer each year in the world; including breast, lung, liver, skin, and brain cancers, etc. So far, the main therapeutic methods for cancer treatment remain surgery and chemotherapy; however, researchers have made enormous efforts in the last years to develop new compounds that are better tolerated by patients during cancer therapy. In order to attack cancer cells and to reduce the same effect in the healthy cells, new treatments methods have been investigated. Among them, transdermal drug delivery systems appear as a promising alternative strategy to carry antineoplastic agents due to certain several advantages such as increased drug solubility, better bioavailability, high stability, controlled drug release, prolonged half-life, selective organ or tissue distribution, and reduction of the total required dose [1]. Enhanced topical delivery of the active principle to the target site, including those for cancer therapy, can be achieved by noninvasive drug delivery systems which can ensure sustained therapy with a single application, thus avoiding first-pass hepatic metabolism, gastric degradation, or frequent dosing and also the inconvenience of parenterals [2].

Carbon dots (CD) are a new star of the carbon nanomaterials family; their unique properties mean they are attractive materials for a wide range of applications such as bioimaging, biosensing, drug delivery, optoelectronics, photovoltaics, and photocatalysis [3,4]. Therefore, more and more researchers are paying significant interest to the synthesis, properties, and applications of these carbon nanostructured materials. However, the investigations toward drug delivery for biomedical applications are still at their beginning [5], along with their putative antitumoral properties. Precursors used to synthesize carbon dots determine their properties, creating a new opportunity for finding new anticancer molecules for certain types of cancer (personalized cancer therapy) [6]. Gels based on natural and/or synthetic polymers represent a good pathway for biomedical applications, due to their hydrophilic properties and their biocompatibility. Physical gels are usually prepared by mixing a polymer and a solvent, with or without thermal treatment, resulting a homogeneous material with remarkable mechanical properties related to the aggregation process of the polymer chains [7]. So far, different types of natural or synthetic polymers have been used to prepare polymer–drug conjugates, produced mainly by physical entrapment of the biologically active principle into macromolecular hydrogels, micro/nanoparticles, or liposomes in therapeutic approaches for the treatment of various cancers.

Polysaccharides are renewable resources found in all living organisms, which have been widely used in different important fields such as the biomedical, pharmaceutical, and food industries and tenvironmental remediation. Their unique properties, such as biocompatibility, biodegradability, nontoxicity, hydrophilicity, stability, and structural variability, allow a high capacity for carrying biological information, making them suitable as a promising natural biomaterial [8]. In recent years, the scientific community has used polysaccharides in a broad range of biomedical applications, such as drug delivery systems for the treatment of inflammation, cell–cell recognition, immune responses, metastasis (tumor cells are spreading from primary tumor to secondary organs e.g., from breast to lungs), and in tissue engineering in scaffolds and wound dressings [9]. Sodium alginate (AS) is a natural polysaccharide, extracted from brown seaweed, used in various biomedical applications such as drug delivery, tissue engineering, and wound healing, due to the attractive properties such as biocompatibility, low toxicity, ease of manipulation, and mild gelation [10]. Another attractive polysaccharide is carboxymethylcellulose (CMC) in its sodium salt form. CMC is the major cellulose ether, also known as cellulose gum, formed from carboxymethyl ether groups. CMC's valuable properties, such as its hydrophilic nature, capacity to form gels at high concentration, and thixotropy, make it suitable for biomedical applications such as drug delivery and tissue engineering [11]. From the synthetic polymers category, carbomers (CARB) are a high-molecular-weight crosslinked polymer of acrylic acid, which are playing an important role in many commercial products such as gels, creams, and lotions, providing viscosity, stabilization, and suspension properties [12].

In this article, we focus on the embedding of *N*-hydroxyphthalimide carbon dots (CD-NHF) in different continuous matrices formed of natural (alginate or carboxymethyl cellulose) or synthetic (CARB) polymers. Due to the recently proven antitumoral activity of NHF [13], CD-NHF were developed and the composite was proven also to induce breast cancer cell apoptosis at doses that only marginally affect normal cell counterparts. The reason for embedding CD-NHF is on one hand, to reduce the aggregation tendency of CD which can affect the treatment efficacy, and on the other hand, to be protected by a polymer environment against chemical modification, which can also affect their properties. The fluorescence analysis demonstrates the CD-NHF presence within the gel.

Appropriate gel properties were investigated from a rheological point of view considering different polymer concentrations; the optimum formulation has been used as the matrix for CD-NHF and further biological investigations.

Cancer is a disease that causes cells to change and grow in an uncontrolled manner. Many—if not all—of the tumor aggressivity behaviors are caused by key molecular defects in multiple biochemical pathways controlling cell survival, proliferation, differentiation, and integration in histological tisular structures and immune interactions. Apoptosis is a cell suicide program that plays an important role in tissue homeostasis by eliminating unnecessary cells [14,15]. The deregulation of the apoptotic pathway results in variety of diseases and has been shown to be involved in cancer cell resistance to conventional anticancer therapy [16]. There are two critical pathways of apoptosis—the extrinsic pathway (death receptor-mediated pathway) and the intrinsic pathway (mitochondria-mediated pathway)—and caspases are fundamental players in both pathways [17]. Mitochondria are vital organelles for energy production and intracellular Ca^{2+} homeostasis and they are involved in a variety of cellular processes, including differentiation, proliferation, and apoptosis. It has been shown that abnormal mitochondrial dynamics have an important role in tumorigenesis. Also, the increased anaerobic glycolysis activity in neoplastic cells was the result of a dysfunction of the mitochondrial activity [18].

We consider that CD-NHF incorporation in gels could be a convenient way to manipulate its local availability (and effects) and its persistence. While the actual mechanism of action is still incompletely understood for many emerging nanoformulations, interactions with extracellular matrix components are plausible objectives in many drug designs. The incorporation of CD-NHF in both gel varieties utilized in our assay was thought to be promising for possible targeted applications in certain malignancies (perhaps including basocellular and spinocellular skin carcinomas). The effect of gels with CD-NHF in 4T1 (mouse mammary breast cancer), MDA-MB-231 (human mammary breast cancer), HDMVECn (primary dermal microvascular endothelial cells), Balb/c-5064 (mouse dermal microvascular endothelial cells), A375 (human malignant melanoma), and B16F10 (mouse malignant melanoma) cells in 2D (cultured cells in normal plastic plates) and 3D (Matrigel) models was evaluated by assessing cell viability, mitochondrial activity, and apoptosis.

2. Materials and Methods

2.1. Materials

The following materials were used for the experiments: the carbomer Ultrez 10 was provided by Antibiotice SA Iasi (Iasi, Romania); *N*-hydroxyphthalimide, medium-viscosity sodium alginate (extracted from *Macrocytis pyrifera*), and glycerin (99.9%) were purchased from Sigma-Aldrich (St. Louis, MO, USA); medium-viscosity carboxymethyl cellulose from Fluka (Buchs, Switzerland); and reagent-grade ethanol (EtOH) and Milli-Q ultrapure distilled water from Merck Chemicals (Darmstadt, Germany). The chemical reagents used in this study were of analytical-grade purity and were used without further purification. MDA-MB-231 cells were purchased from the American Type Culture Collection, Rockville (ATCC), VA, USA; 4T1 cells were from the American Type Culture Collection, Rockville, MD, USA; bovine serum was purchased from Sigma-Aldrich; HDMVECn (primary dermal microvascular endothelial cells) were from ATCC; Balb/c-5064 (mouse dermal microvascular endothelial cells) were from Cell Biologics (Chicago, IL, USA); and Matrigel Matrix from BD Biosciences (356234), Bedford, MA, USA. For biological tests, the Live and Dead Cell Assay (Abcam ab115347), CellTiter-Blue® Cell Viability Assay (Promega, (Madison, WI, USA), MitoTracker™ Red (Molecular Probe, Eugene, OR, USA), Matrigel (Corning, Bedford, MA, USA), and Caspase-3/7 Assay (Promega, Madison, WI, USA) were used according to the manufacturer's recommendations.

2.2. Methods

2.2.1. Preparation of Gel Formulations Based on CARB, AS, and CMC

Antitumoral gels based on CARB, AS, and CMC, respectively, and loaded with CD-NHF were prepared according to the method reported by Hellerbach et al. [19] with slight modification. CD-NHF were prepared according to a protocol reported by Stan et al. [4]. Prior to gel embedment, the carbon dots were prepared through pyrolytic processing of an NHF precursor using an experimental setup presented in Figure S1. Briefly, in a typical synthesis procedure, 0.1 g of NHF is thermally processed in carefully controlled conditions (temperature of the main sequence of the pyrolytic process is 253 °C within 30 min), leading to a typical configuration of the carbon dots consisting of a carbonaceous core with a surface highly decorated by various functional groups. The slightly different processing parameters compared with the previously reported preparation path are due to the different experimental setup, which in this work, was reconfigured and extended in order to increase both the reproducibility of the carbon dots' preparation path and the resultant per batch quantities. The resultant reaction mass is primarily dispersed in high-purity water kept cooled at 4–5 °C. In the next stages, the suspension is centrifuged at 10,000 rpm for 15 min and the supernatant is collected and freeze-dried, with a fine powder of carbon dots being obtained. A series of gel formulations were prepared by mixing different quantities of CARB, AS, or CMC (Table 1) in a solution mixture consisting of 10 mL distilled water and 3 mL ethanol in glass vessels at room temperature. In order to achieve a uniform gel formulation, a T 18 digital ULTRA-TURRAX disperser by IKA was used (10 min, 10,000 rpm). The gelling agent's concentration ranged from 3.8 to 5.8% (Table 1). In order to prepare the gel formulations containing CD-NHF (CARB-F4, AS-F6, or CMC-F6, respectively), the CD-NHF (3.84 mg/mL) was added into the water–ethanol solution and dispersed by sonication for 1 min at room temperature using a sonication bath (Bandelin Sonorex, Berlin, Germany), followed by adding CARB, AS, or CMC. The obtained gel formulations and CD-NHF gel (Figure S3) were stored at 4° C in a refrigerator. The basic molecular formulas for the polymers and CD-NHF used in order to obtain the gel formulations with and without CD-NHF are: sodium alginate (AS)—$(C_6H_8O_6)_n$, carboxymethyl cellulose (CMC)—$[C_6H_7O_2(OH)_2OCH_2COONa]_n$, Ultrez 10 carbomer (CARB)—$(C_3H_4O_2)_n$, and *N*-hydroxyphthalimide (NHF)—$C_8H_5NO_3$.

Table 1. Polymeric composition of various CARB, AS, and CMC gel formulations. CARB: carbomer Ultrez 10; AS: sodium alginate; CMC: carboxymethyl cellulose; GLY: glycerin.

Sample Code	CARB, %	AS, %	CMC, %	GLY, %	Distilled Water (mL)	Ethanol (mL)	CD-NHF (g)
CARB-F1	3.8	-	-	-			-
CARB-F2	4.6	-	-	-			-
CARB-F3	5.8	-	-	-			-
CARB-F4	4.6	-	-	-			0.050
AS-F1	4.6	4.6	-	-			-
AS-F2	4.6	4.6	-	9.2			-
AS-F3	3.5	4.6	-	8.1			-
AS-F4	2.3	4.6	-	6.9	10	3	-
AS-F5	1.2	4.6	-	5.8			-
AS-F6	4.6	4.6	-	5.8			0.050
CMC-F1	2.9	-	2.9	-			-
CMC-F2	2.9	-	2.9	5.8			-
CMC-F3	3.6	-	2.9	6.5			-
CMC-F4	4.3	-	2.9	7.2			-
CMC-F5	5.8	-	2.9	8.7			-
CMC-F6	3.6	-	2.9	6.5			0.050

2.2.2. Characterization of the Prepared NHF Carbon Dots

The morpho-structural characteristics of the prepared NHF carbon dots were essentially similar to those previously reported [4]. In brief, the XPS survey and high-resolution O1s, N1s, and C1s spectra revealed the presence of the defect-rich graphitic core highly decorated with various terminal functional groups, while the recorded FT-IR spectra revealed the modifications/reconfigurations occurring through thermal processing of the NHF precursor and confirmed by XPS results which were revealing the presence of various functional groups attached to the graphitic core, which are essentially responsible for both optical and antitumoral properties of this type of carbon dots. In Figure S2 are presented the recorded spectra of the NHF precursor and resultant carbon dots, and several specific vibrations of various groups are detailed. Interesting details regarding the morphology of the prepared carbon dots were highlighted by the HR-TEM investigation. Figure 1 presents the recorded micrographs of two samples obtained by depositing a highly diluted chloroform-dispersed carbon dot solution on 300-mesh carbon-plated copper grids. Both images suggest a cluster organization of smaller entities, which presumably are individual carbon dots. While their aspect could suggest individual carbon dots, it is also possible to be attributed to even smaller clusters, as can be seen in Figure 1b.

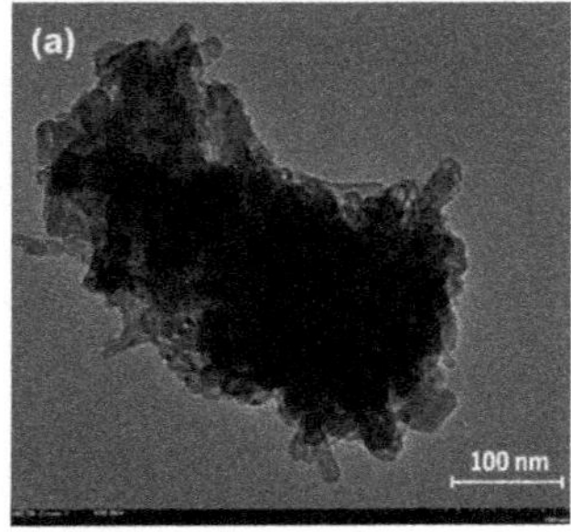

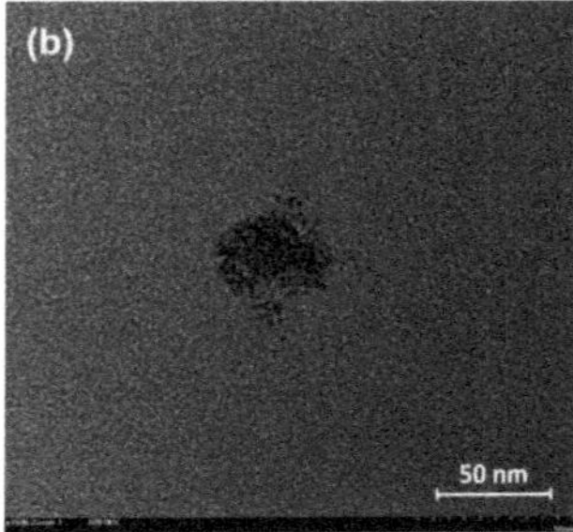

Figure 1. Recorded HR-TEM images of the carbon dots prepared from *N*-hydroxyphthalimide (CD-NHF): (**a**) CD-NHF with an average size of the clusters in 100–300 nm range and (**b**) CD-NHF with an average size of the smaller entities in 2–4 nm range

2.2.3. Rheological Studies

Viscoelastic properties of all gel formulations were evaluated in triplicate by using an Anton Paar Physica MCR 501 (Graz, Austria) rheometer. The modular rheometer is equipped with an electronically commutated synchronous motor (ECMotor, Graz, Austria), allowing rheological testing in controlled stress (CS) and controlled strain (CR) modes. Also, the instrument allows the individual creation of complex real-time tests containing a large number of different intervals under CS or CR control, both in rotational and oscillatory modes. The rheometer is provided with an H-PTD200 system for the temperature control. A parallel-plate geometry specially designed for gels (with serrated plates to avoid slippage) with a diameter of 50 mm was chosen as the measuring system. The freshly prepared, well-homogenized gel formulations were introduced into the measuring cell and analyzed. The rheometer has an intelligent configuration system for automatic identification and configuration—"Tool Master"—which performs automatic transfer of the parameters of the measuring system (constructive characteristics, operating constants, geometry) and temperature control to the Rheoplus program. The chip integrated into the geometry contains all data connected to that and transfers them automatically to the program.

2.2.4. Fluorescence Analysis

The fluorescence analysis of CD-NHF and CD-NHF-loaded gel formulations were recorded on a Horiba Fluoromax 4P (Fluor Essence Version 35.1.20) provided with the Quanta-φ integration sphere. Also, a visual testing of photoluminescence properties was performed using a Philips UVA TL4WBLB lamp (München, Germany) with the emission maximum located in the 370–390 nm range and a 50 mW, 440 nm laser diode.

2.2.5. Cell Culture

MDA-MB-231 cells were cultured in F-12K medium supplemented with 100 U/mL of penicillin and 100 µg/mL of streptomycin and 5% bovine serum; 4T1 cells were cultured in RPMI-1640 supplemented with 100 U/mL of penicillin and 100 µg/mL of streptomycin and 10% bovine serum; HDMVECn (primary dermal microvascular endothelial cells) and Balb/c-5064 (mouse dermal microvascular endothelial cells) were cultured in Vascular Cell Basal Medium supplemented with Microvascular Endothelial Cell Growth Kit-BBE ATCC® PCS-110-040; A375 (human malignant melanoma) and B16F10 (mouse malignant melanoma) were cultured in Dulbecco's Modified Eagle's Medium supplemented with 100 U/mL of penicillin and 100 µg/mL of streptomycin and 10% bovine serum.

2.2.6. Cell Proliferation and Apoptosis Activity

For cell viability estimation, we used the CellTiter-Blue® Cell Viability Assay (Promega). Cells were seeded into a 96-well flat-bottomed tissue culture plate at a density of 2000 cells/well and allowed to adhere to the plate by incubating at 37 °C under 5% CO_2 overnight. Following cell attachment, the cells were incubated with the tested CD-NHF at 5% concentrations (50 µg/mL) for 72 h. For all in vitro experiments, we used the 5% CD-NHF concentration, based on our preliminary results. Control cells were treated with phosphate-buffered saline (PBS), which was equivalent to the amount of PBS used as vehicle. After each of the 72 h treatment time periods, 50 µL of cell viability solution was added to each well and the plate was reincubated for 4 h before fluorescence recording using a multiplate microplate reader (FilterMax F5, Sunnyvale, CA, USA). Apoptosis was assessed using the Caspase-3/7 Assay (Promega) according to the manufacturer's indications.

2.2.7. Mitochondrial Activity

For mitochondrial assay, MitoTracker™ Red (Molecular Probe, Eugene, OR, USA) cells were incubated for 72 h with CD-NHF and then subjected to 1 µg/mL MitoTracker and incubated for 30 min before the fluorescence of the resultant solutions was determined at 590 nm using a multiplate microplate reader.

2.2.8. 3D Matrigel Assays

The 3D Matrigel assays were conducted with 1000 cells seeded in Ibidi plates between 2 layers of Matrigel (BD Matrigel Matrix, Growth Factor Reduced (BD Biosciences)) and cultured for 14 days before microscopy analysis (TissueGnostic rig, Vienna, Austria, Europe). Twelve hours post-seeding, 3D embedded cells began to be treated with gels (CARB-F2, CMC-F3, AS-F5) alone and with gels containing 5% (50 µg/mL) CD-NHF (CARB-F4, CMC-F6, AS-F6) for 72 h. After 72 h, the treatments were removed and replaced with normal 3D Matrigel medium (medium corresponding to every cell type supplemented with 2% fetal bovine serum (FBS) and 1% Matrigel). The Live and Dead Cell Assay (Abcam) was used according to the manufacturer's instructions. Nuclei were counterstained with NucBlue Live Ready Probes Reagent (Thermo Fisher Scientific, Eugene, OR, USA). Fluorescence pictures were acquired at 20× magnification using a Zeiss Axio Observer Z1 Fluorescence Microscope from TissueGnostics rig. Single focal plane images were acquired using Tissue FAXS 4.2 software. The TissueQuest 6.0 software was used for total area segmentation analysis and to quantify the area and sum intensity of fluorescence signal for each spheroid (Figures S4 and S5).

2.2.9. Statistical Analysis

GraphPad Prism was used for statistical analysis using tests stated in the figure legends. Grouped analyses were performed by *t*-test. Significance was established when $p < 0.05$.

3. Results and Discussion

3.1. Preparation of Gel Formulations

Previous studies concerning the preparation of physical gels based on natural or synthetic polymers, namely carbomers, sodium alginate, and carboxymethyl cellulose, have pointed out that these materials are nontoxic and nonhazardous and easily synthesized. We aimed to prepare different topical CD-NHF-loaded gel formulations based on polymers with antitumoral activity. Herein, physical gel formulations based on CARB, AS, and CMC were synthesized according to a slightly modified version of the method reported by Hellerbach et al. [19]. This method offers some advantages compared with other methods reported, such as being simple, rapid, and consistent; performed in the absence of crosslinkers (hence, we eliminated the possibility of toxic chemical contamination commonly associated with methods which use covalent crosslinkers); and the product is a homogeneous soft material formed at room temperature.

Different concentrations of CARB, AS, or CMC were tailored and their effects were observed. First, simple gels were prepared and analyzed before loading with CD-NHF. The gel formulations with or without CD-NHF obtained were stable, odorless, transparent, and highly viscous. Of them, the best formulations presenting a suitable consistency and the best viscoelastic properties, respectively, were considered for further study and loading with CD-NHF, namely samples CARB-F2, AS-F5, and CMC-F3. The carbon-dot-loaded gel samples were denoted as CARB-F4, AS-F6, and CMC-F6.

3.2. Rheology Studies

The viscoelastic nature of polymer gels plays an important role in their adhesion properties; a very important characteristic considering different biomedical applications. Gels based on natural or synthetic polymers exhibit an excellent adhesion property due to both the elastic and viscoelastic properties.

Using the rheological measurements, the mechanical properties of the best formulations were analyzed. All analyzed samples presented rheological characteristics of a gel. The obtained results showed that viscosity was directly dependent on the polymeric content of the formulations. During the rheological measurements, both dynamic moduli were constant. As can be observed from Figure 2, the storage modulus (G′) exhibited higher values than the loss modulus (G″)) over the entire strain range of 0.01 to 100%, indicating the gel-like behavior (Figure 2). The results obtained by rheological analysis confirmed that CARB-F2, AS-F5, and CMC-F3 samples were optimized and further considered for loading with CD-NHF, due to the gel equilibrium modulus. The presence of CD-NHF (for samples CARB-F4, AS-F6, and CMC-F6) in the gel structure has a significant influence over both dynamic moduli. In the case of the AS-F6 sample, the CD-NHF addition induced a slight increase of the dynamic moduli, indicating stiffness of the gel structure, whereas in CARB-F4 and CMC-F6 gels, the effect of CD-NHF addition was exactly the opposite, leading to a slight decrease of the dynamic moduli, indicating a flexible gel structure. Also, the amplitude sweep allows determination of the limits of the linear viscoelastic range for the prepared gels. These values are not affected by the presence of CD-NHF (Table 2).

Table 2. The limits of the linear viscoelastic range (γ_{LVE}) for the tested gel formulations.

Sample Code	γ_{LVE} (%)
CARB-F2	0.25
CARB-F4	0.25
AS-F5	5
AS-F6	5
CMC-F3	10
CMC-F6	10

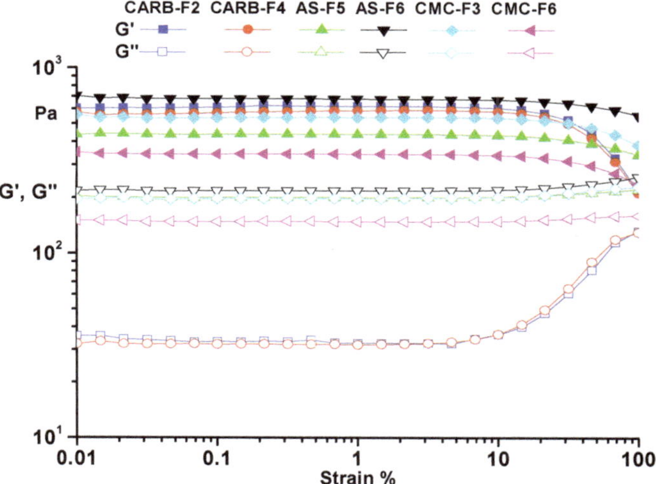

Figure 2. Amplitude sweep for simple gels (CARB-F2, AS-F5, and CMC-F3) and gels with CD-NHF (CARB-F4, AS-F6, and CMC-F6). G′: storage modulus; G″: loss modulus.

All tested gels showed a solid-like behavior, as the storage modulus (G′) was always larger than the loss modulus (G″). As one can observe from Figure 2, the CARB and CMC gel formulations have a more flexible structure than the AS gel, confirmed by storage modulus (G′) values (G'_{CARB} (620 Pa) > G'_{CMC} (540 Pa) > G'_{AS} (437 Pa)). Frequency sweep tests allow observation of the elastic response of gels. The presence of the hydrogen bonding is evidenced by the frequency dependence of the moduli over the entire range of 0.05–500 1/s (γ_{LVE}(CARB) > γ_{LVE}(CMC) > γ_{LVE}(AS)) (Figure 3).

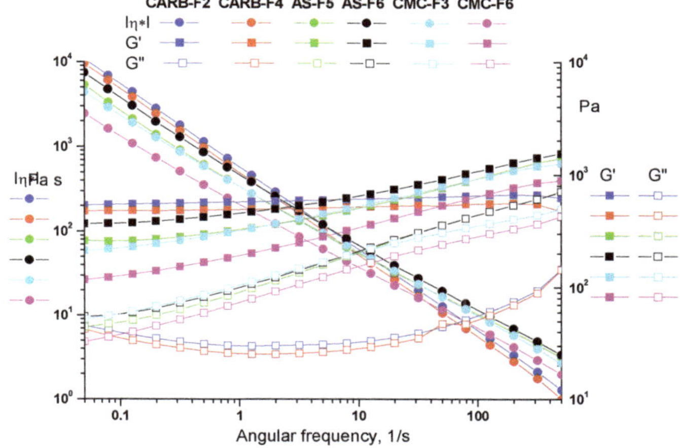

Figure 3. Frequency sweep for simple gels (CARB-F2, AS-F5, and CMC-F3) and gels with CD-NHF (CARB-F4, AS-F6, and CMC-F6).

3.3. Fluorescence Analysis

In order to confirm the presence of CD-NHF in the gel matrices, the formulations were further evaluated through fluorescence analysis. Fluorescence spectra exhibit the maximum emission (λ_{em}) at 424 nm when using an excitation wavelength (λ_{ex}) of 370 nm for CD-NHF (Figure 4 and Table S1).

Also, Figure 4 illustrates the emission profile of CD-NHF-loaded gels (CARB-F4, AS-F6, and CMC-F6) at three different excitation wavelengths ranging from 370 to 410 nm. Moreover, the recorded results (Table S1) revealed that at 370 nm, in the case of CARB, the blue-light emission was not significantly affected by the CARB matrix. The observed difference between CD-NHF blue-light emission and CARB matrix was only 1.5% (Table S1). When AS and CMC were used as a matrix for the entrapment of CD-NHF, an 11% decrease of the blue-light emission was observed, compared to the case of the CARB matrix and CD-NHF. Herein, we can conclude that in this case, the polymer matrix clearly plays an important role for the modification of CD-NHF optical properties.

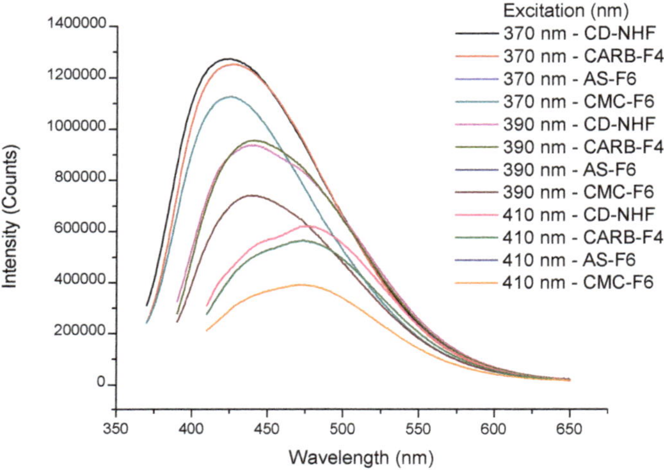

Figure 4. Emission spectra of CD-NHF suspended in H_2O: CARB-F4, AS-F6, and CMC-F6 samples.

3.4. *In Vitro Studies*

Analysis of the cytostatic or cytolytic or growing pattern modulatory activities may be done with end point assays or live cell imaging techniques. 3D cultures create a better similarity between the cultured cells and the living organism. The basic methods to evaluate modulatory activities of putative anticancer agents include reference cancer cell line cultures assays in both classic 2D and 3D cultures.

Viability of HDMVECn cells (primary dermal microvascular endothelial cells) in 2D culture treated with simple gel formulations (CARB-F2, AS-F5, and CMC-F3) and gel formulations containing CD-NHF (conc. 5%; CARB-F4, AS-F6, and CMC-F6) was not affected by CARB-F4 and CMC-F6, whereas, interestingly, AS-F6 had a significant effect (Figure 5).

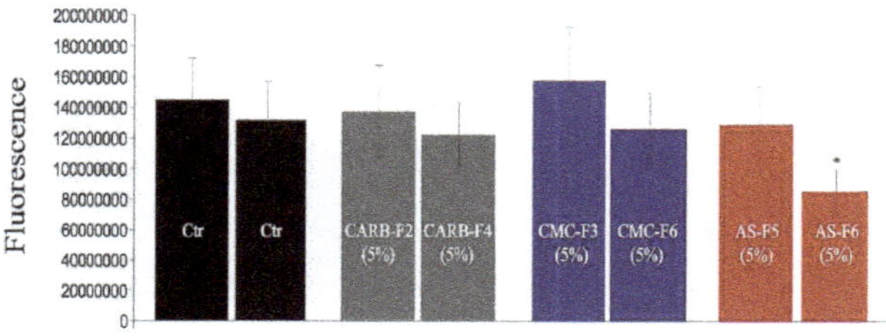

Figure 5. Viability of normal dermal cells (Primary Dermal Endothelial Cells (PDMC)) for gel formulations without and with CD-NHF (N = 15 wells/column from two independent experiments). Ctr (Control).

Proliferation activity of melanoma cell lines (2D system) was not affected the by the presence of CARB-F2, CMC-F3, or AS-F5 formulations (Figure 6a), respectively, while cell viability was affected in a gradual manner in cells treated with CARB-F4, CMC-F6, and AS-F6 formulations (Figure 6b,c).

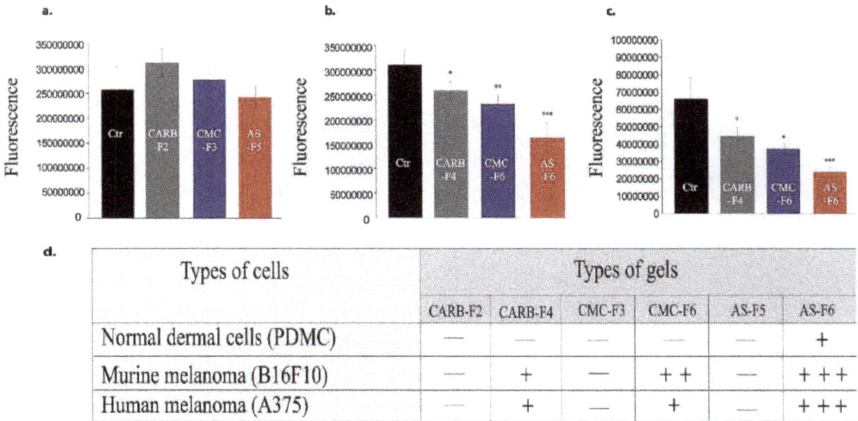

Types of cells	Types of gels					
	CARB-F2	CARB-F4	CMC-F3	CMC-F6	AS-F5	AS-F6
Normal dermal cells (PDMC)	—	—	—	—	—	+
Murine melanoma (B16F10)	—	+	—	+ +	—	+ + +
Human melanoma (A375)	—	+	—	+	—	+ + +

Figure 6. Murine and human melanoma cell viability for gel formulations without (**a**) and with (**b**,**c**) CD-NHF (conc. 5%). $N = 15$ wells/column from two independent experiments; viability of treated groups is expressed relative to the control group. (**a**) Skin cancer; (**b**) murine melanoma (B16F10); (**c**) human melanoma (A375); (**d**) scoring of significance effect, where +++ (***) $p = 0.0003$, ++ (**) $p = 0.01$, + (*) $p = 0.04$, and — = no effect.

In the 3D Matrigel assay, the human melanoma cells form visible and consistently larger colonies compared with the same cells treated with CD-NHF-loaded gel formulations (conc. 5%) (Figure 7 and Figure S4).

By comparing the left column (Matrigel) with the median (simple gel formulations), it can be observed that spheroids are less compact, with decreased cellular uniformity and a tendency to dissociate. In the median column, we find that CMC and AS gel formulations are more unfavorable for maintaining a homogeneity of metabolic activity (the green shade in the L/D kit reflecting live cell population) and induce (or amplify) the dissociation tendency of cells from differentiated proliferation aggregates. The differences found in the analysis of 3D spheroids can be caused by the fact that the different types of gel formulations used either provide different stability conditions for extracellular molecules that favor maintaining a differentiated cellular phenotype, or employ different receptors for adhesion to cell membranes favorable to maintaining both differentiation and inhibition of the cellular dissemination tendency [20,21]. Comparing the median column (single gel formulations) with that on the right (gel formulations supplemented with 5% CD-NHF), we find that exposure to CD was unfavorable to cell proliferation, with global cell counts being significantly impaired and the quantitative weight of cells in invasion reduced. We specify that the shade of the green marker is proportional (L/D kit) to the level of cell metabolic activity. Smaller color intensities can be determined by both a decreased number of cells and more restricted metabolic activity in CD-NHF exposure. Also, the morphological aspect of spheroids under the influence of gel formulations containing CD-NHF is presented in Figure 8.

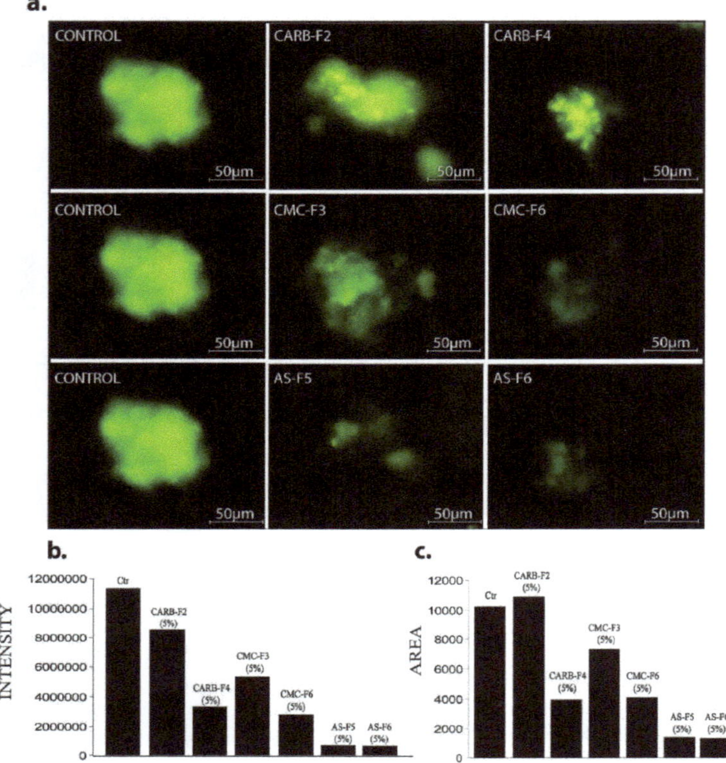

Figure 7. (**a**) 3D Matrigel assay for human melanoma cell cultures in standard conditions (left) and in the presence of added CD-NHF-free gel (middle) versus 5% CD-NHF-loaded gel (right). N = 3D Matrigel cultures per type of gel with or without CD-NHF; 20× microscope objective. (**b**) Sum intensity of fluorescence signal; (**c**) area of spheroids expressed in μm^2.

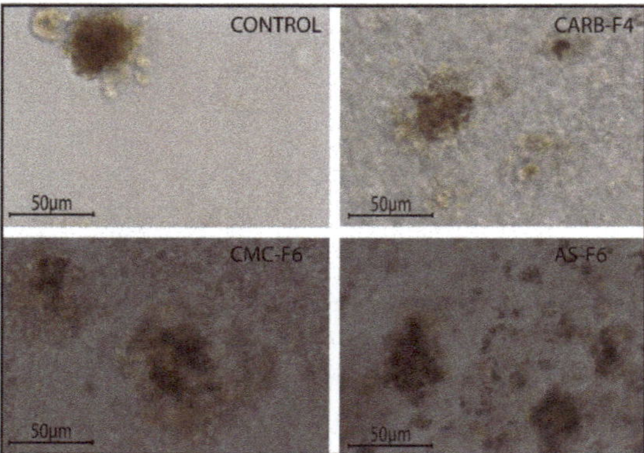

Figure 8. The morphological aspect of spheroids under the influence of gel formulations. N = 3D Matrigel cultures per type of gel with or without CD-NHF; 20× magnification.

Also tested was the viability potential of two different breast cancer cell lines, namely 4T1 (mouse breast cancer) and MDA-MB-231 (human breast adenocarcinoma), in 2D and 3D culture models.

The viability of the breast cancer cell lines 4T1 (mouse breast cancer) and MDA-MB-231 (human breast adenocarcinoma) in the 2D culture system treated with CARB-F4 was affected (Figure 9a). Mitochondrial activity was affected upon CARB-F4 treatment (Figure 9b). In the 3D Matrigel assay, the malignant cells form visible and consistently larger colonies compared with same cells treated with CARB-F4 (Figure 9c). Moreover, the apoptotic foci in 3D cultures treated with gels containing CD-NHF were significantly higher (Figure 9c; white squares in boxes 2 and 4). The apoptotic pattern in the whole spheroid body is also distorted: while in native culturing, inner cells (typically associated with hypoxic conditions inside large spheroids) in spheroids began the apoptotic program, at 5% CD-NHF in exposed spheroids, this form of cell death is initiated in all spheroid compartments. 3D cell cultures displayed visibly larger structure of cancer cells with reduced active phenotype appearance. This suggests that treatment with the CARB-F4 formulation could also either induce cell senescence or cell dormancy—a topic that deserves further clarification. Together, these data suggest that CD-NHF-loaded gels affect the number, size, and cellular organization of spheroids and impacts individual cancer cell ability to proliferate and aggregate in spheroids.

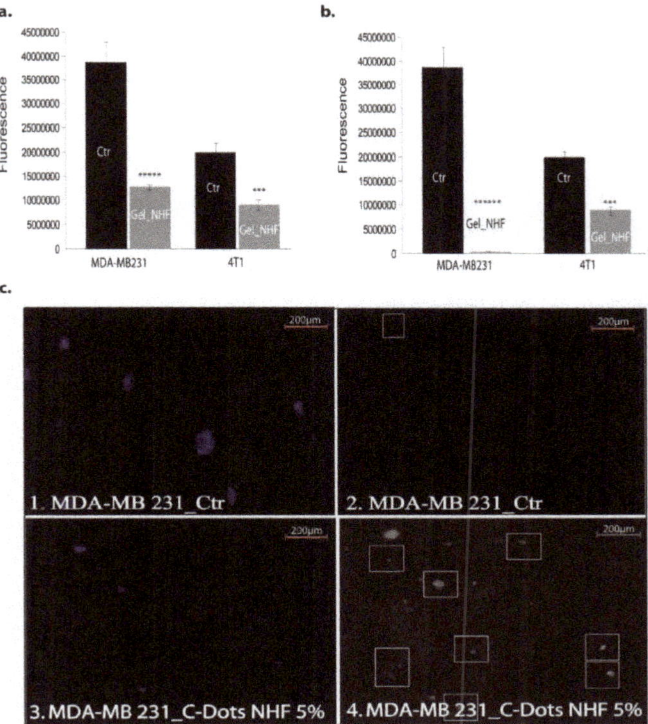

Figure 9. The effects of gels with 5% CD-NHF in 2D and 3D culture. (**a**) Viability of MDA-MB-231 (human breast adenocarcinoma 33.3%) and 4T1 (mouse mammary breast cancer 44.87%) cell lines treated with gels containing 5% CD-NHF ($N = 8$ wells/column from two independent experiments); (**b**) mitochondrial activity in MDA-MB-231 (human breast adenocarcinoma 2%) and 4T1 (mouse mammary breast cancer 49.2%) treated with gels containing 5% CD-NHF ($N = 8$ wells/column from two independent experiments); (**c**) apoptosis evidence in MDA-MB-231 by immunofluorescence microscopy and 3D Matrigel culture treated with gels containing 5% CD-NHF ($N = 6$; 3 Matrigel culture controls and 3 Matrigel cultures treated with 5% NHF). (c1,c3) represent NucBlue nuclei staining; (c2,c4) reflect apoptosis. Picture magnification 5×; where *** $p < 0.0005$, ***** $p < 0.000005$, ****** $p < 0.0000005$.

4. Conclusions

This paper reported the development via a physical method of three topical gel formulations loaded with CD-NHF, which were evaluated from the point of view of the modulatory activities of putative anticancer agents. The obtained results indicate that CD-NHF-loaded gels matrices present complex and interesting cell modulatory activities which are relevant for cancer control in potential clinical applications. The mechanical properties of gels were investigated, concluding that the presence of CD-NHF in the gels induced a slight decrease of the dynamic moduli, indicating a more flexible gel structure. The fluorescence analysis confirmed the presence of the embedded CD-NHF in all gel formulations. Also, worth noting is the potential antitumoral activity of the gel formulations loaded with CD-NHF. The most important finding is that the CD-NHF-loaded gel formulations are able to affect the number, size, and cellular organization of spheroids, while also having a significant impact on the individual tumor cell's ability to proliferate and aggregate in spheroids.

Supplementary Materials: The following are available online at http://www.mdpi.com/1999-4923/11/7/303/s1, Figure S1. Experimental setup used to prepare carbon dots. Figure S2. FT-IR spectra for CD-NHF. Figure S3. Photos of gels without and with CD-NHF under (a) white and (b) UV light. Figure S4. Spheroid fluorescent staining using green/FITC (left column) for live cells, blue/NucBlue for nuclei (middle column), and merged signals (right column) from 3D human melanoma cell cultures. Figure S5. The analytical segmentation procedure for a typical spheroid image. Table S1. Fluorescence results obtained for CD-NHF-loaded gels at different excitation wavelengths: 370–410 nm.

Author Contributions: Gel formulation experimental part, results interpretation, and writing and editing of the manuscript: C.-L.S. and C.A.P.; carbon dot preparation and fluorescence analysis: C.S.S.; rheological analysis and rheology results interpretation: C.A.I.; conceptualization and supervision: B.C.S.; in vitro experiments, biochemical tests, writing, and results interpretation: C.T. and E.C. All authors have given approval to the final version of the manuscript.

Funding: This work has been performed in the frame of the Complex Projects Partnership Program CDI (PCCDI) in priority areas—PN III under UEFISCDI authority, project code PN-III-P1-1.2-PCCDI-2017-0083 (project number 37PCCDI/2018).

Acknowledgments: We wish to thank Florin Zugun-Eloae for his scientific involvement in the in vitro experimental part and the critical review of the manuscript and Adrian Tiron for his involvement in performing the microscopy analysis.

Conflicts of Interest: The authors declare no conflict of interest.

Abbreviations

A375	human malignant melanoma
AS	sodium alginate
B16F10	mouse malignant melanoma cells
Balb/c-5064	mouse dermal microvascular endothelial cells
CARB	Ultrez 10 carbomer
CD	carbon dots
CD-NHF	N-hydroxyphthalimide carbon dots
Conc.	concentration
CMC	carboxymethyl cellulose
CR	controlled strain mode
CS	controlled stress mode
FBS	fetal bovine serum
GLY	glycerin
HDMVECn	primary dermal microvascular endothelial cells
MDA-MB-231	human mammary breast cancer
PBS	phosphate-buffered saline
λ_{em}	emission maximum
λ_{ex}	excitation wavelength

References

1. Vishnubhakthula, S.; Elupula, R.; Durán-Lara, E.F. Recent advances in hydrogel-based drug delivery for melanoma cancer therapy: A mini review. *J. Drug Deliv.* **2017**, *2017*, 7275985. [CrossRef] [PubMed]
2. Ravikumar, P.; Tatke, P. Design of an encapsulated topical formulation for chemoprevention of skin cancer. *Int. J. Pharm. Sci. Res.* **2019**, *10*, 309–319. [CrossRef]
3. Sun, X.; Li, G.; Yin, Y.; Zhang, Y.; Li, H. Carbon quantum dot-based fluorescent vesicles and chiral hydrogels with biosurfactant and biocompatible small molecule. *Soft Matter.* **2018**, *14*, 6983–6993. [CrossRef] [PubMed]
4. Stan, C.S.; Gospei Horlescu, P.; Ursu, L.E.; Popa, M.; Albu, C. Facile preparation of highly luminescent composites by polymer embedding of carbon dots derived from N-hydroxyphthalimide. *J. Mater. Sci.* **2017**, *52*, 185–196. [CrossRef]
5. Peng, Z.; Miyanji, E.H.; Zhou, Y.; Pardo, J.; Hettiarachchi, S.D.; Li, S.; Blackwelder, P.L.; Skromne, I.; Leblanc, R.M. Carbon dots: Promising biomaterials for bone-specific imaging and drug delivery. *Nanoscale* **2017**, *9*, 17533–17543. [CrossRef] [PubMed]
6. Vasimalai, N.; Vilas-Boas, V.; Gallo, J.; Cerqueira, M.F.; Menéndez-Miranda, M.; Costa-Fernández, J.M.; Fernández-Argüelles, M.T. Green synthesis of fluorescent carbon dots from spices for in vitro imaging and tumour cell growth inhibition. *Beilstein J. Nanotechnol.* **2018**, *9*, 530–544. [CrossRef] [PubMed]
7. Djabourov, M. Gelation of physical gels: The gelatin gels. In *Physics of Finely Divided Matter*; Boccara, N., Daoud, M., Eds.; Springer: Berlin, Germany, 1985; pp. 21–23.
8. Desbrieres, J.; Peptu, C.A.; Savin, C.-L.; Popa, M. Chemically modified polysaccharides with applications in nanomedicine. In *Biomass as Renewable Raw Material to Obtain Bioproducts of High-Tech Value*, 1st ed.; Popa, V., Volf, I., Eds.; Elsevier: Amsterdam, The Netherlands, 2018; pp. 351–399. [CrossRef]
9. Xu, H.; Xu, X. Polysaccharide, a potential anti-cancer drug with high efficacy and safety. *Adv. Oncol. Res. Treat* **2016**, *2*, 110.
10. Lee, K.Y.; Mooney, D.J. Alginate: Properties and biomedical applications. *Prog. Polym. Sci.* **2012**, *37*, 106–126. [CrossRef] [PubMed]
11. Aravamudhan, A.; Ramos, D.M.; Nada, A.A.; Kumbar, S.G. Natural polymers. Natural and synthetic biomedical polymers. In *Natural and Synthetic Biomedical Polymers*, 1st ed.; Kumbar, S.G., Laurencin, C.T., Deng, M., Eds.; Elsevier: Burlington, MA, USA, 2014; pp. 67–89. [CrossRef]
12. Karthikeyan, K.; Durgadevi, R.; Saravanan, K.; Shivsankar, K.; Usha, S.; Saravanan, M. Formulation of bioadhesive carbomer gel incorporating drug-loaded gelatin microspheres for periodontal therapy. *Trop. J. Pharm. Res.* **2012**, *11*, 335–343. [CrossRef]
13. Wang, M.; Zhou, A.; An, T.; Kong, L.; Yu, C.; Liu, J.; Xia, C.; Zhou, H.; Li, Y. *N*-Hydroxyphthalimide exhibits antitumor activity by suppressing mTOR signaling pathway in BT-20 and LoVo cells. *J. Exp. Clin. Cancer Res.* **2016**, *35*, 41. [CrossRef] [PubMed]
14. Cotter, T.G. Apoptosis and cancer: The genesis of a research field. *Nat. Rev. Cancer* **2009**, *9*, 501–507. [CrossRef] [PubMed]
15. Amoêdo, N.D.; Valencia, J.P.; Rodrigues, M.F.; Galina, A.; Rumjanek, F.D. How does the metabolism of tumour cells differ from that of normal cells. *Biosci. Rep.* **2013**, *33*, e00080. [CrossRef] [PubMed]
16. Li, J.; Yuan, J. Caspases in apoptosis and beyond. *Oncogene* **2008**, *27*, 6194–6206. [CrossRef] [PubMed]
17. Oakes, S.A.; Korsmeyer, S.J. Untangling the web: Mitochondrial fission and apoptosis. *Dev. Cell* **2004**, *7*, 460–462. [CrossRef] [PubMed]
18. Grandemange, S.; Herzig, S.; Martinou, J.C. Mitochondrial dynamics and cancer. *Semin. Cancer Biol.* **2009**, *19*, 50–56. [CrossRef] [PubMed]
19. Hellerbach, A.; Schuster, V.; Jansen, A.; Sommer, J. MRI phantoms—Are there alternatives to agar? *PLoS ONE* **2013**, *8*, e70343. [CrossRef] [PubMed]
20. Wells, R.G. The role of matrix stiffness in regulating cell behavior. *Hepatology* **2008**, *47*, 1394–1400. [CrossRef] [PubMed]
21. Discher, D.E.; Janmey, P.; Wang, Y.L. Tissue cells feel and respond to the stiffness of their substrate. *Science* **2005**, *310*, 1139–1143. [CrossRef] [PubMed]

Article

Development and Evaluation of Alginate Membranes with Curcumin-Loaded Nanoparticles for Potential Wound-Healing Applications

Mónica C. Guadarrama-Acevedo [1,†], Raisa A. Mendoza-Flores [1,†], María L. Del Prado-Audelo [1,2], Zaida Urbán-Morlán [1], David M. Giraldo-Gomez [3], Jonathan J. Magaña [4], Maykel González-Torres [5,6], Octavio D. Reyes-Hernández [7], Gabriela Figueroa-González [8], Isaac H. Caballero-Florán [1,9], Carla D. Florán-Hernández [4], Benjamín Florán [9], Hernán Cortés [4] and Gerardo Leyva-Gómez [1,*]

1 Departamento de Farmacia, Facultad de Química, Universidad Nacional Autónoma de México, Ciudad Universitaria, Circuito Exterior S/N, Del. Coyoacán, Ciudad de México 04510, México
2 Laboratorio de Posgrado en Tecnología Farmacéutica, FES-Cuautitlán, Universidad Nacional Autónoma de México, Cuautitlán Izcalli 54740, México
3 Departamento de Biología Celular y Tisular, Facultad de Medicina, Universidad Nacional Autónoma de México (UNAM), Edificio "A" 3er piso, Circuito Interior, Avenida Universidad 3000, Ciudad Universitaria, Coyoacán, Ciudad de México 04510, México
4 Laboratorio de Medicina Genómica, Departamento de Genética, Instituto Nacional de Rehabilitación Luis Guillermo Ibarra Ibarra, Ciudad de México 14389, México
5 CONACyT-Laboratorio de Biotecnología, Instituto Nacional de Rehabilitación Luis Guillermo Ibarra Ibarra, Ciudad de México 14389, México
6 Instituto Tecnológico y de Estudios Superiores de Monterrey, Campus Ciudad de México 14380, México
7 Laboratorio de Biología Molecular del Cáncer, UMIEZ, Facultad de Estudios Superiores Zaragoza, Universidad Nacional Autónoma de México, Ciudad de México 09230, México
8 CONACyT-Laboratorio de Genómica, Dirección de Investigación, Instituto Nacional de Cancerología. Av. San Fernando 22, Tlalpan, Sección XVI, Ciudad de México 14080, México
9 Departamento de Fisiología, Biofísica & Neurociencias, Centro de Investigación y de Estudios Avanzados del Instituto Politécnico Nacional, Ciudad de México 07360, México
* Correspondence: gerardoleyva@hotmail.com; Tel.: +52-(55)-5622-3899 (ext. 44408)
† Mónica C. Guadarrama-Acevedo and Raisa A. Mendoza-Flores equally contributed as first authors to the present work.

Received: 6 June 2019; Accepted: 30 July 2019; Published: 3 August 2019

Abstract: Non-biodegradable materials with a low swelling capacity and which are opaque and occlusive are the main problems associated with the clinical performance of some commercially available wound dressings. In this work, a novel biodegradable wound dressing was developed by means of alginate membrane and polycaprolactone nanoparticles loaded with curcumin for potential use in wound healing. Curcumin was employed as a model drug due to its important properties in wound healing, including antimicrobial, antifungal, and anti-inflammatory effects. To determine the potential use of wound dressing, in vitro, ex vivo, and in vivo studies were carried out. The novel membrane exhibited the diverse functional characteristics required to perform as a substitute for synthetic skin, such as a high capacity for swelling and adherence to the skin, evidence of pores to regulate the loss of transepidermal water, transparency for monitoring the wound, and drug-controlled release by the incorporation of nanoparticles. The incorporation of the nanocarriers aids the drug in permeating into different skin layers, solving the solubility problems of curcumin. The clinical application of this system would cover extensive areas of mixed first- and second-degree wounds, without the need for removal, thus decreasing the patient's discomfort and the risk of altering the formation of the new epithelium.

Keywords: wound dressing; polymeric membrane; nanoparticles; curcumin; alginate; pluronic F68; drug skin permeation; Franz cells; tape stripping

---

1. Introduction

Human skin exerts a pivotal function as a protection barrier against diverse exogenous noxious factors; however, it is exposed and undergoes diverse types of injuries, including burns, ulcers, trauma, lacerations, and acute or chronic wounds, which may compromise its integrity [1].

In this regard, when skin is damaged, a specialized and highly regulated dynamic process immediately takes place: wound healing [2]. The main goal of wound healing is to restore tissue integrity and to achieve homeostasis; however, this process may be complicated by distinct intrinsic and extrinsic factors [2]. Thus, in order to accelerate wound healing, a variety of wound dressings have been designed.

Irrespective of the type of wound, the main function of dressings is to aid in the repair of the wound through the reduction of pain and inflammation, to protect the damaged tissue from pathogenic agents, and to enhance cell differentiation and proliferation [3]. Although there several types of dressings which are commercially available, many of these present some drawbacks, such as an inefficient absorption of exudates, poor protection against infections by microbes, the lack of ability to maintain humidity, and the triggering of allergic effects [4]. In addition, several dressings may adhere to the wound and require constant changes, which may interfere with the granulation process and delay the healing course.

Therefore, in recent years, there has been increasing interest in asymmetric membranes as an alternative for designing wound dressings [5]. These types of membranes possess multiple advantages, such as structural similarity with the skin, an ability to absorb exudates due to their porous structure, and improved cell adhesion and proliferation [6,7]. Different polymers have been employed for their development, including chitosan, hyaluronic acid, collagen, poly vinyl alcohol (PVA), polycaprolactone (PCL), and alginate [8]. In particular, alginate is a biopolymer extracted from seaweed and has shown several unique properties, such as biodegradability, good hydrophilicity, and good biocompatibility [4]. Alginate has exhibited potential for improving wound healing due to its hemostatic properties; moreover, it may reduce microbial infections, enhance the absorption of exudates, and decrease allergic reactions [9]. These features render alginate an interesting option for wound dressings. In addition, alginate membranes possess the advantage that they may be functionalized with bioactive compounds that enhance their healing properties.

In this regard, curcumin is a natural compound that possesses a plethora of biological activities, including antimicrobial, antifungal, anti-inflammatory, and antioxidant effects [10–12]. In addition, curcumin improves wound healing, enhances epithelial regeneration, and increases the proliferation of fibroblasts [13,14]. Thus, curcumin could be a suitable pharmacological agent for the elaboration of wound dressings. However, curcumin exhibits low bioavailability and is unstable in neutral and alkaline aqueous solutions, as well as in hydrophilic topical preparations [12,15]. These drawbacks could be overcome by a nanoparticle formulation that permits the controlled and gradual release of the compound.

Therefore, the objective of this study was to design and develop a novel wound dressing comprising an alginate membrane and PCL nanoparticles loaded with curcumin and stabilized with Pluronic® F-68 (CNp) for possible application in wound healing. The wound dressing was physicochemically characterized, and in vivo and ex vivo permeation assays were performed.

2. Materials and Methods

2.1. Materials

For CNp fabrication, we used PCL (Mn 80,000 g/mol), Pluronic® F-68, and trehalose dihydrate, which were purchased from Sigma-Aldrich® (Merck KGaA, Darmstadt, Germany), whereas methanol, ethyl acetate, and curcumin were supplied by Spectrum® (Spectrum Laboratory Products, CA, USA). For membrane elaboration and characterization, we employed sodium alginate [(SA) (300–700 cps 1.0%, RV, 20 rpm, 25 °C], glycerol (Gly), propylene glycol (Prop), and Tween 80, which were purchased from Droguería Cosmopolita (Mexico City, Mexico). PVA, Pluronic® F-127, acetone, and phosphate buffered saline (PBS) solution were supplied by Sigma-Aldrich® (Merck KGaA, Darmstadt, Germany). Kollidon® 30 [Polyvinylpyrrolidone k 30; (PVP k30)] was obtained from BASF (USA). Ethyl acetate was acquired from Distribuidora Química Alvi (State of Mexico, Mexico). Methanol was purchased from Fermont (Nuevo León, Mexico). All other chemicals and reagents were of at least analytical-grade quality.

2.2. Preparation of PCL Nanoparticles Loaded with Curcumin (CNp)

CNp were prepared by the emulsification–diffusion method, as previously described by Quintanar-Guerrero, et al. [16]. Briefly, the organic saturated phase and aqueous saturated phase were obtained by the saturation of an ethyl acetate and distilled water mixture at a 1:1 ratio. After that, 2% (*w*/*v*) solution of PCL was prepared by dissolving 400 mg of PCL in 20 mL of the organic saturated phase; once PCL was dissolved, 100 mg of curcumin was added. At the same time, 5% (*w*/*v*) of Pluronic® F-68 solution was prepared using an aqueous saturated phase as solvent. In order to obtain an emulsion, both solutions were mixed at a 1:2 ratio with a high-speed homogenizer (Ultra Turrax T18; IKA®) at 14,000 rpm for 10 min at room temperature. Then, 160 mL of water was added to the emulsion to generate polymer aggregation, and the system was maintained under the same conditions for 10 min. The organic solvent was evaporated by a rotary vacuum (Heidolph®, Schwabach, Germany), and the nanoparticle suspension obtained was centrifuged at 15,557 *g* for 30 min at 25 °C. Finally, the pellet was dissolved in distilled water. In order to assess the thermal stability and the properties of the nanoparticles, CNp were frozen and lyophilized at −49 °C, 0.05 mBar for 24 h, employing 5% *w*/*v* trehalose dihydrate as a cryoprotectant.

2.3. Physicochemical Characterization of CNp

2.3.1. Particle Size and Zeta Potential Assessment

The particle size and distribution (polydispersity index, PDI) of the CNp were evaluated by dynamic light scattering. On the other hand, to determine the Zeta potential of CNp, laser Doppler velocimetry was employed. CNp dispersions were assessed five times at 25 °C in a Zetasizer (Malvern Instrument ZS90; Malvern, UK).

2.3.2. Atomic Force Microscopy (AFM)

CNp size and geometry were analyzed by atomic force microscopy (AFM) with a scanning probe microscope (JSPM-4210, JEOL®, Tokyo, Japan). In brief, the CNp dispersion was obtained after centrifugation, and the pellet resuspension was diluted to 1:100 with distilled water. A drop was placed on a coverslip, allowing it to dry at room temperature. The coverslip with the drop was held in place with carbon tape, and room-temperature conditions were utilized to assess the samples.

2.3.3. Drug Loading and Entrapment Efficiency of CNp

To calculate entrapment efficiency (EE) and drug loading (DL), the CNp dispersion was centrifuged at 15,557 *g* for 40 min; then, the sediment was resuspended in ethyl acetate and the absorbance was measured by UV–Vis spectrophotometry at 420 nm (DLAB®, SP-UV1000, Beijing, China). The amount of curcumin was obtained by interpolation in a calibration curve (R^2 coefficient = 0.99985).

The percentages of EE and DL were calculated from the equations below:

$$\%\,\text{EE} = \frac{\text{CN}}{\text{IC}} \times 100 \tag{1}$$

$$\%\,\text{DL} = \frac{\text{CN}}{\text{N}} \times 100 \tag{2}$$

where CN = the amount of curcumin in nanoparticles, IC = the initial amount of curcumin, and N = the number of nanoparticles.

2.4. Preparation of Polymer Gels and Membranes

Four membrane formulations based on SA (M1, M2, M3, and M4) (See Table 1) were prepared using the solvent casting method as published by Karki et al. [17]. First, SA and the polymer were dissolved separately in injectable water by stirring at 35 °C. After dissolution, they were mixed with each other by mechanical stirring. Then, the plasticizer (or a plasticizer mixture) was added by stirring at room temperature until a homogeneous gel was obtained. In order to eliminate bubbles from the gel, it was centrifuged at 636 *g* for 20 min at room temperature (BIOBASE, BKC-TH18II, Shandong, China).

Table 1. Formulation of alginate membranes with different polymers as plasticizers. PVA: poly vinyl alcohol; PVP: polyvinylpyrrolidone; CNp: polycaprolactone (PCL) nanoparticles loaded with curcumin.

Formulation Code	**SA (% *w*/*v*)**	**Polymer**		**Plasticizer**	
		PVA (%*w*/*v*)	**PVP (%*w*/*v*)**	**Gly (%*w*/*v*)**	**Prop (%*v*/*v*)**
	4	2	—	10	—
M2	4	2	—	10	12
M3	4	—	2	10	—
M4	4	—	2	10	12
CNp-M4	4	—	2	10	12

In order to obtain the membranes, 10 g of each gel, prepared with the previously mentioned methodology, was poured into a Teflon cast 12 cm in diameter and left to dry into an oven (OAKTON Stable Temp, IL, USA) at 75.0 ± 0.5 °C for 3 h.

2.4.1. Preparation of Nanoparticle-Coated Alginate Membranes (CNp-M4)

The methodology described previously was followed to prepare our nanoparticle-coated alginate membrane (CNp-M4), but the vehicle utilized was a dispersion of CNp to obtain 0.01% *w*/*v* of curcumin instead of injectable water. In order to prepare the membranes, the gel obtained was poured into a Teflon cast and left to dry in an oven at 40.0 ± 0.5 °C for 4 h.

2.5. Physicochemical Characterization of Membranes

2.5.1. Swelling Test

Samples of M1, M2, M3, M4, and CNp-M4 membranes were cut to a size of 1 × 1 cm and weighed on pre-weighed aluminum trays. The samples were divided into five blocks, corresponding to different times as follows: 5, 10, 20, 30, and 60 min. The membranes on the trays were placed on a flat surface, and 750 µL of PBS 1X pH 7.4 was added to each sample.

Once the established time had elapsed, the trays were turned vertically for 2 min on absorbent paper, allowing the draining and absorbption of the excess of PBS. After that time, the membranes in the trays were weighed again [18]. The assessment was performed in triplicate for each of the different times.

The swelling percentage (%S) was calculated using the following formula [19]:

$$\%S = \frac{M_s - M_d}{M_d} \times 100 \quad (3)$$

where M_s and M_d are the weight of the swollen membrane and dried membrane, respectively.

2.5.2. Mechanical Test

Tensile strength (TS) and the elongation at a break (%E) of M1, M2, M3, M4, and CNp-M4 were determined using a Sintech $\frac{1}{2}$ testing machine (MTS, USA), which was equipped with a 100-N load cell at a crosshead speed of 2.4 mm s^{-1}. Three samples of each formulation were cut into a dumbbell shape with a width of 10 mm and an effective length of 40 mm between the clamps at the beginning of the measurement. The thickness of each sample was measured using a Vernier at five different points before testing, and the average of these was employed for TS calculation. The load (Lb) and displacement (mm) of each film were recorded during the stretching. TS and %E were calculated by Equations (4) and (5), as published by Karki et al. [17]:

$$TS = \frac{F}{A} \quad (4)$$

$$\%E = \frac{D}{L} \times 100 \quad (5)$$

where TS is reported in MPa, F is the maximum load (N) required to break the film, and A is the initial cross-sectional area in mm^2. In Equation (5), D is the displacement of the film elongation at the rupture and L is the initial length.

2.5.3. Thermogravimetric Analysis (TGA)

The thermal stability of the membranes (M4, CNp-M4), CNp, and curcumin was evaluated through TGA, employing a Hi-Res TGA 2950 Thermogravimetric Analyzer (Modulated TA Instruments, New Castle, DE, USA). In brief, 5 mg of each sample was analyzed starting at room temperature and increasing to 500 °C at a heating rate of 10 °C/min under a nitrogen atmosphere.

2.5.4. Differential Scanning Calorimetry (DSC)

The thermal properties of the membranes (M4, CNp-M4), CNp, and curcumin were determined with the DSC 2910 (Modulated TA Instruments, DE, USA). Lyophilized samples were placed in hermetic aluminum cells and evaluated starting at room temperature and increasing to 250 °C at a heating rate of 10 °C/min under a nitrogen atmosphere.

2.5.5. pH Values

The pH values of M1, M2, M3, M4, and CNp-M4 polymer gels was measured through a pH meter. The samples were put into contact with the electrode until a constant value was obtained. The pH meter was previously calibrated against standard solutions to ensure the highest level of accuracy.

2.5.6. Structure and Morphology of M4 and CNp-M4 Membranes

In order to analyze the structure and morphology of the membranes, samples were cut in a circular shape 1 cm in diameter and in a longitudinal section and were analyzed using a scanning electron microscope (Jeol-JCM 6000, MA, USA) at 100× and 220× magnification; then, photographs were taken in four different fields.

The pore diameter and membrane width were measured using a software package (ImageJ, MD, USA), and the average was calculated. Pore number was counted field-by-field, and the average was determined.

In order to evaluate the transparency of M4 and CNp-M4 membranes, samples were cut into a circular shape 3 cm in diameter and these were placed over an image (before swelling); then, they were observed and photographed. Afterward, 1 mL of PBS 1X was added to the samples, and after 20 min (swelling process), photographs were taken of the samples.

2.5.7. In Vitro Release Study of Drug Dispersion, CNp and CNp-M4 Membrane

To evaluate the curcumin release profile from the CNp and CNp-M4 membrane, the direct dispersion method was applied. For the drug and CNp dispersion, a certain amount of curcumin (and the equivalent in nanoparticles) was dissolved into PBS solution (pH 7.4, Pluronic® F-127 2% *w*/*v*), divided into sets of three tubes each, and placed in a shaker incubator, maintaining this at 37 °C. For the CNp-M4 membrane, disks 1 cm in diameter were placed in tubes with the PBS solution under the conditions previously described. At the defined times of 0, 0.25, 0.5, 1, 2, 4, 6, 24, and 48 h, one set of tubes was removed from shaking and centrifuged at 18,514 *g* for 30 min. The curcumin released in the supernatant was quantified by UV-Vis spectrophotometry at 420 nm.

The results of the release tests of CNp and CNp-M4 membrane were analyzed by mathematical models such as zero-order, first-order, the Higuchi model, and the Korsmeyer–Peppas model to predict the drug release mechanism.

2.6. Permeation Assays

2.6.1. Ex Vivo Permeation Assay

Porcine skin was obtained from the back of pig ears within 12 h after slaughter. The pig ears were cut into circular sections 3 cm in diameter. The excess of fat was removed with surgical scissors, and the samples were washed with saline solution. The samples were divided into three groups: drug dispersion with 0.01% *w*/*v* of curcumin (2 mL; $n = 3$), CNp dispersion with 0.01% *w*/*v* of curcumin (2 mL; $n = 3$), and CNp-M4 membranes (3 cm in diameter; $n = 3$).

The experiments were conducted in 12 independent Franz cells with a diffusion area of 7.07 cm^2. The freshly excised skin was placed into Franz cells, and the stratum corneum remained in contact with the donor compartment, with the dermis facing the receptor compartment. This was filled with 30 mL of 0.1 M PBS solution (pH 7.4) with 2.5% of Tween 80 and maintained under constant stirring at 400 rpm.

Franz cells were immersed in a water bath at a constant temperature of 37.0 ± 0.5 °C. At predetermined times (1, 2, 3, 4, 5, 6, 7, 8, 22, 24, 26, 28, and 30 h), 1000 µL of medium was removed from the receptor compartment and replaced with 1000 µL of fresh receptor medium.

At the end of the test, the skin samples were carefully removed from the Franz cells to conduct the tape-stripping test, following the in vivo permeation assay methodology (described in Section 2.6.2). All tapes were placed in a flask with 40 mL of acetone and mechanically stirred for 15 h. Subsequently, each skin sample was fragmented into small pieces using surgical scissors. The curcumin was extracted with 25 mL of ethyl acetate:methanol at a ratio of 9:1. The extract was centrifuged at 10,174 *g* for 10 min. All samples were analyzed by UV-visible spectrophotometry at a wavelength of 420 nm.

2.6.2. In Vivo Permeation Assay

To evaluate the in vivo permeation of the CNp-M4 membrane through the stratum corneum, the tape-stripping technique was employed. Four healthy Mexican males aged 22–36 years were recruited as volunteers. The individuals had neither a history of skin disorders nor had used cosmetic products on their forearms 24 h prior to the test. Written informed consent was obtained from each volunteer before each study.

Four sites (3 × 3 cm) were demarcated: two on the right forearm and two on the left forearm. Before administering the treatment, the sites were cleaned with cotton impregnated with water. Three treatments were administered to each volunteer. The CNp dispersion with 0.01% of curcumin (2 mL) and the drug dispersion at the same concentration were contained within a glass cylinder on the left forearm. The CNp-M4 membrane (3 cm in diameter) was put on the right forearm, and 1.75 mL of distilled water was added onto the surface of the membrane. The last site on the right forearm was used for the blank. The treatments were in contact with the skin for 6 h [20]. Once the time had elapsed, the membrane was removed with steel nippers. The skin was subjected to 15 successive tape strips (Scotch® 3M®), previously cut into 3 × 3 cm squares. In each case, the site was pressed uniformly by sliding a spatula over the surface of the tape five times; then, it was removed by pulling it with steel nippers from the lower right to the upper left end. The blank site was subjected to the same procedure. Each tape was immersed in 15 mL of acetone in a different amber glass bottle with a lid. All of the bottles were placed under mechanical agitation for 15 h; subsequently, the samples were filtered to remove the glue particles. The quantification of curcumin was performed by UV–visible spectrophotometry at a wavelength of 420 nm. Each sample was analyzed in triplicate.

3. Results and Discussion

3.1. Physicochemical Characterization of CNp

3.1.1. Particle Size and Zeta Potential Assessment

CNps were obtained, and their mean particle size and PDI were 148.3 ± 1.9 nm and 0.044 ± 0.020, respectively. These values were expected to improve dermal permeation [21], since nanoparticles below 500 nm have a larger surface area-to-volume ratio, which ensures direct contact with the stratum corneum and skin appendages [6,7]. Moreover, the PDI value was below 0.1, which indicates that the small size measured in the sample is reliable and monodisperse [22].

On the other hand, CNp exhibited a zeta potential value of −7.32 ± 0.03 mV. In this regard, zeta potential is commonly employed to measure the charge in the nanoparticles and/or electrostatic repulsion [1], and the literature indicates that nanoparticles are stable in suspension with a zeta potential above ±30 mV [23]. Despite the CNp zeta potential value not being in this range, it should be considered that Pluronic® F-68 was added as a stabilizer of nanoparticles, which provides them with stability by means of a repulsion effect through a steric mechanism [24]. Likewise, CNp possesses a negative charge, which is related to the carboxylic end group of PCL [25]; thus, negatively charged nanoparticles could permeate adequately in conjunction with the negative charges existing on the skin [26].

3.1.2. Atomic Force Microscopy (AFM)

The morphology and size of CNp were evaluated by AFM. In agreement with the particle-size and zeta-potential assessments, the nanoparticles demonstrated a spherical shape and a size of approximately 200 nm with no agglomeration (Figure 1) [27]. These results support the idea that the small size of CNp could improve the dermal permeation of curcumin, which would increase its anti-inflammatory, antimicrobial, and wound-healing activities [6,12].

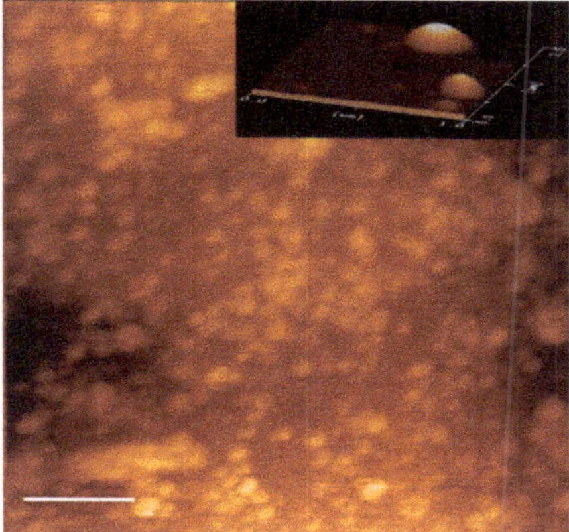

Figure 1. Atomic force microscopy (AFM) topography images of the CNp. The size bar is 1 μm.

3.1.3. Drug Loading and Entrapment Efficiency of Nanoparticles

In order to determine the amount of curcumin entrapped inside the CNp, the DL and EE were determined. The CNp showed DL and EE values of 4.9 ± 0.7% and 96.01 ± 0.95%, respectively. It is known that the emulsification diffusion method ensures high encapsulation efficiencies (generally >70%) [28]. On the other hand, DL depends on nanoparticle structure and methodology [29], whereas both DL and EE depend on the interactions between the drug, the matrix, and the medium [29].

3.2. Physicochemical and Mechanical Characterization of Nanoparticle-Coated Alginate Membranes as Wound Dressings

3.2.1. Swelling Test

From a practical point of view, the membranes should absorb the exudate from the wound and at the same time provide a moist environment that promotes healing. For this reason, the percentage of swelling of five different formulations was determined by weight difference (Figure 2).

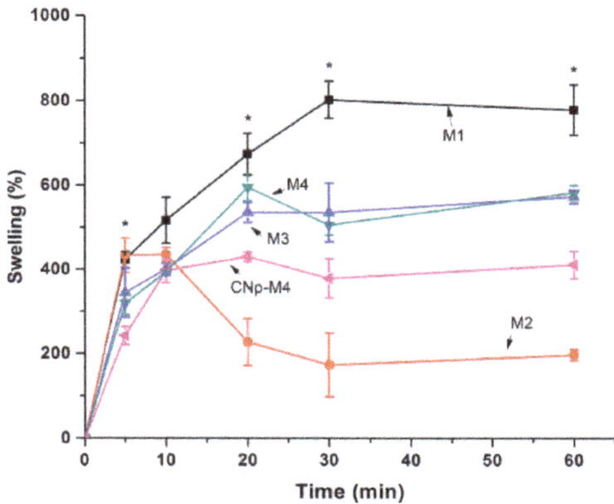

Figure 2. Percentage of swelling in alginate membranes as a function of time. Effect on swelling capacity by the addition of PVA + glycerin (**M1**), PVA + glycerin + propylene glycol (**M2**), PVP + glycerin (**M3**), PVP + glycerin + propylene glycol (**M4**), and CNp + PVP + glycerin + propylene glycol (**CNp-M4**) (mean ± SE; $n = 3$). There were significant differences when ANOVA was applied at 5, 20, 30, and 60 min ($p < 0.05$).

Figure 2 depicts a difference at a time of 20 min, when propylene glycol is added to the formulation with PVA (M2) in comparison to M1; however, there is no significant difference between using or not using propylene glycol in the formulation combined with PVP (M3 and M4, respectively). In order to determine the differences in the percentage of swelling according to each formulation as a function of time, an ANOVA was performed. It was demonstrated that both the formulation and the time during which the membrane was exposed to PBS exerted a statistically significant effect on swelling with a 95% confidence level.

The formulation with the highest swelling value was M1, at 30 min (802.8 ± 76.0%), while the addition of propylene glycol (M2) decreased the swelling percentage after 10 min. On the other hand, M2 showed the lowest value in the swelling test before dissolving. The addition of propylene glycol to the polyvinyl alcohol mixture could increase the intermolecular interactions between both excipients via hydrogen bonds in the –OH groups, favored by a wide steric disposition, but could decrease the entry capacity of water molecules and their interaction with alginate by saturation, decreasing the swelling ability. In the case of the interaction of propylene glycol and PVP, the situation could be the opposite. The interaction of both excipients is lower, resulting in a lower swelling capacity of the alginate; i.e., the excipient that determines the majority of the response. This phenomenon is confirmed with the profile observed for CNp-M4 in relation to M4.

Namely, when water enters the polymer matrix, the chains begin to relax, giving rise to the opening of the polymer networks. This promotes the penetration of more water; however, in the last stages of swelling, the diffusion coefficient is diminished because the chains are completely relaxed and near equilibrium [30].

The membranes began to dissolve after being exposed to PBS for a longer period of time. According to the composition of the medium, polymers undergo degradation and erosion processes. When a polymer degrades, the chains are cleaved into oligomers and subsequently monomers. The continuous loss of monomers will eventually lead to the phenomenon of erosion, which progressively changes the microstructure of the membrane through the formation of pores [31]. The combination of these processes could favor the possible application of our M4 and CNp-M4 membranes as wound dressings, because it would not be necessary to remove them from the application site, avoiding harm through injury and discomfort to the patient.

Moreover, the level of exudate from a wound (for example, an ulcer) can vary from absent (dry ulcer) to minimally exuding (<5 mL fluid per 24 h), to moderately exuding (5 to 10 mL fluid per 24 h), and finally to highly exuding (>10 mL fluid per 24 h) [32]. In this regard, the measurement of the swelling capacity of a wound dressing developed with alginate could be classified as dressings of low absorbance (alginate wound dressing that absorbs less than 6 g of liquid per g of dressing, or less than 12 g/100 cm^2), and dressings of high absorbency (an alginate wound dressing that absorbs 6 g or more liquid per g of dressing, or 12 g or more/100 cm^2) [33]. With this consideration, CNp-M4 possesses a value of 17.48 g/100 cm^2, corresponding to high absorbency and similar to several commercial products.

3.2.2. Mechanical Test

Wound dressings must be resistant and flexible for ease of handling [17]. Thus, the mechanical properties of the membranes are depicted in Figure 3. Formulations with PVA (M1 and M2) did not demonstrate a significant difference in %E with 73.54 ± 0.87% and 74.90 ± 1.67%, respectively. TS was similar for M1, M2, and M3 samples with 1.32 ± 0.02, 1.34 ± 0.01, and 1.96 ± 0.05 MPa, respectively. By way of comparison, M1 and M2 membranes exhibited a lower TS and %E than the remaining formulations, probably due to PVA being a polymer that has been characterized as possessing poor elasticity, a rigid membrane, and low hydrophilic characteristics [34].

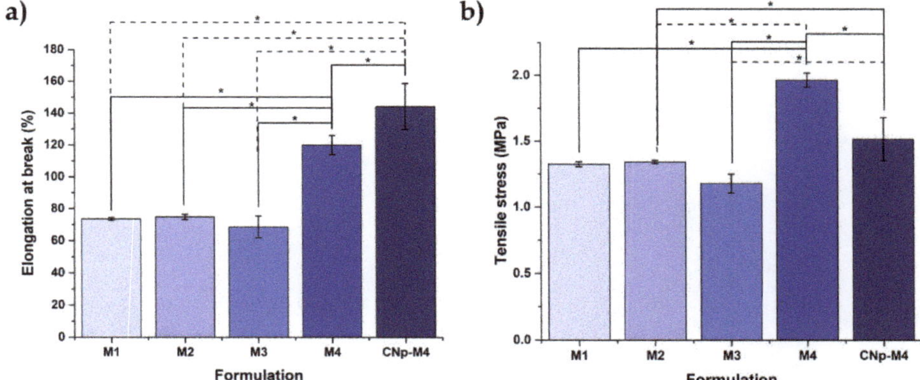

Figure 3. Effect of plasticizers on the mechanical properties of alginate membranes. Effect on mechanical properties by the addition of PVA + glycerin (M1), PVA + glycerin + propylene glycol (M2), PVP + glycerin (M3), PVP + glycerin + propylene glycol (M4), and CNp + PVP + glycerin + propylene glycol (CNp-M4), respectively. (**a**) Elongation at break; (**b**) tensile stress, (mean ± SD; $n = 3$). * indicates $p < 0.05$ as statistically significant.

The %E and TS of M3 were 68.63 ± 6.75% and 1.18 ± 0.07 MPa, respectively; these values significantly increased with the addition of propylene glycol (M4), to 120.01 ± 5.97% and 1.96 ± 0.05 MPa, respectively. M4 exhibited the highest values in the assay and showed a significant difference with respect to the remaining formulations without CNp. This may be explained, at least in part, by the properties of propylene glycol, which is a plasticizer with a small molecular weight that is able to create multiple H-bonds with PVP and SA chains into a package and, consequently, aid in the formation of cross-linked networks [35].

Therefore, because of its greater swelling capacity and better mechanical properties, the M4 membrane (a mixture of sodium alginate, PVP, and propylene glycol) was chosen for the incorporation of CNp.

On the other hand, the mechanical properties of M4 were modified when CNp dispersion was added to the formulation. The %E of CNp-M4 showed the highest value, with 144.39 ± 14.52%; in contrast, TS decreased to 1.52 ± 0.16 MPa. This could be due to the addition of CNp dispersion to the formulation decreasing the number of hydrogen bonds between the polymer molecular chains; as a result, less strength is necessary to break the membrane [36]. In comparison, CNp-M4 showed the highest %E, which could be due to the effect of Pluronic® F-68. The latter is a surfactant that decreases the pore number, providing a membrane with a homogeneous structure; thus, it is more resistant to changes, rendering higher elasticity properties.

3.2.3. Thermogravimetric Analysis (TGA)

It is important to determine the thermal properties of a substance, because these provide useful information for their identification and the characterization of materials. In Figure 4, thermograms of curcumin, CNp, M4, and CNp-M4 membranes are presented. For curcumin, mass loss was observed at 173 °C by TGA (Figure 4a); due to the degradation of turmeric powder, water loss was not observed, possibly due to its high hydrophobicity [37]. The weight loss of CNp started at approximately 90 °C, corresponding to dehydration, and there was a second plateau from 280 to 350 °C, suggesting a better thermal stability for curcumin when it is inside PCL nanoparticles than when alone. However, CNp lost more weight in a smaller temperature range.

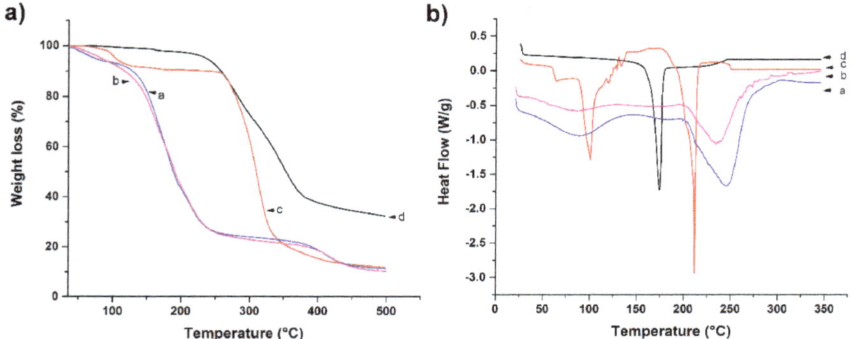

Figure 4. Thermal analysis of nanoparticle-coated alginate membrane using (**a**) thermogravimetric analysis (TGA) and (**b**) differential scanning calorimetry (DSC). The thermal properties of the M4 membrane, CNp-M4 membrane, CNp, and curcumin are represented as a, b, c, and d, respectively.

On the other hand, in the M4 membrane, the first mass loss occurred between 90 and 240 °C, whereas for the CNp-M4 sample, weight loss began between 100 and 240 °C. This could be due to the evaporation of water traces, the degradation of propylene glycol, PVP, glycerin (150-220 °C), and turmeric powder [38]. In the case of the CNp-M4 membrane, it presented a slighter weight loss compared to the M4 membrane. The second mass loss of both samples was between 270 and 425 °C; in this stage, the decomposition of the functional groups of SA polymer chains is presented. Finally, the last plateau in M4 and CNp-M4 membranes started at 425 °C, which corresponds to the degradation of the SA backbone [19]. In the same manner, the CNp-M4 membrane thermogram revealed a lower weight-loss temperature compared with that of CNp; this is probably because membrane formulation is a mixture that contains more substances than CNp. This is similar to the thermal behavior exhibited by curcumin and CNp.

3.2.4. Differential Scanning Calorimetry (DSC)

DSC is a technique used to determine the quantity of heat either absorbed or released when substances undergo physical or chemical changes [39]. In Figure 4b, DSC thermograms of curcumin, CNp, M4 membrane as vehicle, and CNp-M4 membrane are presented. The melting point of curcumin was found to be 174 °C (Table 2), which was expected with regard to the literature [40]. Furthermore, three thermal events were observed in the CNp: at 63.5; 101, and 212 °C. The first thermal event could correspond to the melting point of PCL (61 °C) [41], while events at 101 and 212 °C may indicate the presence of trehalose, which was employed as a cryoprotectant to lyophilize the CNp [42]. Interestingly, the melting point of curcumin was not detected in the CNp sample; this could be due to the high EE of curcumin inside PCL nanoparticles as a molecular dispersion.

Table 2. Thermal events of curcumin, CNp, and alginate membranes by DSC.

Sample	T_{m1} (°C)	T_{m2} (°C)	T_{m3} (°C)
Curcumin	174	—	—
CNp	63.5	101	212
M4	87	249	—
CNp-M4	87	233	—

On the other hand, both M4 and CNp-M4 membranes revealed two thermal events. The first peak was around 87 °C for both formulations, whereas the second peak was around 233 °C for the M4 membrane and 249 °C for the CNp-M4 membrane. The latter peaks were due to the presence of SA in the formulation. A thermal peak prior to 100 °C was observed for both samples, possibly due to the presence of water in the membranes.

3.2.5. pH Determination

The pH values of all the polymer gels were 5.78 ± 0.06, 5.76 ± 0.03, 5.65 ± 0.01, 5.97 ± 0.05 and 5.68 ± 0.03 for M1, M2, M3, M4, and CNp-M4, respectively. All of these values are acceptable because pH wound dressings must be neither acid nor alkaline in order to avoid skin irritation. Moreover, membrane pH is important to regulate the wound-healing process. The natural pH of the skin is within a range of 5–6, depending on the person, while the pH of the chronic wound oscillates in an alkaline range between pH 7 and 8, which increases susceptibility to wound infection.

3.2.6. Structure and Morphology of M4 and CNp-M4 Membranes

An ideal scaffold is expected to have a suitable microstructure (number of pores and pore size controlled) in order to transport nutrients, cells, metabolites, gases, and signaling molecules [43]. In this respect, pores were observed in the top of the membrane structure, which did not span the membrane (Figure 5a). However, it should be expected that the addition of water to the membranes (for example, from the wound exudate) promotes the total formation of pores through these. This would allow skin transpiration and an optimal environment for the wound.

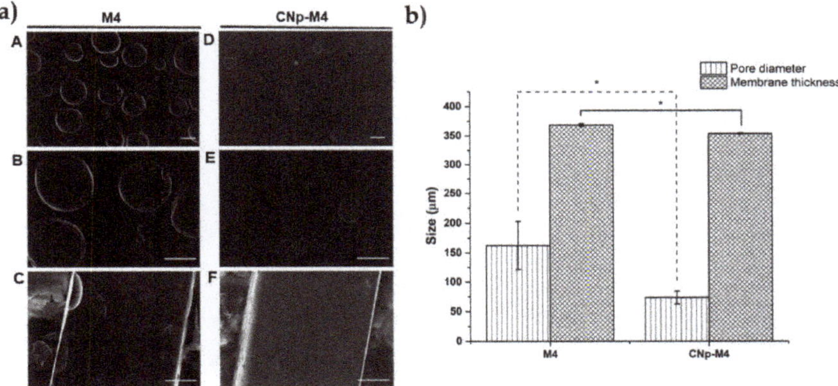

Figure 5. Morphology and porosity of alginate membranes. (**a**) Micrographs by scanning electronic microscopy of the alginate membrane surface (A, B, D, and E) and membrane thickness (C and F). Magnification of 100× for A, D; 220× in B, C, E and F; the scale bar is 100 µm; (**b**) pore diameter and membrane thickness of M4 and CNp-M4 membranes, mean ± SE, $n = 3$. * indicates that $p < 0.05$ is statistically significant.

On the other hand, the mean pore numbers in M4 and CNp-M4 membranes found in each 0.199 mm^2 were 4 and 2, respectively. The pore diameter in the M4 membrane was 162.25 ± 40.75 µm, while for CNp-M4 membranes, this was 73.43 ± 11.04 µm (Figure 5b). Moreover, M4 membranes were significantly thicker and more homogenous in structure than CNp-M4 membranes. These features could be due to the fact that CNp-M4 membranes have CNp dispersion in their formulation, which contains Pluronic® F-68, a nonionic surfactant used as a nanoparticle coating that decreases tensile surface in the polymer gel, thus reducing the number and size of the pores formed in the membranes [43].

Finally, transparency is an expected feature in our alginate membranes used as wound dressings, in order to observe the possible wound-healing process without removing the dressing. M4 and CNp-M4 membranes revealed a transparent feature prior to swelling, while CNp-M4 membranes demonstrated a translucent feature during the swelling process (Figure 6). Although the latter was not completely transparent, it was possible to see through it.

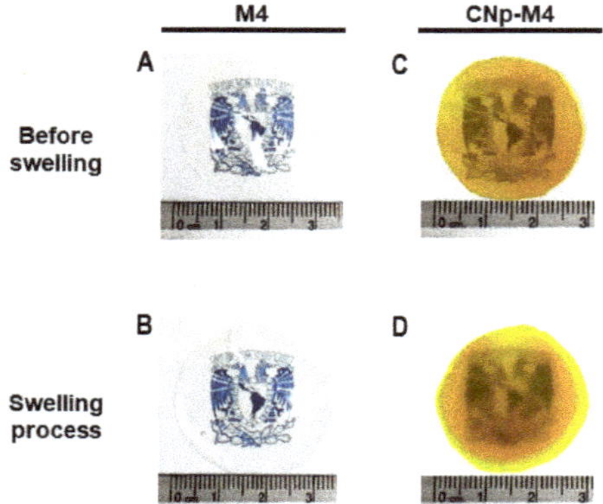

Figure 6. Alginate membranes (M4) before swelling and during the swelling process in PBS medium (**A**,**B**, respectively), and alginate membranes with curcumin nanoparticles (CNp-M4) before swelling and during the swelling process in PBS medium (**C**,**D**, respectively). Scale in centimeters.

3.2.7. In Vitro Release Study of Drug Dispersion, CNp and CNp-M4

To analyze the mechanism of drug release from the nanoparticles (CNp) and from the nanoparticles inside the membrane (CNp-M4), an in vitro release study was performed via the dispersion method and is presented in Figure 7. The curcumin release from CNp (red line, circle symbol) showed a low burst effect at 2 h; this behavior could be related to the presence of the curcumin released from the nanoparticles, as well as to the curcumin outside the nanoparticle, inside the border-zone matrix-stabilizer. After that, the release profile exhibited a linear behavior, with nearly 60% of the curcumin released after 48 h of study. Interestingly, in the release profile for CNp-M4 (blue line, triangle symbol), the burst effect was not evident, and a faster release than CNp (80% of curcumin released at 48 h) was found. These behaviors may be due to the interaction among the membrane excipients, the solvents, and the nanoparticles. To elaborate the CNp-M4 membrane (Section 2.4.1), the nanoparticles are in contact with water, and the hydrolysis of CNp could be stimulated. In addition, some excipients, such as glycerin and propylene glycol, are co-solvents that could improve the prior solubilization of curcumin. As can be observed, the release from both the CNp and CNp-M4 membrane was considerably slower than the drug dispersion release (black line, square symbol).

In order to investigate the mechanism of curcumin release from CNp and CNp-M4, different mathematical models were applied (Table 3). The CNp data were fixed with the Higuchi model (A = 0.0852, B = −0.0459, R^2 = 0.9551), according to those previously reported [44]. This model describes the release of the drug by diffusion from the nanoparticle core into the external solution. On the other hand, the release from CNp-M4 could be explained with the Korsmeyer–Peppas model, due to the highest squared-correlation-coefficient value being obtained with this method (A = 0.3119, B = −0.5609, R^2 = 0.9536). This model combines the diffusion and erosion mechanisms of the nanoparticles as the explanation for drug release.

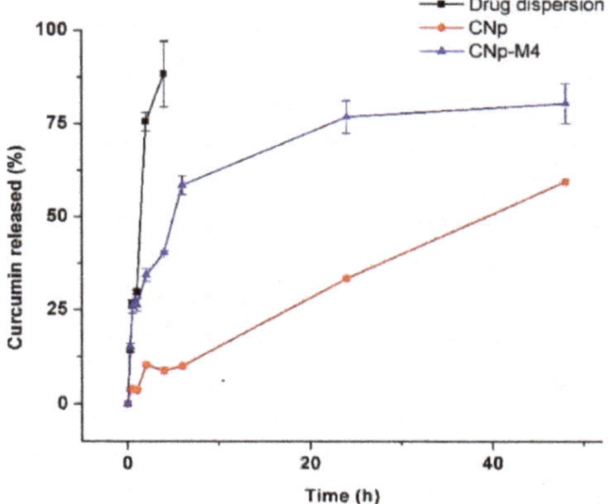

Figure 7. Release profile of curcumin from the drug dispersion, CNp and CNp-M4 membrane in PBS pH 7.4 (0.1 M, Pluronic® F-127 2% *w*/*v*) at 37 °C. Each point represents the mean ± SE, *n* = 3.

Table 3. Squared of the correlation coefficient (R^2) and coefficients obtained after the linear regression of release data from CNp and CNP-M4 utilizing four mathematical models.

Mathematical Model	Equation	CNp			CNp-M4		
		R^2	A	B	R^2	A	B
Zero-order	$Q_t = Q_0 + K_0 t$	0.842	0.5264	4.7385	0.7358	1.2362	31.432
First-order	$\ln Q_t = \mathrm{lo}Q_0 + K_1 t$	0.8436	−0.0066	−2.3991	0.8453	−0.0303	−3.1975
Higuchi	$Q_t = K_H t^{1/2}$	0.9551	0.0852	−0.0459	0.8939	0.1017	0.194
Korsmeyer–Peppas	$\frac{Q_t}{Q_\infty} = K_K t^n$	0.9037	0.5422	−1.2655	0.9536	0.3119	−0.5609

Q_t: amount of drug released in time t; Q_0: initial amount of drug in the dosage form; Q_∞: total amount of drug dissolved when the dosage form is exhausted; K_0, K_1, K_H, K_K: release rate constants; R^2: squared correlation coefficient. y = ax ± b is an equation obtained after regression: a, slope and b, linear coefficient.

3.3. *Curcumin Permeation Assays*

3.3.1. Ex Vivo Permeation

An ideal system for potential use in chronic diseases with a slow healing process, such as wounds or psoriasis, should exhibit sustained drug release in order to allow permeation through the skin [20,45]. In this regard, the alginate membranes developed in our study possess polymeric networks as a structure, as well as the curcumin encapsulated in PCL nanoparticles dispersed within these networks. These features will allow the slow release of curcumin.

An ex vivo permeation study was conducted to determine the distribution of the drug and CNp throughout the stratum corneum, epidermis, and dermis, and to verify whether curcumin can pass through the skin and reach blood circulation after the administration of CNp-M4 and CNp formulations. With respect to the aqueous dispersion of curcumin (Figure 8), due to the high lipophilic character of the drug, a high accumulation was observed in some superficial layers of the stratum corneum (10.04 ± 1.73 μg/cm^2). Interestingly, permeation comprises a considerable amount, even from the application of a curcumin dispersion in water, which involves solid particle clusters. This means that the drug particles engage in a dissolution process with the oily components of the stratum corneum, leading to their brief permeation in the superficial region. Although the values are high, there is a higher efficiency of permeation with CNp (14.80 ± 1.61 μg/cm^2). In the region of the dermis, with a hydrophilic character, the permeated amount of the drug dispersion decreases considerably due to its highly limited solubility in aqueous media (2.40 ± 0.46 μg/cm^2); this is nearly one third in relation to the CNp value (6.99 ± 0.27 μg/cm^2). No detection was recorded for the dispersion of the drug that could completely permeate the skin. According to Figure 8, a significant difference was observed between CNp dispersion and CNp-M4 membrane treatments, with the highest permeation values observed for CNp. The CNp-M4 membrane treatment revealed the highest amount of curcumin retained in the epidermis and dermis—that is, 5.7 μg/cm^2 (1.620 ± 0.051% of the total concentration of curcumin)—and the lowest concentration was found in the stratum corneum (SC), at 0.65 μg/cm^2 (0.140 ± 0.006%). On the other hand, in the CNp dispersion treatment, curcumin was found mostly in the SC; that is, 14.8 μg/cm^2 (2.62 ± 0.49%) [26], the highest curcumin value in the entire assay.

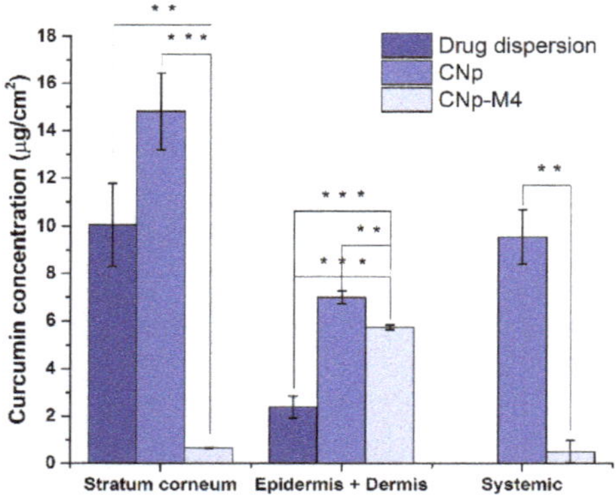

Figure 8. Ex vivo permeation of curcumin after 30 h of treatment with drug dispersion, CNp dispersion, or CNp-M4 membrane. Stratum corneum-bound particles (obtained from 15 tape strips), epidermis + dermis (surface on which dosed skin was handled after 30 h), and systemic (receptor compartment), mean ± SE, $n = 4$; ** indicates $p < 0.01$ and *** indicates $p < 0.001$ as statistically significant.

There was also a significant difference in the amount of curcumin that crossed through the skin (which, in an in vivo model, means reaching the systemic circulation), since the concentration derived from the CNp-M4 membrane was significantly lower than CNp (0.32% and 2.04% respectively; p-value = 0.0019).

The aqueous system of CNp dispersion permitted the curcumin to permeate through the dermis and completely cross the skin. Moreover, CNps have a negative charge; negatively charged nanoparticles permeate the skin more rapidly than positively charged nanoparticles. The skin is predominately negatively charged, and the electrostatic interaction of positive particles with the negatively charged molecules in the skin matrix slows particle diffusion [26]. However, curcumin from CNp-M4 membranes diffuses more slowly through the skin than that from CNp. These observations may be due to the fact that, in the CNp dispersion, water was used as the medium. Water affects the absorption rates of different substances through the stratum corneum, which is in a constant state of partial hydration under normal conditions. Thus, when immersed in water, dead keratinocytes quickly absorb it, resulting in the pruning effect of the skin [45]. Furthermore, water in contact with skin creates a flow gradient toward the skin's inner layers, since the inside of the stratum corneum is more hydrated than the surface [46]. On the other hand, the CNp-M4 membrane does not have a liquid medium in the interface that allows the nanoparticles to flow easily into the deep layers, such as water in the case of the CNp dispersion. Despite this, for the CNp-M4 membrane, a modulated release is expected because of the degree of swelling of the membrane in response to the presence of exudate. The swelling would permit the relaxation of the polymer chains and the release from the CNp. Otherwise, diffusion from a solid state (CNp-M4) into a semi-solid state (the skin) would be expected. Therefore, the CNp-M4 membrane represents a prolonged release system.

In addition, the skin possesses furrows, in which a considerable amount of curcumin is retained, which could not be extracted with the application of adhesive tapes [47]. This curcumin was quantified until mechanical disaggregation. This could explain why the epidermis and dermis had the highest concentration of the drug.

On the other hand, particle size is an important factor in obtaining the desired therapeutic effect, because nanoparticles with a small size can more easily permeate the physiological barriers; moreover, due to their greater surface, release of the drug is favored [48]. Thus, it should be expected that CNps, with their small size (148.3 nm), and the surfactant effect of Pluronic® F-68 can permeate intercellularly and through hair follicles, favoring accumulation for several hours. In the same way, due to the contact of nanoparticles with the corneocytes of the skin, as well as the prolonged release thereof, a large amount of curcumin was found to be retained in the dermis [24,27]. Although after 30 h the majority of curcumin remained in the stratum corneum when CNp dispersion was applied, the monitoring of the permeation at longer times could allow the observation of a prolonged release system [49]. Therefore, the CNp-M4 membrane is proposed as a functional prolonged release system for drug delivery in chronic diseases; however, it would be necessary to perform a more prolonged test to observe the diffusion of the drug at a greater proportion.

3.3.2. Permeation Assay in Vivo

In order to evaluate the in vivo skin permeation of curcumin from the CNp-M4 membrane, CNp dispersion [20], and drug dispersion, a quantification of the curcumin deposited in the stratum corneum by UV–Vis spectrophotometry was performed. The results are presented in Figure 9, according to the work of Goto et al. [50].

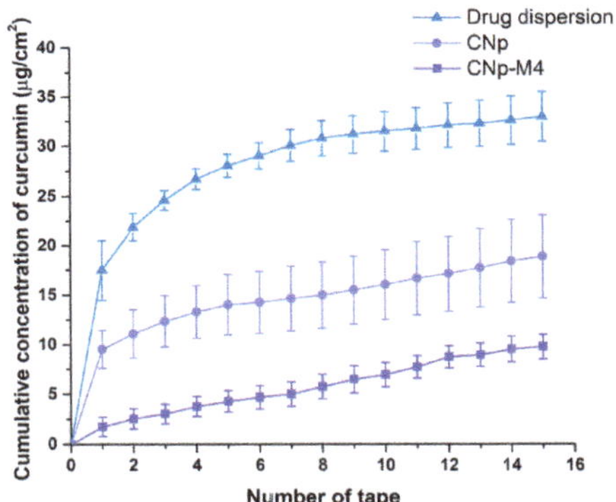

Figure 9. Cumulative concentration of curcumin quantified in the stratum corneum of healthy volunteers. Drug extraction from 15 adhesive tapes applied to the treatment site after placing a CNp-M4 membrane, CNp dispersion, or drug dispersion for 6 h (mean ± SE; $n = 4$).

Higher permeation values were observed for the drug dispersion in water compared to the CNp and CNp-M4 membrane, at least in the superficial layers of the stratum corneum, where the tape stripping technique is applied. The higher values of drug permeation could correspond to the high lipophilicity value of the drug, structural symmetry, and low molecular weight. These values coincide with Figure 8: greater drug deposition on the stratum corneum surface, a lower proportion in the dermis due to an inadequate hydrophilic–lipophilic balance, and no recorded quantity that completely permeates the skin. At 6 h after the application of the drug dispersion, CNp-M4 membrane, and CNp dispersion, the curcumin measured in the stratum corneum reached 33.08 ± 2.57, 9.82 ± 4.23, and 18.96 ± 1.25 μg/cm^2, respectively [25]. CNp-M4 remained well adhered during the study, even up to 48 h in other volunteers (data not shown). As can be noted, a greater amount of curcumin was observed in the stratum corneum when the CNp dispersion treatment was applied, compared with the

CNp-M4 membrane. This is because, in the CNp-M4 membrane, the nanoparticles must be released from the polymeric matrix. This result suggests our formulation as a system of prolonged release that may be useful for long treatments, such as that for a wound (for example, for 7–14 days of application until closure of the lesion). It is noteworthy that the smallest variation observed in permeation values with the CNp-M4 treatment reveals a system that allows better gradual control release. Moreover, these nanoparticles also possess a stabilizer on the outside, Pluronic® F-68, which interacts through hydrogen bonds with the -OH groups of plasticizers and polymers used in the formulation. Thus, when the water makes contact with the membrane, the external part of the polymers swells and promotes the drug—in this case, CNp—to flow outward, permitting its release [51].

With respect to the CNp dispersion, the nanoparticles are free in the medium; thus, they interact more easily with the stratum corneum. Curcumin is encapsulated and uniformly distributed in the PCL nanoparticles, forming nanospheres [52]. The release of curcumin from these latter will depend on the solubility of the drug, the diffusion of curcumin through the matrix of the nanoparticles, thedesorption of curcumin from PCL, the erosion or degradation of the matrix of the nanoparticles, and on the combination of the erosion and diffusion processes [23]. It is also known that, under physiological conditions, a random cleavage of PCL ester bonds occurs, which produces a destabilization of the polymer matrix of the CNp, inducing the release of curcumin [25].

Finally, many studies have reported an accumulation of nanoparticles in hair follicles and the pilosebaceous glands in ex vivo skin experiments [49]. In the same manner, it has been reported that there is better permeation of the drug into the skin, as well as greater absorption in areas with high hair-follicle density. The hair follicle can become a reservoir of substances comparable to the stratum corneum [53]. This means that drugs will penetrate better through the skin of a person with greater amounts of hair follicles. This may explain slight variations in the results obtained, together with variability among individuals (age, body mass, color, and skin moisturization).

4. Conclusions

In this study, the development of a new wound dressing for possible application in wound healing was demonstrated. The wound dressing comprises an alginate membrane and PCL nanoparticles stabilized with Pluronic® F-68 with curcumin inside. This new system was designed with a mixture of aqueous plasticizers to confer high-strength mechanical properties. The characterization of the formulation exhibited that it possesses a high absorbency capacity for the removal of possible exudates, transparency for monitoring the wound bed, the presence of pores with controlled dimensions that could facilitate the transpiration of the wound as a synthetic skin substitute, and the gradual release of the drug according to the ex vivo and in vivo studies. In addition, the high adherence of the wound dressing should be noted even in dry skin, as well as its degradation at later times. Therefore, all of these advantages reveal our wound dressing as a good option for wound healing without the need for its removal from the patient.

Author Contributions: Conceptualization, G.L.-G.; methodology, M.C.G.-A., R.A.M.-F., and M.L.D.P.-A.; validation, I.H.C.-F., C.D.F.-H., and B.F.; formal analysis, M.C.G.-A., R.A.M.-F., M.L.D.P.-A., and G.L.-G.; investigation, M.C.G.-A., R.A.M.-F., and G.L.-G.; resources, M.C.G.-A., R.A.M.-F., M.L.D.P.-A., and G.L.-G.; data curation, M.C.G.-A., R.A.M.-F., M.L.D.P.-A., Z.U.-M., H.C., and G.L.-G.; writing—original draft preparation, M.C.G.-A., R.A.M.-F., M.L.D.P.-A., Z.U.-M., H.C., and G.L.-G.; writing—review and editing, M.C.G.-A., R.A.M.-F., Z.U.-M., D.M.G.-G., J.J.M., M.G.-T., O.D.R.-H., G.F.-G., I.H.C.-F., C.D.F.-H., B.F., H.C., and G.L.-G.; visualization, G.L.-G.; supervision, M.L.D.P.-A., Z.U.-M., H.C., and G.L.-G.; project administration, M.L.D.P.-A. and G.L.-G.; funding acquisition, G.L.-G.

Funding: This research was funded by Dirección General de Asuntos del Personal Académico, Universidad Nacional Autónoma de México (Becas Posdoctorales, PAPIIT TA 200318) and CONACYT A1-S-15759.

Conflicts of Interest: The authors declare no conflict of interest.

References

1. Eming, S.A.; Brachvogel, B.; Odorisio, T.; Koch, M. Regulation of angiogenesis: Wound healing as a model. *Prog. Histochem. Cytochem.* **2007**, *42*, 115–170. [CrossRef] [PubMed]
2. Demidova-Rice, T.N.; Hamblin, M.R.; Herman, I.M. Current Methods for Drug Delivery, Part 1: Normal and Chronic Wounds: Biology, Causes, and Approaches to Care. *Adv. Skin Wound Care* **2013**, *25*, 304–314. [CrossRef] [PubMed]
3. Axibal, E.; Brown, M. Surgical Dressings and Novel Skin Substitutes. *Dermatol. Clin.* **2019**, *37*, 349–366. [CrossRef] [PubMed]
4. Aderibigbe, B.A.; Buyana, B. Alginate in wound dressings. *Pharmaceutics* **2018**, *10*, 42. [CrossRef] [PubMed]
5. Morgado, P.I.; Aguiar-Ricardo, A.; Correia, I.J. Asymmetric membranes as ideal wound dressings: An overview on production methods, structure, properties and performance relationship. *J. Memb. Sci.* **2015**, *490*, 139–151. [CrossRef]
6. Mi, F.L.; Wu, Y.B.; Shyu, S.S.; Chao, A.C.; Lai, J.Y.; Su, C.C. Asymmetric chitosan membranes prepared by dry/wet phase separation: A new type of wound dressing for controlled antibacterial release. *J. Memb. Sci.* **2003**, *212*, 237–254. [CrossRef]
7. Priya, S.G.; Gupta, A.; Jain, E.; Sarkar, J.; Damania, A.; Jagdale, P.R.; Chaudhari, B.P.; Gupta, K.C.; Kumar, A. Bilayer Cryogel Wound Dressing and Skin Regeneration Grafts for the Treatment of Acute Skin Wounds. *ACS Appl. Mater. Interfaces* **2016**, *8*, 15145–15159. [CrossRef]
8. Miguel, S.P.; Moreira, A.F.; Correia, I.J. Chitosan based-asymmetric membranes for wound healing: A review. *Int. J. Biol. Macromol.* **2019**, *127*, 460–475. [CrossRef]
9. Sood, A.; Granick, M.S.; Tomaselli, N.L. Wound Dressings and Comparative Effectiveness Data. *Adv. Wound Care* **2014**, *3*, 511–529. [CrossRef]
10. Aggarwal, B.B.; Harikumar, K.B. Potential therapeutic effects of curcumin, the anti-inflammatory agent, against neurodegenerative, cardiovascular, pulmonary, metabolic, autoimmune and neoplastic diseases. *Int. J. Biochem. Cell Biol.* **2009**, *41*, 40–59. [CrossRef]
11. Augustyniak, A.; Bartosz, G.; Čipak, A.; Duburs, G.; Horáková, L.; Łuczaj, W.; Majekova, M.; Odysseos, A.D.; Rackova, L.; Skrzydlewska, E.; et al. Natural and synthetic antioxidants: An updated overview. *Free Radic. Res.* **2010**, *44*, 1216–1262. [CrossRef]
12. Del Prado-Audelo, M.L.; Caballero-Florán, I.H.; Meza-Toledo, J.A.; Mendoza-Muñoz, N.; González-Torres, M.; Florán, B.; Cortés, H.; Leyva-Gómez, G. Formulations of curcumin nanoparticles for brain diseases. *Biomolecules* **2019**, *9*, 56. [CrossRef]
13. Sidhu, G.S.; Singh, A.K.; Thaloor, D.; Banaudha, K.K.; Patnaik, G.K.; Srimal, R.C.; Maheshwari, R.K. Enhancement of wound healing by curcumin in animals. *Wound Repair Regen.* **1998**, *6*, 167–177. [CrossRef]
14. Thangapazham, R.L.; Sharad, S.; Maheshwari, R.K. Skin regenerative potentials of curcumin. *Biofactors* **2013**, *39*, 141–149. [CrossRef]
15. Yallapu, M.M.; Jaggi, M.; Chauhan, S.C. Curcumin nanoformulations: A future nanomedicine for cancer. *Drug Discov. Today* **2012**, *17*, 71–80. [CrossRef]
16. Quintanar-guerrero, D.; Allémann, E.; Fessi, H.; Doelker, E.; Allémann, E.; Fessi, H.; Doelker, E. Preparation Techniques and Mechanisms of Formation of Biodegradable Nanoparticles from Preformed Polymers. *Drug Dev. Ind. Pharm.* **1998**, *24*, 1113–1128. [CrossRef]
17. Karki, S.; Kim, H.; Na, S.; Shin, D.; Jo, K.; Lee, J. Thin films as an emerging platform for drug delivery. *Asian J. Pharm. Sci.* **2016**, *11*, 559–574. [CrossRef]
18. Golafshan, N.; Rezahasani, R.; Tarkesh Esfahani, M.; Kharaziha, M.; Khorasani, S.N. Nanohybrid hydrogels of laponite: PVA-Alginate as a potential wound healing material. *Carbohydr. Polym.* **2017**, *176*, 392–401. [CrossRef]
19. Song, Y.; Jiang, Z.; Gao, B.; Wang, H.; Wang, M.; He, Z.; Cao, X.; Pan, F. Embedding hydrophobic MoS2nanosheets within hydrophilic sodium alginate membrane for enhanced ethanol dehydration. *Chem. Eng. Sci.* **2018**, *185*, 231–242. [CrossRef]
20. Mao, K.; Fan, Z.; Yuan, J.; Chen, P.; Yang, J.; Xu, J. Skin-penetrating polymeric nanoparticles incorporated in silk fibroin hydrogel for topical delivery of curcumin to improve its therapeutic effect on psoriasis mouse model. *Colloids Surf. B Biointerfaces* **2017**, *160*, 704–714. [CrossRef]

21. Kamar, S.S.; Abdel-Kader, D.H.; Rashed, L.A. Beneficial effect of Curcumin Nanoparticles-Hydrogel on excisional skin wound healing in type-I diabetic rat: Histological and immunohistochemical studies. *Ann. Anat.* **2019**, *222*, 94–102. [CrossRef]
22. Clayton, K.N.; Salameh, J.W.; Wereley, S.T.; Kinzer-Ursem, T.L. Physical characterization of nanoparticle size and surface modification using particle scattering diffusometry. *Biomicrofluidics* **2016**, *10*, 1–14. [CrossRef]
23. Singh, R.; Lillard, J.W., Jr. Nanoparticle-based targeted drug delivery. *Exp. Mol. Pathol.* **2009**, *86*, 215–223. [CrossRef]
24. Moghimi, S.M.; Hunter, A.C. Good Review on Poloxamers and Poloxamines in Pharma. *Elsevier* **2000**, *18*, 412–420.
25. Ramanujam, R.; Sundaram, B.; Janarthanan, G.; Devendran, E.; Venkadasalam, M.; John Milton, M.C. Biodegradable Polycaprolactone Nanoparticles Based Drug Delivery Systems: A Short Review. *Biosci. Biotechnol. Res. Asia* **2018**, *15*, 679–685. [CrossRef]
26. Kraeling, M.E.K.; Topping, V.D.; Keltner, Z.M.; Belgrave, K.R.; Bailey, K.D.; Gao, X.; Yourick, J. In vitro percutaneous penetration of silver nanoparticles in pig and human skin. *Regul. Toxicol. Pharmacol.* **2018**, *95*, 314–322. [CrossRef]
27. Quintanar-Guerrero, D.; de la Luz Zambrano-Zaragoza, M.; Gutierrez-Cortez, E.; Mendoza-Munoz, N. Impact of the emulsification-diffusion method on the development of pharmaceutical nanoparticles. *Recent Pat. Drug Deliv. Formul.* **2012**, *6*, 184–194. [CrossRef]
28. Pinto Reis, C.; Neufeld, R.J.; Ribeiro, A.J.; Veiga, F.; Nanoencapsulation, I. Methods for preparation of drug-loaded polymeric nanoparticles. *Nanomed. Nanotechnol. Biol. Med.* **2006**, *2*, 8–21. [CrossRef]
29. Judefeind, A.; de Villiers, M.M. *Nanotechnology in Drug Delivery*; de Villiers, M.M., Aramwit, P.S., Kwon, G., Eds.; Springer US: New York, NY, USA, 2009; ISBN 9780387776675.
30. Singh, B.; Pal, L. Radiation crosslinking polymerization of sterculia polysaccharide-PVA-PVP for making hydrogel wound dressings. *Int. J. Biol. Macromol.* **2011**, *48*, 501–510. [CrossRef]
31. Göpferich, A. Mechanisms of polymer degradation and erosion. *Biomaterials* **1996**, *17*, 103–114. [CrossRef]
32. GD, M. Quantifying wound fluids for the clinician and researcher. *Ostomy Wound Manag.* **1994**, *40*, 66–69.
33. Parikh, D.V.; Fink, T.; Delucca, A.J.; Parikh, A.D. Absorption and swelling characteristics of silver (I) antimicrobial wound dressings. *Text. Res. J.* **2011**, *81*, 494–503. [CrossRef]
34. Kamoun, E.A.; Kenawy, E.S.; Chen, X. A review on polymeric hydrogel membranes for wound dressing applications: PVA-based hydrogel dressings. *J. Adv. Res.* **2017**, *8*, 217–233. [CrossRef]
35. Shi, S.; Peng, X.; Liu, T.; Chen, Y.N.; He, C.; Wang, H. Facile preparation of hydrogen-bonded supramolecular polyvinyl alcohol-glycerol gels with excellent thermoplasticity and mechanical properties. *Polymer* **2017**, *111*, 168–176. [CrossRef]
36. Wang, R.-M.; Zheng, S.-R.; Zheng, Y.-P. Matrix materials. In *Polymer Matrix Composites and Technology*; Elsevier: Amsterdam, The Netherlands, 2011; pp. 101–548.
37. Sun, B.; Tian, Y.; Chen, L.; Jin, Z. Food Hydrocolloids Linear dextrin as curcumin delivery system: Effect of degree of polymerization on the functional stability of curcumin. *Food Hydrocoll.* **2018**, *77*, 911–920. [CrossRef]
38. Campus, K.; Kanchanaburi, M.L.; Road, S. Nanocomposites Based on Cassava Starch and Chitosan-Modified Clay: Physico-Mechanical Properties and Biodegradability in Simulated Compost Soil. *J. Braz. Chem. Soc.* **2017**, *28*, 649–658.
39. Aghazadeh, M.; Karim, R.; Abdul Rahman, R.; Sultan, M.T.; Johnson, S.K.; Paykary, M. Effect of Glycerol on the Physicochemical Properties of Cereal Starch Films. *Food Technol. Econ. Eng. Phys. Prop.* **2018**, *36*, 403–409. [CrossRef]
40. Shrotriya, S.; Ranpise, N.; Satpute, P.; Vidhate, B. Skin targeting of curcumin solid lipid nanoparticles-engrossed topical gel for the treatment of pigmentation and irritant contact dermatitis. *Artif. Cells Nanomed. Biotechnol.* **2018**, *46*, 1471–1482. [CrossRef]
41. Kossack, W.; Kremer, F. Banded spherulites and twisting lamellae in poly-ε-caprolactone. *Colloid Polym. Sci.* **2019**, *297*, 771–779. [CrossRef]
42. Verhoeven, N.; Neoh, T.L.; Furuta, T.; Yamamoto, C.; Ohashi, T.; Yoshii, H. Characteristics of dehydration kinetics of dihydrate trehalose to its anhydrous form in ethanol by DSC. *Food Chem.* **2012**, *132*, 1638–1643. [CrossRef]

43. Bueno, C.Z.; Moraes, Â.M. Development of Porous Lamellar Chitosa-Alginate Membranes: Effect of Different Surfactants on Biomaterial Properties. *J. Appl. Polym. Sci.* **2011**, *122*, 624–631. [CrossRef]
44. Del Prado-Audelo, M.L.; Magaña, J.J.; Mejía-Contreras, B.A.; Borbolla-Jiménez, F.V.; Giraldo-Gomez, D.M.; Piña-Barba, M.C.; Quintanar-Guerrero, D.; Leyva-Gómez, G. In vitro cell uptake evaluation of curcumin-loaded PCL/F68 nanoparticles for potential application in neuronal diseases. *J. Drug Deliv. Sci. Technol.* **2019**, *52*, 905–914. [CrossRef]
45. Blattner, C.M.; Coman, G.; Blickenstaff, N.R.; Maibach, H.I. Percutaneous absorption of water in skin: A review. *Rev. Environ. Health* **2014**, *29*, 175–180. [CrossRef]
46. Tagami, H.; Kanamaru, Y.; Inoue, K.; Suehisa, S.; Iwatsuki, K.; Yoshikuni, K.; Yamada, M. Water sorption-desorption test of the skin in vivo for functional assessment of the stratum corneum. *J. Invest. Dermatol.* **1982**, *78*, 425–428. [CrossRef]
47. Van der Molen, R.; Spies, F.; van´t Noordende, J.M.; Boelsma, E.; Mommaas, A.M.; Koerten, H.K. Tape stripping of human stratum corneum yields cell layers that originate from various depths because of furrows in the skin. *Arch. Dermatol. Res.* **1997**, *289*, 514–518. [CrossRef]
48. Kucukturmen, B.; Oz, U.C.; Bozkir, A. In Situ Hydrogel Formulation for Intra-Articular Application of Diclofenac Sodium-Loaded Polymeric Nanoparticles. *Turk. J. Pharm. Sci.* **2017**, *14*, 56–64. [CrossRef]
49. Zhang, N.; Said, A.; Wischke, C.; Kral, V.; Brodwolf, R.; Volz, P.; Boreham, A.; Gerecke, C.; Li, W.; Neffe, A.T.; et al. Poly[acrylonitrile-co-(N-vinyl pyrrolidone)] nanoparticles—Composition-dependent skin penetration enhancement of a dye probe and biocompatibility. *Eur. J. Pharm. Biopharm.* **2017**, *116*, 66–75. [CrossRef]
50. Goto, N.; Morita, Y.; Terada, K. Deposits from Creams Containing 20% (w/w) Urea and Suppression of Crystallization (Part 2): Novel Analytical Methods of Urea Accumulated in the Stratum Corneum by Tape stripping and Colorimetry. *Chem. Pharm. Bull.* **2016**, *64*, 1092–1098. [CrossRef]
51. Langer, R. New Methods of Drug Delivery. *Science* **1990**, *249*, 1527–1533. [CrossRef]
52. Urrejola, M.C.; Soto, L.V.; Zumarán, C.; Peñaloza, P.; Álvarez, B.; Fuentevilla, I.; Haidar, Z.S. Sistemas de Nanopartículas Poliméricas II: Estructura, Métodos de Elaboración, Características, Propiedades, Biofuncionalización y Tecnologías de Auto-Ensamblaje Capa por Capa (Layer-by-Layer Self-Assembly). *Int. J. Morphol.* **2018**, *36*, 1463–1471. [CrossRef]
53. Lademann, J.; Richter, H.; Teichmann, A.; Otberg, N.; Blume-Peytavi, U.; Luengo, J.; Weiß, B.; Schaefer, U.F.; Lehr, C.M.; Wepf, R.; et al. Nanoparticles—An efficient carrier for drug delivery into the hair follicles. *Eur. J. Pharm. Biopharm.* **2007**, *66*, 159–164. [CrossRef]

Article

Evaluation of Drug Delivery and Efficacy of Ciprofloxacin-Loaded Povidone Foils and Nanofiber Mats in a Wound-Infection Model Based on Ex Vivo Human Skin

Fiorenza Rancan [1,*], Marco Contardi [2], Jana Jurisch [1], Ulrike Blume-Peytavi [1], Annika Vogt [1], Ilker S. Bayer [2] and Christoph Schaudinn [3]

1 Clinical Research Center for Hair and Skin Science, Department of Dermatology and Allergy, Charité—Universitätsmedizin Berlin, Corporate Member of Freie Universität Berlin, Humboldt-Universität zu Berlin, and Berlin Institute of Health, 10117 Berlin, Germany; jana.jurisch@gmx.de (J.J.); ulrike.blume-peytavi@charite.de (U.B.-P.); annika.vogt@charite.de (A.V.)

2 Smart Materials, Istituto Italiano di Tecnologia, 16163 Genova, Italy; marco.contardi@iit.it (M.C.); ilker.bayer@iit.it (I.S.B.)

3 Advanced Light and Electron Microscopy, ZBS4, Robert Koch Institute, 13353 Berlin, Germany; schaudinnc@rki.de

* Correspondence: fiorenza.rancan@charite.de; Tel.: +49-30-450518347

Received: 16 September 2019; Accepted: 10 October 2019; Published: 12 October 2019

Abstract: Topical treatment of wound infections is often a challenge due to limited drug availability at the site of infection. Topical drug delivery is an attractive option for reducing systemic side effects, provided that a more selective and sustained local drug delivery is achieved. In this study, a poorly water-soluble antibiotic, ciprofloxacin, was loaded on polyvinylpyrrolidone (PVP)-based foils and nanofiber mats using acetic acid as a solubilizer. Drug delivery kinetics, local toxicity, and antimicrobial activity were tested on an ex vivo wound model based on full-thickness human skin. Wounds of 5 mm in diameter were created on 1.5 × 1.5 cm skin blocks and treated with the investigated materials. While nanofiber mats reached the highest amount of delivered drug after 6 h, foils rapidly achieved a maximum drug concentration and maintained it over 24 h. The treatment had no effect on the overall skin metabolic activity but influenced the wound healing process, as observed using histological analysis. Both delivery systems were efficient in preventing the growth of *Pseudomonas aeruginosa* biofilms in ex vivo human skin. Interestingly, foils loaded with 500 μg of ciprofloxacin accomplished the complete eradication of biofilm infections with 1×10^9 bacteria/wound. We conclude that antimicrobial-loaded resorbable PVP foils and nanofiber mats are promising delivery systems for the prevention or topical treatment of infected wounds.

Keywords: wound infection; biofilm; pseudomonas aeruginosa; antimicrobial delivery; polyvinylpyrrolidone; nanofibers

1. Introduction

The number of antibiotic resistant bacteria, as well as the number of immune deficient patients, is increasing. For this reason, infections of post-operative and chronic wounds are becoming a concern for many patients and health care providers [1]. Age-related immune deficiency, diabetes mellitus type 2, venous insufficiency, or immobility are the major conditions leading to chronic wounds [2–4]. Chronic wounds are often associated with microbial biofilms, i.e., organized communities of one or more microorganism species encased and shielded by extracellular polymeric substances [5,6]. It was shown that significantly higher concentrations of antibiotics are required to treat biofilm-associated infections [7,8]. One of the reasons for biofilm drug-resistance is the extracellular polymeric matrix, which acts as a shield protecting bacteria from the external environment. In addition, the low concentrations of antibiotics that reach the bacteria in the biofilm favor the formation of persisters, i.e., physiologically inactive dormant cells that are less responsive to antibiotics [9]. Thus, for an efficient treatment of a biofilm-associated infection, a high and sustained concentration of antimicrobial drugs should be achieved at the site of infection and within the biofilm. This is often not fulfilled by most of the available antimicrobial formulations because of unfavorable physicochemical properties of drugs like instability in different biological environments, low solubility, or high molecular weight. New solubilization strategies and innovative pharmaceutical formulations, like polymer conjugates, nanocarriers, and membranes, have the potential to improve drug delivery and thus increase their concentration at the site of infection [10,11].

In this study, we investigated the drug delivery properties of polyvinylpyrrolidone (PVP)-based foils and nanofiber mats. A ciprofloxacin base was chosen as model drug because of its low water-solubility, detectability using fluorescence spectroscopy, and quorum sensing inhibiting properties in *Pseudomonas aeruginosa* (*P. aeruginosa*) at concentrations below the minimal inhibitory concentration (MIC) [12]. It is a broad-spectrum antibiotic that has been approved by the U.S. Food and Drug Administration (FDA) for the treatment of urinary tract and other infections, including skin and skin-structure infections. Nevertheless, it has severe side effects, and in 2016, the FDA recognized the existence of a rare side effect, a potentially permanent syndrome called fluoroquinolone-associated disability [13]. Because of the low water solubility, ciprofloxacin is formulated into tablets for oral use as monohydrochloride monohydrate salt. In this form, its absolute bioavailability is of 60% [14]. Alternatively, a ciprofloxacin base has been formulated in oil-based syrups [15] with a better bioavailability of up to 70% [16]. Formulations for topical use (e.g., ciprofloxacin ophthalmic ointment) are made by dissolving the drug in mineral oil. However, it was shown that drug crystals can still form and even cause the blockage of the bottle nozzle [17]. In general, antibiotics are not used for the topical treatment of chronic wounds because of insufficient drug bioavailability and possible local side effects. However, topical drug delivery would be an advantage in those cases, like an infected diabetic foot, where blood circulation is reduced and systemic therapies are often inefficient. In addition, it has been shown that locally applied antibiotics can reduce the risk of surgical site infections [18]. Finally, topical therapies reduce systemic toxicity and spare the gut microbiome. Thus, a more efficient topical drug delivery would improve both the management and prevention of wound infections.

In a previous study, we used a pH modification strategy to solubilize a ciprofloxacin base and increase the amount loaded on PVP foils and nanofiber mats. Interestingly, the residual acetic acid bound to PVP conferred a peculiar transparency and elasticity to the foils. These delivery systems were shown to be safe and to have a good anti-microbial activity in vitro [19]. In this study, we tested further properties of these materials using a wound infection model based on full-thickness human ex vivo skin. Different models are available to test the antimicrobial activity of a drug. In vitro grown biofilms give the possibility to study drug efficacy toward bacteria grown in communities. Nevertheless, in vitro biofilms are often very different from those found in in vivo infections. In vivo animal models have diverse advantages, including a tissue scaffold, host immune response, and wound healing processes. However, besides ethical reasons, animal studies are expensive and results are not always reproducible in humans. To reduce the number of animal studies, ex vivo models have been proposed for preliminary studies prior to animal studies. Ex vivo human skin has been used for several years to test the skin penetration of chemicals and drugs [20]. Yet, in the last few years, ex vivo porcine or human skin has been used to develop wound infection models [21–27]. Human full-thickness skin represents not only a three-dimensional scaffold where bacteria can grow, but also a complex environment with extracellular enzymatic activity, antimicrobial peptides, and several different cell populations, including immune-active cells. Thus, human skin infection models are realistic animal-free systems; even if they cannot completely replace in vivo studies, they are useful tools to screen antimicrobial formulations [26].

Using this model, we could measure the drug delivery kinetics, monitor eventual toxic effects, and measure the antimicrobial activity of the tested materials.

2. Materials and Methods

2.1. Preparation of Ciprofloxacin-Loaded PVP Foils and Nanofiber Mats

Transparent films and nanofiber mats were prepared as described by Contardi et al. [15]. Briefly, films were produced using a solvent casting method starting from aqueous solutions of PVP (3% *w/v*) with a molecular weight (MW) of 360,000 g/mol (Sigma-Aldrich, Milan, Italy), monohydrochloride monohydrate free ciprofloxacin (≥ 98.0% measured by high performance liquid chromatography (HPLC), Sigma-Aldrich, Milan, Italy) and acetic acid (≥ 99.7%, Sigma-Aldrich, Milan, Italy). Three different initial quantities of ciprofloxacin (1.2, 30, and 60 mg) were combined with the polymer and dissolved in acetic acid 30% (*v/v*) to a final volume of 30 mL to reach different drug concentrations (2.2, 44, and 88 mmol). The solutions were cast on Petri dishes (diameter 8.75 cm) for 3 days under an aspirated hood under ambient conditions (16–20 °C and 40–50% r.h.). Then, the films were placed in a vacuum desiccator for 3 more days to complete the removal of excess acetic acid. The nanofiber mats were fabricated by using a vertical electrospinning set-up. Starting solutions of PVP, acetic acid (30% *v/v*), and ciprofloxacin were prepared (final volume 6.2 mL). A higher concentration of the polymer (25% *w/v*) was used with respect to the films to allow for the electrospinning process. Three different concentrations of ciprofloxacin (2.2, 44, and 88 mM) were also prepared for the nanofiber mats. Syringes with a stainless-steel needle (18 gauge) were filled with the three different solutions and connected to a syringe pump (NE-1000, New Era Pump Systems, Inc., New York, NY, USA) working at a constant flow rate (500 µL/h). The needles were clamped to the positive electrode of a high-voltage power supply generating 26 kV at a distance of 24 cm from an aluminum disk used as a collector (diameter of 8.75 cm). Only 2 mL of each solution were electrospun in order to obtain the same amount of ciprofloxacin on both the films and nanofibers.

The morphology of foils and nanofiber mats was analyzed using SEM with a variable pressure JSM-649 microscope (JEOL, Milan, Italy) equipped with a tungsten thermionic electron source working in high vacuum mode and an acceleration voltage of 10 kV. The cross-section of the films was obtained by cutting slices with a UCS ultramicrotome (Leica Microsystems, Wetzlar, Germany) equipped with a glass knife. The specimens were coated with a 10-nm thick film of gold using the sputter coater 208 HR (Cressington Scientific Instruments, Watford, U.K.).

2.2. Skin Samples and the Creation of Superficial Wounds

Abdominal skin was obtained after getting informed consent from healthy donors undergoing plastic surgery. The study was conducted according to the Declaration of Helsinki guidelines and after approval by the Ethics Committee of the Charité—Universitätsmedizin Berlin (approval EA1/135/06, renewed on January 2018). Skin explants were used within 2–4 h after surgery. Subcutaneous fat tissue was partially removed, keeping a layer of approximately 5 mm and skin pieces (1.5 × 1.5 cm) were stretched and fixed on a Styrofoam block covered with Parafilm (Bemis Company, Neenah, WI, USA) using needles. The surface of the ex vivo skin (free of injuries or redness) was cleaned with saline solution (0.9% NaCl). The epidermis was then removed with a ball-shaped milling cutter 6 mm in size (No. 28725, Proxxon, Föhren, Germany) mounted on a micro motor handpiece (Marathon N7, TPC Advanced Technology, Inc. Diamond Bar, CA, USA) and rotating at 16,000 rpm. In this way, superficial wounds of approximately 5 mm in diameter were produced [21].

2.3. Drug Penetration Kinetics

Using a punch biopsy cutter, discs of 8 mm in diameter were cut out from foils, as well as nanofiber mats containing 44 mmol ciprofloxacin, so that each disc contained approximately 250 µg of ciprofloxacin. Wounds of 5 mm in diameter were produced on 1.5 × 1.5 cm pieces of skin. The disks were applied on the top of the wounds on skin blocs of 1.5 × 1.5 cm that were stretched on a Styrofoam block, placed in a humid chamber, and incubated at 37 °C, 5% CO_2, and 100% humidity for different time points. Thereafter, non-penetrated material was removed with a cotton swab and the treated wound was removed from the rest of the tissue by means of an 8 mm punch biopsy tool. The tissue was chopped into small pieces and placed in 2-mL tubes filled with HCl (0.1 N, 1.5 mL) to extract the ciprofloxacin. The samples were gently mixed on a shaker for 24 h at room temperature. After centrifugation for 5 min at 300× *g*, the supernatant was collected and placed in triplicate in a 96-well plate (100 µL/well). Ciprofloxacin fluorescence (excitation wavelength: 275 nm, emission wavelength: 480 nm) was measured with an EnSpire® Multimode plate reader (Perkin Elmer, Akron, OH, USA). A standard curve was prepared by dissolving ciprofloxacin in HCl (0.1 N) and preparing dilutions (0.5–10 µg/mL). The amount of penetrated drug was calculated on the basis of the standard curve. Results are presented as the means and standard deviations of three independent experiments.

2.4. Metabolic Activity of Skin Cells after the Topical Application of Ciprofloxacin on Ex Vivo Skin Wounds

Wounds were treated with 8 mm discs from PVP-foils loaded with 44 mmol of drug. Wounds treated with NaCl solution (0.9%, 20 µL) served as negative controls, and wounds treated with 50 µg/cm^2 PVP-coated silver nanoparticles (50 nm size, nanoComposix, San Diego, CA, USA) served as positive controls. Samples and controls were incubated in six-well plates for 20 h in 2 mL RPMI-1640 medium (Gibco, Darmstadt, Germany) without phenol red supplemented with fetal calf serum (FCS) (10%, Gibco, Darmstadt, Germany), glutamine (2mM, Gibco, Darmstadt, Germany), streptomycin (100 µg/mL, Gibco, Darmstadt, Germany), and penicillin (100 I.E./mL, Sigma-Aldrich, Hamburg, Germany). Thereafter, the old medium was replaced with fresh medium added with 500 µL of 2,3-bis-(2-methoxy-4-nitro-5-sulfophenyl)-2H-tetrazolium-5-carboxanilide (XTT) (Roche Diagnostic, Berlin, Germany). After 4 h, the medium from each sample and control was collected (3 × 100 µL) and placed in a 96-well microplate. The optical density at 450 nm was read with the EnSpire® Multimode plate reader using 650 nm as the reference wavelength. Results are presented as a percentage with respect to the values from the control wounds. The averages and standard deviations of values from three independent experiments are reported.

2.5. Histological Analysis of Wound Tissue after the Topical Application of Ciprofloxacin-Loaded Foils

Wounds created on ex vivo skin were treated with ciprofloxacin in solution or loaded on foils and nanofiber mats (250 µg/wound). Wounds left untreated served as controls. Skin explant cultures were grown in supplemented RPMI-1640 medium as described in Section 2.4. Every 2 days, the medium (500 mL) was removed and fresh medium (500 mL) was added. After eight days of incubation, the skin was plunge frozen in Tissue Freezing Medium (Leica Microsystems, Wetzlar, Germany) and cryosections were prepared. Hematoxylin and eosin (H&E) staining was performed following the manufacturer's manual (Roth, Karlsruhe, Germany) and images were taken using optical microscopy with an Olympus IX 50 (OLYMPUS, Hamburg, Germany).

2.6. Bacteria Inoculation and Characterization of the PAO1 Wound Infection

An overnight culture of the *P. aeruginosa* strain PAO1 (ATCC 15692) was diluted with tryptic soy broth. The suspension (5 µL, 1 × 10^7 bacteria) was injected with a 10 µL syringe (26 gauge) with a tapered tip (SGE Analytical Science, Ringwood, VIC, Australia) from the edge of the wound into the dermis. As a control, sterile saline (5 µL, 0.9% NaCl) was injected in an uninfected control wound. Wound samples were incubated in a humidified chamber at 37 °C for 20 h. Biopsies (8 mm) were taken and fixed in a solution of formaldehyde (4%) and glutaraldehyde (0.5% in 50 mmol (4-(2-hydroxyethyl)-1-piperazineethanesulfonic acid) (HEPES) for 48 h at room temperature. Skin samples were then washed in HEPES (50 mmol), and dehydrated in 30, 50, 70, 90, 95, and 100% ethanol. Samples were infiltrated first with a LR White/ethanol solution (1:1, 10 min), and then with pure LR White (2 × 15 min). Successively, samples were transferred to polyallomer centrifuge tubes

(5 × 20 mm, Beckman Coulter, Inc., Brea, CA, USA) containing LR White with an accelerator (5 µL/mL monomer). The centrifugation tubes were capped with a gelatin capsule and the samples were left to polymerize for 1 h on ice, and then at 60 °C overnight. Five hundred nanometer sections were cut using a ultramicrotome (EM UC7, Leica, Wetzlar, Germany), mounted on poly-L-lysine slides, and incubated for 10 min on an 80 °C thermo-plate. Sections were stained for 4 min with Richardson's stain (1% Azure II, 1% Methyleneblue, 1% Borax, Sigma-Aldrich, Hamburg, Germany, then washed with double distilled water (ddH_2O), and imaged using a microscope (Axiophot, Carl Zeiss Microscopy GmbH, Jena, Germany). For analysis with a scanning electron microscope (SEM), after fixation, one of the sample duplicates was cut with a scalpel in order to reveal the skin profile. The samples were then dehydrated as described above, critical-point dried, mounted on aluminum stubs, sputter-coated with a 12-nm layer of gold-palladium, and finally examined with an SEM (ZEISS 1530 Gemini, Carl Zeiss Microscopy GmbH, Jena Germany) operating at 3 kV using the in-lens electron detector. Images have been cropped, adjusted for optimal brightness, and contrasted using Photoshop Lightroom (version 6.0, Adobe Systems, San Jose, CA, USA).

2.7. Antimicrobial Activity of Ciprofloxacin-Loaded Foils and Nanofiber Mats

Each tested setting was done in triplicates with a total of at least three runs. Three different dosages of ciprofloxacin were tested (11, 250, and 500 µg/wound). The treatments with PVP foils and nanofiber mats started either 1 h or 20 h after bacteria inoculation and lasted for 20 h. Thereafter, an 8 mm punch biopsy was used to collect the wound tissue, including some of the surrounding intact skin. The tissue was placed in a 1.5 mL microcentrifuge tube containing saline (0.2 mL) and homogenized for 3 min with a sterile steel pistil mounted on a digital overhead stirrer at 150 rpm (DSL, VELP Scientifica Srl, Usmate, MB, Italy). Thereafter, samples were sonicated for 10 min in an ultrasonic bath (BactoSonic1, Bandelin, Berlin, Germany) at 40 kHz using 200 Weff to detach the bacteria. Volumes of each sample were transferred to the wells of 96-well microplates and diluted in 1:10 steps (20 µL sample + 180 µL saline) by using a multichannel pipette. A volume (5 µL) of each well was plated on square tryptic soy agar plates. After incubation overnight at 37 °C, spotting areas with 5 to 50 colony forming units (CFU) were counted. Mean values of the triplicates were calculated and bacteria number per wound was calculated considering the used dilutions. In the diagrams, bacteria counts/wound are presented as the mean and standard deviation of three independent experiments.

2.8. Data Analysis

Data are reported as arithmetic means and standard deviations of at least three experiments. Calculations, data processing, and graphics were prepared with Excel 2018 (Microsoft, Redmond, WA, USA).

3. Results and Discussion

3.1. Preparation and Characterization of Drug-Loaded Foils and Nanofiber Mats

Starting from aqueous solutions of PVP, acetic acid, and ciprofloxacin at different concentrations, transparent films and nanofiber mats were fabricated using the solvent casting and electrospinning methods, respectively (Figure 1).

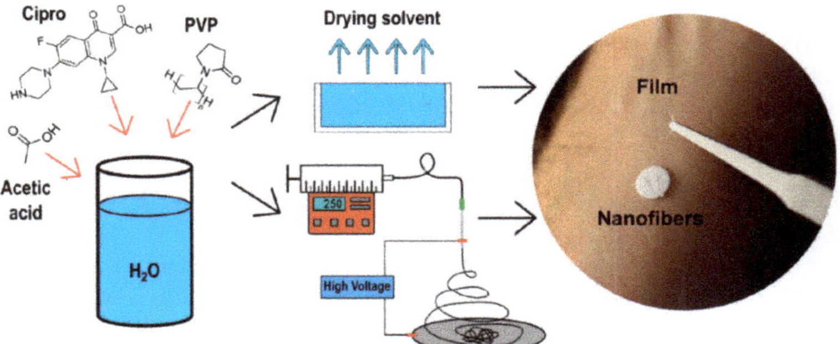

Figure 1. Schematic representation of the methods used for the film and nanofiber mat fabrication, and pictures of the discs used in the experimental set up. Cipro—ciprofloxacin; PVP—polyvinylpyrrolidone.

The final dry transparent foils had a thickness of 150–180 µm and the absence of crystal formation was verified using SEM (Figure A1, Appendix A). As recently demonstrated, the obtained PVP-based foils represent suitable wound dressings due to their flexibility, adhesion, and resorption properties [19,28]. The nanofiber had an average diameter of 360 ± 80 nm and did not show any defects or beads (Figure A2, Appendix A).

3.2. Foils and Nanofiber Mats Had Different Drug Delivery Profiles

Foils and nanofiber mats (discs of 8 mm in diameter) were applied on the top of the 5 mm in diameter wounds created on the 1.5 × 1.5 cm ex vivo skin tissue to provide an applied drug dosage of approximately 250 µg/wound. The skin explants were incubated in humidified chambers in an incubator with 100% humidity. This maintained the wetness of the skin blocks, which favored the dissolution of foils and nanofibers. After different incubation times, the non-penetrated material was removed, 8 mm biopsies were taken, and extracts were prepared to measure the amount of penetrated drug (Figure 2).

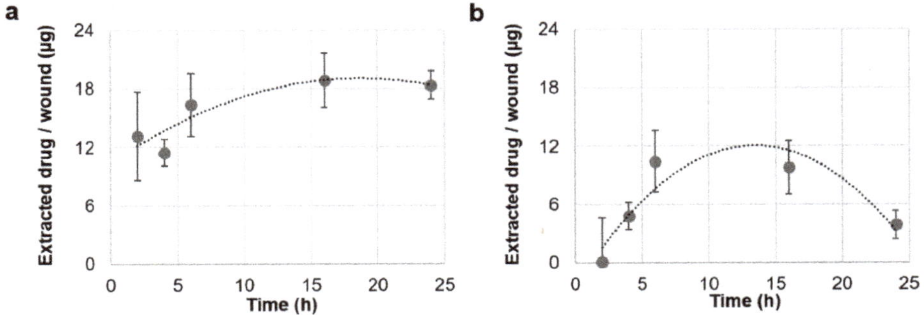

Figure 2. Time-dependent concentration of ciprofloxacin into wound tissue after the topical application of ciprofloxacin-loaded foils (**a**) and nanofibers (**b**). Ciprofloxacin in wound tissue extracts was measured by means of a fluorescence microplate reader and drug concentration was calculated from standard curves. Means and standard deviations from three independent experiments using skin from three different donors are reported.

Both penetration kinetics could be well-fitted to a second-degree polynomial trend line. Amounts ranging between 12 and 18 µg of ciprofloxacin/wound were detected for foils, whereas amounts ranging between 5 and 10 µg/wound were measured for nanofiber discs. Foils delivered high amounts of ciprofloxacin very quickly (approximately 12 µg after only 2 h), whereas nanofiber mats reached similar concentrations later, after 6 h. Interestingly, foils maintained the reached drug concentration in the wound over 24 h. On the contrary, in samples treated with nanofiber mats, the maximal concentration was maintained for a shorter time range (6–16 h). The decreased amounts of drug in the wound tissue observed for nanofiber mats (Figure 2b) were due to drug diffusion to the neighboring tissue (Figure A3, Appendix B). We suppose that for nanofiber mats, a lower amount of drug penetrated the wound tissue and that, due to diffusion to the nearby skin tissue, a reduction of local drug concentration in the wound was measured after 24 h. On the contrary, foils delivered a higher amount of drug to the wound and could saturate both the wound and the nearby tissue.

In a previous study, the dissolution rate and drug release for both materials was investigated in a phosphate buffer and on a mice full-thickness wound model [19]. Results showed that nanofibers dissolved and released the loaded drug very quickly, within the first two hours, whereas foils took longer. The fast dissolution of nanofibers might result in a high local concentration causing drug precipitation or crystallization, which in turn can hinder drug penetration. Another important aspect to be considered is that the investigated PVP-foils contained higher amounts (1.4%) of residual acetic acid than the nanofiber mats (0.3%) [19], which might also have had an influence on drug solubility and penetration.

In summary, drug penetration experiments using the ex vivo wound model showed that both drug delivery systems could deliver the loaded antibiotic to the wound tissue in a controlled manner. The ability of PVP-foils to maintain a high drug concentration in the wound over 24 h is of significance, especially for the treatment of biofilm infections, which need high antibiotic concentrations to be resolved. On the other side, the more controlled delivery profile of the nanofiber mats might be more useful for the prevention of wound infections.

3.3. Local Toxicity of Ciprofloxacin in Full-Thickness Ex Vivo Skin

The wound model was also used to test the potential cytotoxicity of the drug and investigated materials after topical application. For these experiments, tissue blocks were cultured in supplemented RPMI 1640 medium at the air–liquid interface. After 20 h of treatment, tissue viability was tested by means of an XTT test (Figure 3a).

Many cells in skin explants were still metabolically active and an evident formation of the red formazan product was observed. PVP foils did not show any toxicity. This was somehow expected with PVP being a well-tolerated, FDA-approved polymer with many uses, such as a food additive, binder in tablets, and plasma volume expander [29]. Also, foils prepared from a PVP solution in 30% acetic acid resulted in having no effect on skin viability due to the fact that only residual amounts of acetic acid molecules remained within the PVP polymers after solvent evaporation during the preparation process [19]. Finally, the PVP foils loaded with the three concentrations of ciprofloxacin also resulted in having no influence on the overall viability of skin cells after 20 h of incubation. Wounds treated with silver nanoparticles, which release toxic silver ions, served as positive controls. The evident reduction of formazan formation was indicative of silver toxicity toward wound cells. Thus, even if this test gave no information about the type of cells being affected by the tested substances, it was a useful method to detect overall acute toxic effects.

Previous studies have shown that ciprofloxacin has toxic effects on fibroblasts [30] and keratinocytes [31]. Toxicity was shown to be time- and concentration-dependent. For this reason, we tested the effect of ciprofloxacin delivered by the investigated foils with regard to re-epithelialization. After 9 days of incubation in supplemented RPMI-1640 medium, a partial re-epithelialization had occurred in controls, with the re-growing of a keratinocyte layer on the edges and in the wound bed (Figure 3b,d,e). Cell nuclei and the collagen network presented a normal morphology in samples treated with empty PVP foils with residual acetic acid. Only a delayed re-epithelialization was observed, probably due to oxygen deprivation (data not shown). On the contrary, in wounds treated with ciprofloxacin, only a thin epithelial layer was observed. Cells' nuclei appeared small with an irregular morphology typical of necrotic cells and the collagen matrix was less organized than in the controls (Figure 3c,f,g). This effect was visible in all samples treated with ciprofloxacin at different concentration and was independent of the formulation (data not shown).

In recent years, it has been recognized that fluoroquinolones, despite being well tolerated by a broad portion of patients, can have rare and very disabling side effects like a tendon rupture and irreversible nerve damage [13]. There are several hypotheses on the mechanism of toxicity and on the reason why some persons develop these side effects more than others do. One theory is that there might be a gene variant responsible for a disrupted quinolone metabolism. Our results show that, despite ciprofloxacin inducing no changes in whole skin metabolic activity, it negatively influenced the re-epithelialization of ex vivo wounds. Even if this side effect is tolerable in the case of severe wound infections, these results stress the importance of a more local and controlled release of drugs like ciprofloxacin that have narrow therapeutic windows.

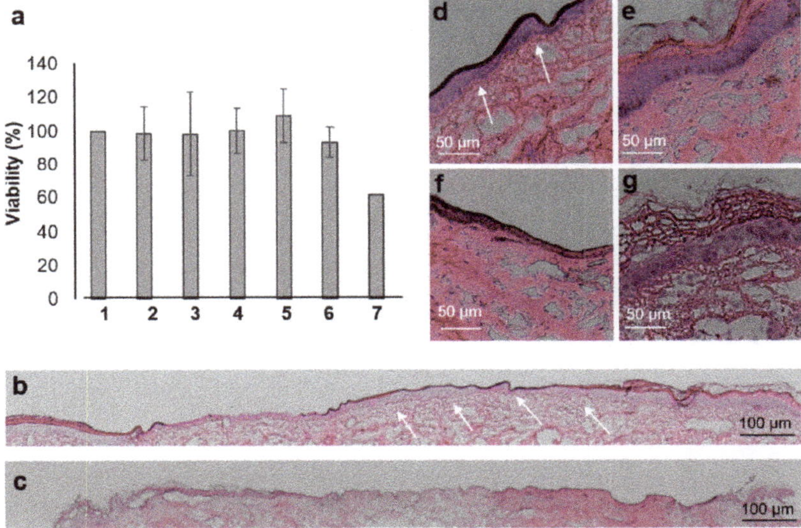

Figure 3. Local toxicity of ciprofloxacin delivered using PVP-foils. (**a**) The XTT assay was run with a skin biopsy previously treated for 20 h with: 0.9% NaCl (1); PVP foils prepared from solutions in ddH_2O (2); PVP foils prepared from solutions in 30% acetic acid (3); PVP foils loaded with 11 µg (4), 250 µg (5), and 500 µg (6) of ciprofloxacin; and 50 µg of PVP-coated silver nanoparticles (7). Means and standard deviations from three independent experiments are reported. (**b**,**c**) Picture collage of H&E stained sections after 9 days of tissue culture of (**b**) an untreated wound and (**c**) a wound treated with a PVP foil loaded with 250 µg ciprofloxacin. (**d–g**) Details from the centre (**d**,**f**) and the edges (**e**,**g**) of control (**d**,**e**) and treated (**f**,**g**) wounds. Arrows show a newly formed epithelial layer.

3.4. Ciprofloxacin-Loaded Foils and Nanofiber Mats Efficiently Reduced P. aeruginosa Infections

Next, we investigated the antimicrobial efficacy of the investigated drug delivery systems using the ex vivo skin infection model (Figure 4). The PAO1 strain was used, which possesses several proteolytic enzymes, among those collagenases, and was shown to build biofilm-like infections on ex vivo skin [21]. Bacteria inoculated in the wounds grew from 1×10^7 to 1×10^9 bacteria per wound after 20 h of incubation. Macroscopically, the surface of the PAO1-infected wound appeared shiny and yellowish (Figure 4c). Scanning electron microscopy images of the wound surface showed bacteria conglomerates typical of a biofilm. These three-dimensional structures are made of bacteria, extracellular materials, and probably of degraded collagen material (Figure 4d,e). The microscopic pictures of H&E-stained skin sections (Figure 4f) revealed an approximately 10-µm-thick bacteria film on the surface of the wound, but also groups of bacteria deep in the wound tissue (circles). The scanning electron microscopic analysis of the wound profile (Figure 4g,h) confirmed the formation of a bacterial biofilm within the superficial wound tissue, as well as the presence of small agglomerates or scattered bacteria deep in the wound.

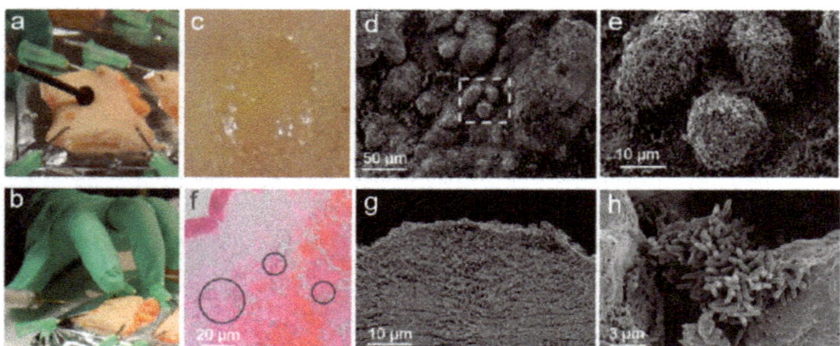

Figure 4. Wound infection model with the *P. aeruginosa* strain PAO1. (**a**) A superficial wound of approximately 5 mm in diameter was produced on skin explants. (**b**) Bacteria (1×10^7 per wound) were inoculated and skin explants, which were incubated for 20 h in the previous analysis. (**c**) Microscopic appearance of a representative wound 20 h after the PAO1 injection. (**d**,**e**) SEM images of the wound surface at two different magnifications showing the typical morphology of bacterial communities in the biofilm. (**f**) Wound section stained with H&E showing a bacteria layer on the surface of the wound and bacteria agglomerates deep into the tissue (circles). (**g**,**h**) SEM images of wound sections at different magnifications confirming bacterial growth on the surface (**g**) and deep in the wound tissue (**h**).

In the ex vivo model, most of bacteria grew on the wound's surface, despite the fact that they were inoculated deep in the connective tissue. *P. aeruginosa* can grow in both aerobic and anaerobic conditions. However, the less favorable conditions and shortage of oxygen in the deeper wound layers resulted in a slower proliferation of bacteria. Nevertheless, these small colonies can be the reason for recurrent infections if not eradicated during a treatment. For this reason, the penetration of adequate drug concentrations deep in the wound tissue and the eradication of all bacterial colonies is a crucial factor for a successful therapy.

Pseudomonas aeruginosa has already been cultured on ex vivo skin [24,32–35]. In this study, we used the ex vivo wound infection model as a three dimensional set up to test the capacity of ciprofloxacin-loaded PVP drug delivery systems to eradicate bacteria located not only on the surface, but also deep in the wound's tissue. The investigated foils and nanofiber mats were loaded with three different concentrations of ciprofloxacin, corresponding to final doses of 11, 250, and 500 µg per wound. The treatments were applied on the top of the wounds 1 h or 20 h after bacteria inoculation and kept for a further 20 h. After 1 h of inoculation, bacteria were still planktonic and thus less resistant to antibiotics. After 20 h of inoculation, the number of bacteria had increased and bacteria were organized in a biofilm, which was more difficult to treat. Samples treated 1 h after bacteria inoculation served to test the drug efficacy toward a moderate infection with approximately 1×10^7 planktonic bacteria, whereas samples treated 20 h after PAO1 inoculation served to test the efficacy of foils and nanofiber mats on a severe infection with approximately 1×10^9 bacteria primarily organized into a biofilm. The experiment terminated after a total of 40 h.

Macroscopically, a thick, yellow biofilm was visible on the surface of untreated wounds, whereas no bacterial film was visible on the treated samples (Figure 5a).

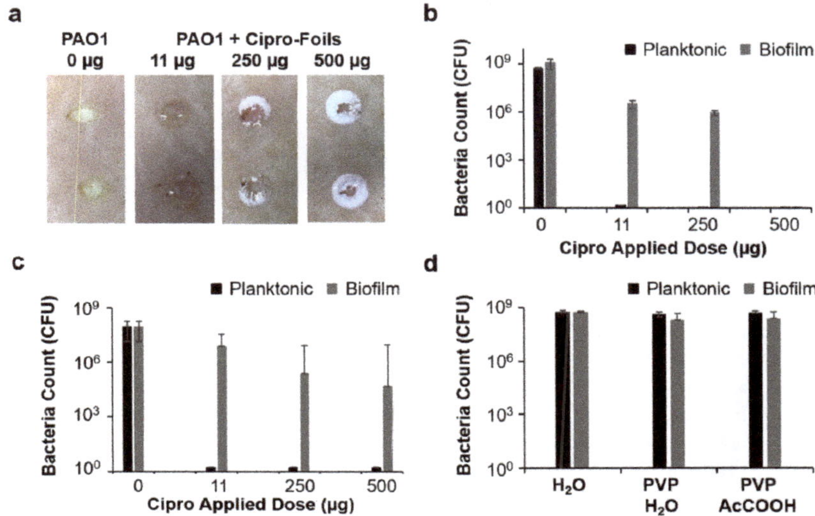

Figure 5. Antimicrobial efficacy of PVP foils and nanofiber mats with different ciprofloxacin payloads applied on wounds infected with planktonic bacteria in a biofilm. (**a**) Representative pictures of untreated and treated wounds after 40 h of incubation. (**b**,**c**) Bacteria count in tissue extracts of wounds treated with foils (**b**) or nanofiber mats (**c**). (**d**) Bacteria counts in control wounds treated with plain PVP foils prepared from solutions in water or 30% acetic acid. Means and standard deviations of three independent experiments (three donors) run in triplicate are reported. Cipro—ciprofloxacin; AcCOOH—acetic acid.

Drug crystals were visible on the wound edges, where higher concentrations of the drug were reached after the foils or nanofibers had dissolved. When applied 1 h after bacterial inoculation, both foils and nanofiber mats resulted in a complete eradication of the bacterial infection independent from the amount of drug that was applied. This result shows the potential of these drug delivery materials for the prevention of wound infections. In particular, nanofibers should be tailored to obtain a more controlled drug delivery in order to have effective antimicrobial concentrations and limit adverse side effects.

In contrast, the 20-h-biofilm infection was more difficult to treat. PVP foils had the best performance (Figure 5b). Foils loaded with 500 μg ciprofloxacin achieved a complete eradication of the bacteria, and the two lower concentrations resulted in a 3-log reduction of bacteria concentration with respect to the untreated controls. Nanofiber mats had a lower antimicrobial activity with a maximal 3-log reduction achieved by the highest concentration. No effect on bacterial growth was measured for plain PVP foils or foils with residual acetic acid (Figure 5d).

The most striking result was the complete eradication of the biofilm infection achieved with the 500-μg PVP foils. Foils have a higher percentage of residual acetic acid (1.4%) than nanofiber mats (0.1%) [19]. Because acetic acid also has antimicrobial activity, it might have acted synergistically with ciprofloxacin. Nevertheless, the superior efficacy of the foils compared to the nanofiber mats correlated well with the delivery kinetics. Foils delivered higher amounts of the drug and over a longer time than nanofibers did. Phillips et al. assessed the efficacy of several commercially available treatments in a porcine ex vivo wound model and found that time-release silver gel and cadexomer iodine dressings were the most effective in reducing a mature biofilm with a reduction between 5 and 7 logs out of a 7-log total [24]. Using an in vivo mouse model, Roy et al. found that ciprofloxacin-loaded keratin-based hydrogels with a sustained drug release profile could reduce the amount of *P. aeruginosa* in the wound bed by 99.9% without interfering with the key processes of wound healing [36]. Thus, these and our results underline the importance of drug delivery systems with a sustained release profile for the efficient treatment of *P. aeruginosa* wound infections.

4. Conclusions

In this study, we used full-thickness ex vivo human skin and a wound infection model to investigate the efficacy and tolerability of ciprofloxacin-loaded PVP foils and nanofiber mats. The model allowed us to test the antimicrobial efficacy of these materials and to correlate it to their delivery properties. Ciprofloxacin was representative of a poorly water-soluble drug to be loaded in PVP-based drug delivery systems. The use of a solubilizer (acetic acid) increased the loading capacity and drug delivery properties of nanofibers and foils. This was in turn a crucial point for the accomplishment of high local drug concentrations that were required for the successful prevention and eradication of the PAO1 biofilm infections.

This strategy can be used to load PVP-based delivery systems with different types of drugs or disinfectants, including high molecular weight moieties. Besides their sustained drug release properties, such matrixes allow for a more precise dosing of the active ingredient in comparison to ointments or creams. Foils and nanofiber mats are flexible and can be easily applied in less accessible skin areas, e.g., the lower back or between the toes, which is an advantage, especially for old and disabled patients. Thus, we conclude that the combination of drug delivery systems and solubilizing agents is a promising strategy to create attractive new pharmaceutical forms for topical drug delivery to treat or prevent wound infections.

Author Contributions: Conceptualization, F.R., M.C., I.S.B. and C.S.; methodology, F.R. and C.S.; formal analysis, F.R., J.J. and M.C.; investigation, M.C., J.J. and C.S.; resources, U.B.-P., A.V. and C.S.; writing—original draft preparation, F.R.; writing—review and editing, F.R., M.C., I.S.B., U.B.-P., A.V. and C.S.; visualization, F.R. and M.C.; funding acquisition, F.R. and C.S.

Funding: This research was partially funded by the German Federal Ministry of Economics (grants: KF2928204MD4 and KF2088119MD4). We acknowledge support from the German Research Foundation (DFG) and the Open Access Publication Fund of Charité—Universitätsmedizin Berlin.

Conflicts of Interest: The authors declare no conflict of interest.

Appendix A

Morphological Analysis of Films and Fibers

To verify the absence of crystals in the drug-loaded membranes, SEM analysis was performed. In Figure A1, representative images of the top-view and cross-section of ciprofloxacin-loaded foils (88 mmol) are reported.

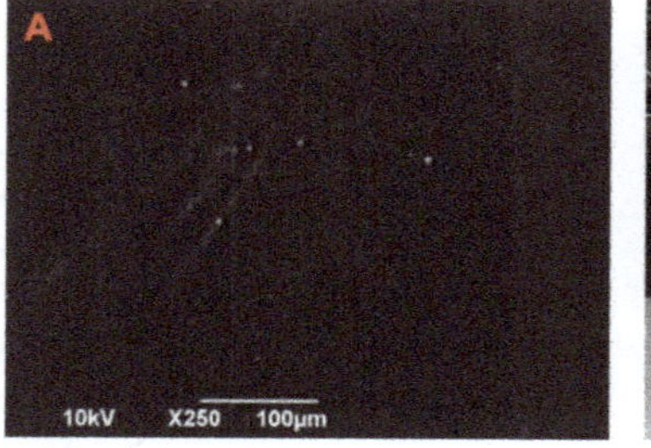

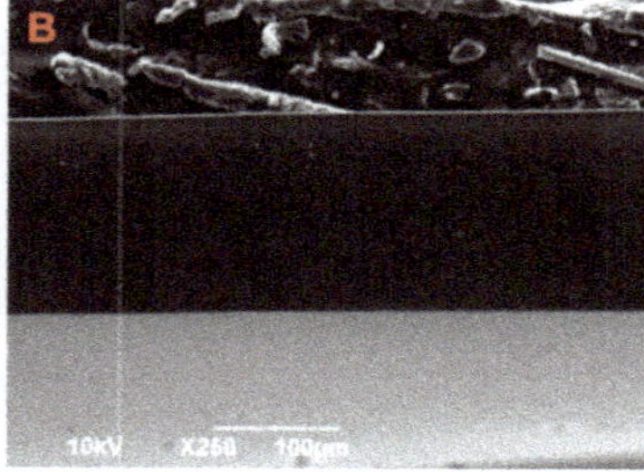

Figure A1. SEM images of the top-view (**A**) and cross-section (**B**) of a ciprofloxacin-loaded PVP foils.

Figure A2 shows representative SEM images of the ciprofloxacin-loaded nanofibers (2.2, 44, and 88 mmol). The diameter of the fibers was determined using ImageJ software. Approximately 100 measurements were taken to obtain the diameter distribution of each type of fiber mat.

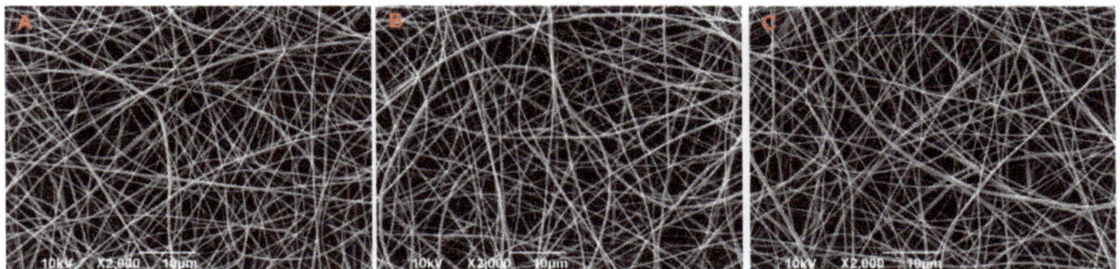

Figure A2. (**A–C**) Top-view SEM images of the PVP nanofibers loaded with (**A**) 1.2, (**B**) 30, and (**C**) 60 mg of ciprofloxacin respectively.

Appendix B

Drug Diffusion to the Surrounding Skin Tissue

In order to estimate the amount of drug that diffused to the nearby tissue, skin blocks were treated with ciprofloxacin loaded on PVP-foils or a solution of the drug in water. The wound tissue, as well as the surrounding tissue, were extracted and analyzed (Figure A3). The kinetics showed that the drug accumulated first in the wound tissue and diffused successively to the nearby tissue. After approximately 16 h, similar concentrations were reached in both areas of the ex vivo tissue block.

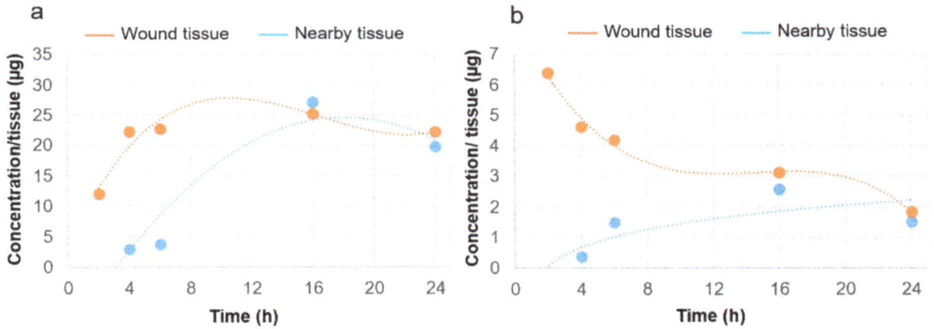

Figure A3. Penetration of ciprofloxacin into the wound and surrounding tissue after the topical application of (**a**) ciprofloxacin-loaded PVP foils (250 μg/wound) and (**b**) a solution of ciprofloxacin base in water (13 μg/wound).

References

1. Tang, S.S.; Apisarnthanarak, A.; Hsu, L.Y. Mechanisms of β-lactam antimicrobial resistance and epidemiology of major community-and healthcare-associated multidrug-resistant bacteria. *Adv. Drug Deliv. Rev.* **2014**, *78*, 3–13. [CrossRef] [PubMed]
2. Pawelec, G. Age and immunity: What is "immunosenescence"? *Exp. Gerontol.* **2018**, *105*, 4–9. [CrossRef] [PubMed]
3. Han, G.; Ceilley, R. Chronic wound healing: A review of current management and treatments. *Adv. Ther.* **2017**, *34*, 599–610. [CrossRef] [PubMed]
4. Bjarnsholt, T.; Kirketerp-Møller, K.; Jensen, P.Ø.; Madsen, K.G.; Phipps, R.; Krogfelt, K.; Høiby, N.; Givskov, M. Why chronic wounds will not heal: A novel hypothesis. *Wound Repair Regen.* **2008**, *16*, 2–10. [CrossRef] [PubMed]
5. Malone, M.; Bjarnsholt, T.; McBain, A.J.; James, G.A.; Stoodley, P.; Leaper, D.; Tachi, M.; Schultz, G.; Swanson, T.; Wolcott, R.D. The prevalence of biofilms in chronic wounds: A systematic review and meta-analysis of published data. *J. Wound Care* **2017**, *26*, 20–25. [CrossRef] [PubMed]

6. Percival, S.L.; Hill, K.E.; Williams, D.W.; Hooper, S.J.; Thomas, D.W.; Costerton, J.W. A review of the scientific evidence for biofilms in wounds. *Wound Repair Regen.* **2012**, *20*, 647–657. [CrossRef] [PubMed]
7. Wu, H.; Moser, C.; Wang, H.-Z.; Høiby, N.; Song, Z.-J. Strategies for combating bacterial biofilm infections. *Int. J. Oral Sci.* **2015**, *7*, 1–7. [CrossRef]
8. Høiby, N.; Bjarnsholt, T.; Givskov, M.; Molin, S.; Ciofu, O. Antibiotic resistance of bacterial biofilms. *Int. J. Antimicrob. Agents* **2010**, *35*, 322–332. [CrossRef]
9. Harms, A.; Maisonneuve, E.; Gerdes, K. Mechanisms of bacterial persistence during stress and antibiotic exposure. *Science* **2016**, *354*, aaf4268. [CrossRef]
10. Kalepu, S.; Nekkanti, V. Insoluble drug delivery strategies: Review of recent advances and business prospects. *Acta Pharm. Sin. B* **2015**, *5*, 442–453. [CrossRef]
11. Goyal, R.; Macri, L.K.; Kaplan, H.M.; Kohn, J. Nanoparticles and nanofibers for topical drug delivery. *J. Control. Release* **2016**, *240*, 77–92. [CrossRef] [PubMed]
12. Skindersoe, M.E.; Alhede, M.; Phipps, R.; Yang, L.; Jensen, P.O.; Rasmussen, T.B.; Bjarnsholt, T.; Tolker-Nielsen, T.; Høiby, N.; Givskov, M. Effects of antibiotics on quorum sensing in Pseudomonas aeruginosa. *Antimicrob. Agents Chemother.* **2008**, *52*, 3648–3663. [CrossRef] [PubMed]
13. Marchant, J. When antibiotics turn toxic. *Nature* **2018**, *555*, 431–433. [CrossRef] [PubMed]
14. Drusano, G.; Standiford, H.; Plaisance, K.; Forrest, A.; Leslie, J.; Caldwell, J. Absolute oral bioavailability of ciprofloxacin. *Antimicrob Agents Chemother.* **1986**, *30*, 444–446. [CrossRef]
15. Johnson, C.E.; Wong, D.V.; Hoppe, H.L.; Bhatt-Mehta, V. Stability of ciprofloxacin in an extemporaneous oral liquid dosage form. *Int. J. Pharm. Comp.* **1998**, *2*, 314–317.
16. Shah, A.; Liu, M.-C.; Vaughan, D.; Heller, A.H. Oral bioequivalence of three ciprofloxacin formulations following single-dose administration: 500 mg tablet compared with 500 mg/10 mL or 500 mg/5 mL suspension and the effect of food on the absorption of ciprofloxacin oral suspension. *J. Antimicrob. Chemother.* **1999**, *43*, 49–54. [CrossRef]
17. John, T. Scanning electron microscopic study of a Ciloxan bottle blocked by ciprofloxacin crystals. *Eye* **2001**, *15*, 786. [CrossRef]
18. Heal, C.F.; Banks, J.L.; Lepper, P.D.; Kontopantelis, E.; van Driel, M.L. Topical antibiotics for preventing surgical site infection in wounds healing by primary intention. *Cochrane Database Syst. Rev.* **2016**, *11*, CD011426. [CrossRef]
19. Contardi, M.; Heredia-Guerrero, J.A.; Perotto, G.; Valentini, P.; Pompa, P.P.; Spanò, R.; Goldoni, L.; Bertorelli, R.; Athanassiou, A.; Bayer, I.S. Transparent ciprofloxacin-povidone antibiotic films and nanofiber mats as potential skin and wound care dressings. *Eur. J. Pharm. Sci.* **2017**, *104*, 133–144. [CrossRef]
20. Flaten, G.E.; Palac, Z.; Engesland, A.; Filipović-Grčić, J.; Vanić, Ž.; Škalko-Basnet, N. In vitro skin models as a tool in optimization of drug formulation. *Eur. J. Pharm. Sci.* **2015**, *75*, 10–24. [CrossRef]
21. Schaudinn, C.; Dittmann, C.; Jurisch, J.; Laue, M.; Günday-Türeli, N.; Blume-Peytavi, U.; Vogt, A.; Rancan, F. Development, standardization and testing of a bacterial wound infection model based on ex vivo human skin. *PLoS ONE* **2017**, *12*, e0186946. [CrossRef] [PubMed]
22. Maboni, G.; Davenport, R.; Sessford, K.; Baiker, K.; Jensen, T.K.; Blanchard, A.M.; Wattegedera, S.; Entrican, G.; Tötemeyer, S. A novel 3D skin explant model to study anaerobic bacterial infection. *Front. Cell. Infect. Microbiol.* **2017**, *7*, 404. [CrossRef] [PubMed]
23. Steinstraesser, L.; Sorkin, M.; Niederbichler, A.; Becerikli, M.; Stupka, J.; Daigeler, A.; Kesting, M.; Stricker, I.; Jacobsen, F.; Schulte, M. A novel human skin chamber model to study wound infection ex vivo. *Arch. Dermatol. Res.* **2010**, *302*, 357–365. [CrossRef] [PubMed]
24. Phillips, P.L.; Yang, Q.; Davis, S.; Sampson, E.M.; Azeke, J.I.; Hamad, A.; Schultz, G.S. Antimicrobial dressing efficacy against mature Pseudomonas aeruginosa biofilm on porcine skin explants. *Int. Wound J.* **2015**, *12*, 469–483. [CrossRef] [PubMed]
25. Yang, Q.; Larose, C.; Della Porta, A.; CSchultz, G.S.; Gibson, D.J. A surfactant-based wound dressing can reduce bacterial biofilms in a porcine skin explant model. *Int. Wound J.* **2017**, *14*, 408–413. [CrossRef] [PubMed]
26. Ramirez, H.A.; Pastar, I.; Jozic, I.; Stojadinovic, O.; Stone, R.C.; Ojeh, N.; Gil, J.; Davis, S.C.; Kirsner, R.S.; Tomic-Canic, M. Staphylococcus aureus triggers induction of miR-15B-5P to diminish DNA repair and deregulate inflammatory response in diabetic foot ulcers. *J. Investig. Dermatol.* **2018**, *138*, 1187–1196. [CrossRef]

27. Alhusein, N.; Blagbrough, I.S.; Beeton, M.L.; Bolhuis, A.; Paul, A. Electrospun zein/PCL fibrous matrices release tetracycline in a controlled manner, killing Staphylococcus aureus both in biofilms and ex vivo on pig skin, and are compatible with human skin cells. *Pharm. Res.* **2016**, *33*, 237–246. [CrossRef]
28. Contardi, M.; Russo, D.; Suarato, G.; Heredia-Guerrero, J.A.; Ceseracciu, L.; Penna, I.; Margaroli, N.; Summa, M.; Spanò, R.; Tassistro, G. Polyvinylpyrrolidone/hyaluronic acid-based bilayer constructs for sequential delivery of cutaneous antiseptic and antibiotic. *Chem. Eng. J.* **2019**, *358*, 912–923. [CrossRef]
29. Ravin, H.A.; Seligman, A.M.; Fine, J. Polyvinyl pyrrolidone as a plasma expander: Studies on its excretion, distribution and metabolism. *N. Engl. J. Med.* **1952**, *247*, 921–929. [CrossRef]
30. Gürbay, A.; Garrel, C.; Osman, M.; Richard, M.; Favier, A.; Hincal, F. Cytotoxicity in ciprofloxacin-treated human fibroblast cells and protection by vitamin E. *Hum. Exp. Toxicol.* **2002**, *21*, 635–641. [CrossRef]
31. Kautzky, F.; Hartinger, A.; Köhler, L.D.; Vogt, H.J. In vitro cytotoxicity of antimicrobial agents to human keratinocytes. *J. Eur. Acad. Dermatol. Venereol.* **1996**, *6*, 159–166. [CrossRef]
32. Schmidtchen, A.; Holst, E.; Tapper, H.; Björck, L. Elastase-producing Pseudomonas aeruginosa degrade plasma proteins and extracellular products of human skin and fibroblasts, and inhibit fibroblast growth. *Microb. Pathog.* **2003**, *34*, 47–55. [CrossRef]
33. Werthen, M.; Davoudi, M.; Sonesson, A.; Nitsche, D.; Mörgelin, M.; Blom, K.; Schmidtchen, A. Pseudomonas aeruginosa-induced infection and degradation of human wound fluid and skin proteins ex vivo are eradicated by a synthetic cationic polymer. *J. Antimicrob. Chemother.* **2004**, *54*, 772–779. [CrossRef] [PubMed]
34. Vieira, A.; Silva, Y.; Cunha, A.; Gomes, N.; Ackermann, H.-W.; Almeida, A. Phage therapy to control multidrug-resistant Pseudomonas aeruginosa skin infections: In vitro and ex vivo experiments. *Eur. J. Clin. Microbiol. Infect. Dis.* **2012**, *31*, 3241–3249. [CrossRef]
35. Björn, C.; Mahlapuu, M.; Mattsby-Baltzer, I.; Håkansson, J. Anti-infective efficacy of the lactoferrin-derived antimicrobial peptide HLR1r. *Peptides* **2016**, *81*, 21–28. [CrossRef]
36. Roy, D.C.; Tomblyn, S.; Burmeister, D.M.; Wrice, N.L.; Becerra, S.C.; Burnett, L.R.; Saul, J.M.; Christy, R.J. Ciprofloxacin-loaded keratin hydrogels prevent Pseudomonas aeruginosa infection and support healing in a porcine full-thickness excisional wound. *Adv. Wound Care* **2015**, *4*, 457–468. [CrossRef]

Article

Erythrocyte Membrane-Coated Arsenic Trioxide-Loaded Sodium Alginate Nanoparticles for Tumor Therapy

Yumei Lian, Xuerui Wang, Pengcheng Guo, Yichen Li, Faisal Raza, Jing Su * and Mingfeng Qiu *

School of Pharmacy, Shanghai Jiao Tong University, Shanghai 200240, China; lym-0517@sjtu.edu.cn (Y.L.); wangxuerui0303@163.com (X.W.); gpcedu@sjtu.edu.cn (P.G.); liyichen592@sjtu.edu.cn (Y.L.); faisalraza@sjtu.edu.cn (F.R.)

* Correspondence: jingsu@sjtu.edu.cn (J.S.); mfqiu@sjtu.edu.cn (M.Q.); Tel.: +86-21-34204052 (M.Q.)

Received: 22 November 2019; Accepted: 18 December 2019; Published: 24 December 2019

Abstract: Arsenic trioxide (ATO) has a significant effect on the treatment of acute promyelocytic leukemia (APL) and advanced primary liver cancer, but it still faces severe side effects. Considering these problems, red blood cell membrane-camouflaged ATO-loaded sodium alginate nanoparticles (RBCM-SA-ATO-NPs, RSANs) were developed to relieve the toxicity of ATO while maintaining its efficacy. ATO-loaded sodium alginate nanoparticles (SA-ATO-NPs, SANs) were prepared by the ion crosslinking method, and then RBCM was extruded onto the surface to obtain RSANs. The average particle size of RSANs was found to be 163.2 nm with a complete shell-core bilayer structure, and the average encapsulation efficiency was 14.31%. Compared with SANs, RAW 264.7 macrophages reduced the phagocytosis of RSANs by 51%, and the in vitro cumulative release rate of RSANs was 95% at 84 h, which revealed a prominent sustained release. Furthermore, it demonstrated that RSANs had lower cytotoxicity as compared to normal 293 cells and exhibited anti-tumor effects on both NB4 cells and 7721 cells. In vivo studies further showed that ATO could cause mild lesions of main organs while RSANs could reduce the toxicity and improve the anti-tumor effects. In brief, the developed RSANs system provides a promising alternative for ATO treatment safely and effectively.

Keywords: red blood cells membrane; arsenic trioxide; sodium alginate nanoparticles; reduce toxicity; anti-tumor

1. Introduction

Arsenic trioxide (ATO) is the main active ingredient of traditional Chinese medicine (TCM) Arsenic. In the 1970s, it was first applied to acute promyelocytic leukemia (APL) with significant efficacy [1] and was approved by the National Medical Products Administration (NMPA) and Food and Drug Administration (FDA) as a first-line treatment for APL in 1999 and 2000, respectively [2,3]. ATO can induce cell differentiation, inhibit apoptosis, and exert anti-tumor effect [4]. In recent years, research studies have confirmed the significant growth inhibition and apoptosis induction effect of ATO in solid tumors, such as liver cancer, breast cancer, stomach cancer, glioma and lung cancer [5–11]. At present, ATO injection has been employed clinically in the treatment of APL and advanced primary liver cancer. However, the unique physicochemical properties of ATO allow it to be rapidly cleared from blood, and it requires daily administration during clinical treatment. At the same time, the uptake of the reticuloendothelial system (RES) makes only a slight amount of ATO reach the tumor site. Nevertheless, considering the potent toxicity of ATO, increasing the dose of ATO will increase the systemic toxicity and cause damage to the liver, kidney, heart, and peripheral nerve [12–14].

Based on the size advantage, nanoparticles (NPs) can exude through the tumor vasculature and effectively delivery the drugs to cells through enhanced permeability and retention (EPR) effects [15].

Therefore, it is considered to be a kind of formulation with low toxicity and high stability. Different ATO delivery systems (DDS) have been developed, including magnetic nanoparticles [16], chitosan nanoparticles [17], microspheres [18], liposomes [19], and mesoporous silica nanoparticles [10,20]. These formulations can achieve sustained release of ATO, which can reduce the transient plasma concentration and toxicity of drugs to a certain extent. However, they are still deficient in biocompatibility, and the safety of these systems needs to be verified [21].

Sodium alginate (SA) is a sort of polyanionic polysaccharide alginic acid salt found in brown algae that is water-soluble and has the advantages of anti-tumor effect, immune regulation, non-toxic, biodegradability, and excellent biocompatibility [22,23]. It has been approved by FDA for the pharmaceutical industry as an excipient [24,25]. For the past few years, NPs prepared from SA as drug carrier systems have also attracted more attention [26,27]. Red blood cells membrane (RBCM) will be formed into vesicles (RVs) using extrusion or sonication methods [28]. As a drug carrier, it can be attached to the surface of NPs to sustain the release of drugs, avoid elimination by the immune system, increase drug stability, improve biocompatibility and thus prolong drug circulation in vivo [29,30].

Moreover, RBCM coating nanotechnology already has excellent precedents. Che-Ming J et.al [31] demonstrated the synthesis of an RBCM coated polymeric nanoparticle for long-circulating cargo delivery, Jinghan Su et al. [32–34] extensively studied the effectiveness of RBCM-camouflaged NPs for treating metastatic breast cancer. In addition, if the non-toxic SA nanoparticles can be encapsulated by natural RBCM and combine the superiorities of sustained release and prolonging residence time, the nano-system can achieve the purpose of maintaining efficacy and reducing toxicity. In addition, this can provide a new possibility for safe application of ATO.

In this study, RBCM-camouflaged ATO-loaded sodium alginate nanoparticles (RBCM-SA-ATO-NPs, RSANs) were prepared as shown in Figure 1. ATO-loaded sodium alginate nanoparticles (SA-ATO-NPs, SANs) were prepared by the ion crosslinking method, followed by coating of RBCM to obtain RSANs. It was then systematically characterized and evaluated for its efficacy and toxicity. The results indicated that the system might become a promising delivery system for the safe, effective, and sustained release of ATO.

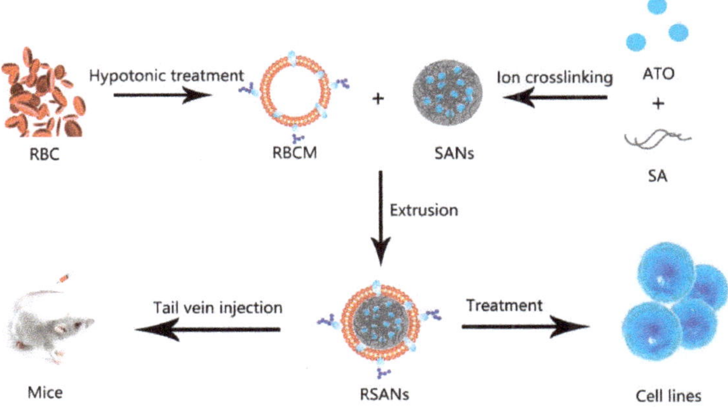

Figure 1. Preparation and Characterization of RSANs.

2. Materials and Methods

2.1. Materials

SA (200–500 Pa·s) was purchased from Shanghai Taitan Technology Co., Ltd., Shanghai, China, anhydrous calcium chloride ($CaCl_2$) was purchased from Shanghai Lingfeng Chemical Reagent Co., Ltd., Shanghai, China and Polosham 188 (F-188) was purchased from Shaanxi

Zhengyi Pharmaceutical Accessories Co., Ltd. Carbon support copper mesh (230 mesh) and phosphotungstic acid were obtained from Beijing Zhongjing Keyi Technology Co., Ltd., Beijing, China. 1,1′-dioctadecyl-3,3,3′,3′-tetramethylindocarbocyanine perchlorate (Dil, cell membrane green fluorescent probe), Hoechst 33342, 4% paraformaldehyde fix solution, antifade mounting medium and Cell Counting Kit-8 (CCK-8), were all purchased from Biyuntian Biotechnology Co., Ltd., (sShanghai, China) 5(6)-aminofluorescein was bought from Nanjing Xinfan Biotechnology Co., Ltd., Nanjing, China. Polycarbonate film was bought from Whatman Company, City, UK. Dialysis bag (Cut-off molecular weight = 3500Da) was obtained from United States for carbonization. Fetal bovine serum (FBS), RPMI 1640 medium, and DMEM medium were ordered from the Shanghai Chenyi Biotechnology Company, Shanghai, China. Trypsin and penicillin-streptomycin were purchased Yingjie Jieji (Shanghai) Trading Co., Ltd., Shanghai, China.

2.2. Cells and Animals

RAW264.7 cells (mice macrophages) and HEK-293 cells (normal human embryonic kidney cells) were bought from the Shanghai Cell Bank of the Chinese Academy of Sciences, Shanghai, China, and cultured with DMEM complete medium. SMMC-7721 cells (human liver cancer cells) and NB4 cells (APL cells) were obtained from Shanghai Jihe Biotechnology Co., Ltd., Shanghai, China and cultured with RPMI 1640 complete medium. Culture of the cells was performed in an incubator kept at 5% CO_2 and 37 °C. The male BALB/c nude mice (SCXK 2017-0005) were obtained from Shanghai Slack Laboratory Animals Co., Ltd., Shanghai, China and kept in the Specific Pathogen Free (SPF) animal room of School of Pharmacy, Shanghai Jiao Tong University. Guidelines for care and use of laboratory animals of Shanghai Jiao Tong University were used to perform animal studies and these studies were duly approved by the animal ethics committee of Shanghai Jiao Tong University (No: A2019046, Date: 5 July 2019).

2.3. Preparation of SANs

SANs were prepared by the ion crosslinking method. In brief, 1.5 mL of 2 mg/mL $CaCl_2$ was slowly added into 10 mL of 0.3 mg/mL SA (pH = 5) under stirring. After sonication for 5 min at 250 W, 0.1 mL of 10 mg/mL F-188 was added and then stirred for 30 min. To obtain SANs, 0.2 mL of 8 mg/mL ATO was added and stirred for other 30 min.

2.4. Preparation of RBCM

RBCM was extracted by hypotonic rupture method. In brief, whole blood of SD rats (bought from Shanghai Jiesijie Experimental Animal Co., Ltd., Shanghai, China) was collected through abdominal aorta. The blood was centrifuged (2000 rpm, 5 min, 4 °C), to obtain red blood cells (RBCs), and then washed with 1× phosphate buffer saline (1× PBS) for 3 times. To collect RBCM, 900 μL EDTA (0.2 mM) was added to disrupt the RBCs, followed by centrifugation (13,200 rpm, 10 min, 4 °C), and the above steps were repeated until the supernatant turned colorless. The obtained RBCM was resuspended in EDTA, and then stored in −80 °C refrigerator.

2.5. Preparation of RSANs

The prepared RBCM was sonicated at 250 W for 3 min to obtain RVs, and then sequentially extruded through polycarbonate films of 800 nm, 400 nm, and 200 nm by LF-50 extruder (Avestin Inc, agented by Shanghai Narujie Biotechnology Co., LTD, Shanghai, China) for at least 15 times respectively. The solutions of RVs and SANs were mixed at a ratio of 1:8(*v*/*v*). The prepared mixture was then extruded through polycarbonate films of 400 nm and 200 nm at least ten times, respectively, to obtain RSANs.

2.6. Characterization and Stability Test

The particle size and polydispersity index (PDI) of SANs and RSANs were determined by Malvern Zetasizer (He-Ne, 4.0 Mw, λo = 633 nm, Marvin instruments Ltd., Marvin, United Kingdom. The stability test was investigated simultaneously. The particle size and PDI of SANs and RSANs were measured for 15 days consecutively at both 4 °C and 37 °C.

2.7. Morphological Observation

Transmission electron microscopy (TEM, Thermo Fisher Scientific, Shanghai, China) was performed to evaluate the morphology of RSANs. 10 µL sample solutions were dropped on carbon-supported copper, air-dried and then rinses with 10 µL ultrapure water. Drip 10 µL 2% phosphotungstic acid to stain the samples, and the excess dye was removed with filter paper from the edge. After baking 30 min under an infrared baking lamp, the morphology was observed by TEM.

2.8. Drug Loading Capacity and In Vitro Release Study

The release study of ATO encapsulated in nanoparticles was performed in 1×PBS (pH 7.4). 2 mL ATO solution (final concentration of 135.6 µg/ mL), SANs, and RSANs solution were placed into dialysis bags, respectively (M_w = 3500). The dialysis bags were fastened at both ends and maintained under sink conditions at 37 °C using 50 mL PBS, then magnetically stirred at 100 rpm. At time point of 3 h, encapsulation efficiency (EE) and drug loading capacity (DL) of the nanoparticles were analyzed by extracting 1 mL release solution. The optimal DL and EE were investigated by adding different concentrations of ATO. For the determination of in vitro release, 1 mL release sample was taken at 0.5, 1, 2, 4, 8, 12, 24, 36, 48, 72, and 84 h, and then replaced with 1 mL of fresh PBS. The aliquots were filtered through 0.22 µm microfiltration membrane and diluted up to 50 times. Inductively coupled plasma-atomic emission spectroscopy (ICP-AES) was used to detect the concentration of ATO. In addition, the following formulas were used to calculated DL and EE.

$$\text{Encapsulation Efficiency (\%)} = (M_1 - M_2)/M_1 \times 100\%$$

$$\text{Drug Loading Capacity (\%)} = (M_1 - M_2)/(M_1 - M_2 + M_3) \times 100\%$$

where M_1 means the total amount of used ATO, M_2 is the amount of ATO in the dialysis solution, and M_3 is the amount of used SA.

2.9. Hemolysis Test

Since it is administered by intravenous injection (iv), it is necessary to ensure that the nano-preparation will not cause hemolysis or cell aggregation. RBCs from the blood of SD rats were collected and normal saline was added to obtain a 2% (*v*/*v*) RBC suspension. The water, normal saline, RBC suspension, and RSANs were mixed into 12 tubes according to the ratio shown in Table 1. Tube 1 (RSANs) replaced with ultrapure water served as a positive control, while tube 2 (RSANs) replaced with saline as negative control. The results after incubation of 3 h and 24 h at 37 °C were recorded. Meanwhile, the absorbances of supernatant were determined at 3 h and 24 h through an ultraviolet spectrophotometer at the wavelength of 540 nm, and then the percentage of hemolysis was calculated according to formula:

$$\text{Percentage of Hemolysis (\%)} = (A_{\text{sample}} - A_{\text{tube 2}})/(A_{\text{tube 1}} - A_{\text{tube 2}})$$

Table 1. Hemolysis test of RSANs.

Tube No.	1	2	3	4	5	6	7	8	9	10	11	12
2% RBC Suspension (mL)	2.5	2.5	2.5	2.5	2.5	2.5	2.5	2.5	2.5	2.5	2.5	2.5
Saline (mL)	0	2.5	2.4	2.3	2.2	2.1	2.0	1.9	1.8	1.7	1.6	1.5
Ultrapure Water (mL)	2.5	0	0	0	0	0	0	0	0	0	0	0
Samples (mL)	0	0	0.1	0.2	0.3	0.4	0.5	0.6	0.7	0.8	0.9	1.0

2.10. Macrophage Uptake Study

The immune escape ability of nanoparticles was investigated through RAW 264.7 cells. SA was labeled with 5(6)-aminofluorescein, and SANs and RSANs were prepared with labeled SA. A 12-well plate (2×10^5 cells per well) was used for 24 h incubation. Serum-free medium was used to starve cells for 1 h and then replaced by complete medium containing SANs or RSANs (SA concentration was 50 μg/mL). A complete medium without nanoparticles was used as blank group. The cells were observed with laser scanning confocal microscope (LSCM) after incubation for 2 h. Before being photographed, the cells were fixed with 4% paraformaldehyde. The excitation wavelength of 5(6)-aminofluorescein was 493 nm. Also, three parallel groups were set up to detect the uptake quantitatively. After incubation with the drugs, each group was first digested with trypsin and then washed with PBS for 3 times, then resuspended in 500 μL PBS. The detection was performed through the FITC channel of a flow cytometer.

2.11. In Vitro Cellular Uptake

To investigate the uptake of nano-formulations by tumor cells and the structural integrity of NPs during this process, the test was implemented on NB4 and 7721 cells. The determination method was described in Section 2.10. For the qualitative detection, RSANs were prepared with labeled SANs and RVs (the RVs was labeled with Dil). The density of 7721 cells was 2×10^5 cells per well, and the cells were cultured for 24 h. The cells were then treated in the same way as Section 2.10. After an incubation of 2 h, Hoechst 33342 was used to stain the nuclei for 15 min, and then the uptakes were observed with LSCM. The cells were fixed with 4% paraformaldehyde before being photographed. Since NB4 cells is a kind of suspension cells, the density of inoculation was doubled to avoid loss during the experiment, and finally immobilized on a glass slide coated with polylysine. The other steps were the same as the 7721 cells. The excitation wavelengths of 5(6)-aminofluorescein, Dil and Hoechst 33342 were 493 nm, 549 nm, and 405 nm, respectively.

2.12. Cytotoxicity Test

To investigate the toxicity of blank carrier material, the prepared SA nanoparticles (SNs) and RBCM-SA nanoparticles (RSNs) without ATO were diluted with DMEM complete medium to obtain SA concentrations of 8, 12, 20, 30, 40, 50, 60 μg/mL respectively. To investigate the toxicity of the nanoparticles after drug loading, free ATO solution, SANs, and RSANs were diluted to obtain different concentrations of ATO (1, 2, 4, 6, 8, 10, 12, 20 μg/mL). The above groups were administration groups (AG). 100 μL 293 cells (5×10^4 cells/mL) were placed into a 96-well culture plate for 24 h. The nutrient medium was then replaced with 100 μL AG. At 24 h time interval, all the groups were cultured with 100 μL CCK-8 for 2 h. Only DMEM medium was set as a blank group (BG) and only 293 cells were set as a control group (CG). The optical density (OD) of samples were determined at 450 nm by a microplate reader and the following formula was used to calculate the cell viability.

$$\text{Cell viability (\%)} = (\text{ODAG} - \text{ODBG})/(\text{ODCG} - \text{ODBG}) \times 100\%$$

2.13. *In Vitro Efficacy Study*

NB4 and 7721 cells were selected to evaluate efficacy. 100 μL NB4 and 7721 cells (5×10^4 cells/mL) were inoculated in a 96-well culture plate overnight, respectively. Then free ATO solution, SANs, and RSANs were administered to the cells, respectively. The concentrations of ATO were diluted to 1, 2, 4, 6, 8, 10, 12, 20 μg/mL. After 24 h, 100 μL CCK-8 were administered to evaluate the cell viability. To further investigate the inhibitory effect, the concentration of the ATO of each group was fixed at 1 μg/mL. The OD was measured after incubation of 4, 8, 12, 24, 36, 48, and 60 h, and the cell viability was determined. The blank and control groups were same as mentioned in Section 2.12.

2.14. *In Vivo Toxicity and Safety Test*

Due to the potent toxicity of ATO, the administration concentration at a safe level should be determined at first. For 2 weeks, ATO with high (40 μg/mL), medium (20 μg/mL) and low concentration (10 μg/mL) was administered through the tail vein respectively once a week at a dose volume of 0.2 mL per mouse. The mental state and death of nude mice were registered during the period.

The aim of the safety trial was to investigate whether the continuous intravenous injection of RSANs would cause lesions on systemic, hematological, and major organs. Afterward, the healthy nude mice were divided into the saline group, ATO group, SANs group, and RSANs group. The drugs were administered once every 2 days at dose volume of 0.2 mL per mouse and repeated seven times. The weight of the mice was recorded at 1, 3, 5, 7, 9, 11, 13 days after administration. Meanwhile, the mental state and death of mice were observed during the process. 2 days after the last administration, orbital blood was collected into the tubes, pre-mixed with heparin sodium, and white blood cells (WBC), glutamate pyruvic transaminase (ALT), aspartate aminotransferase (AST) were analyzed. In addition, after the mice were sacrificed by CO_2 asphyxiation, the principal organs were excised. The viscera coefficients were calculated after weighing. Furthermore, the tissues were fixed with paraffin solution for immunohistochemical analysis (H & E) to examine its structure and morphology.

$$\text{Visceral coefficient} = \text{weight of organ/body weight}$$

2.15. *In Vivo Anti-Tumor Studies*

Male nude mice 6–8 weeks old with 7721 cells were used to investigate the anti-tumor effects of our nanoparticles. First we established xenograft tumor model. For the establishment of tumor model, 7721 cell suspension (2×10^7 cells / mL) in a volume of 0.2 mL was injected subcutaneously in the armpit of the upper limb. In addition, then the formula $\text{width}^2 \times \text{length}/2$ was used to calculate the tumor volume every other day. At tumor volume of 100–250 mm^3, then mice were grouped into four treatment groups ($n = 5$): (1) saline, (2) ATO, (3) SANs, and (4) RSANs. The drugs were administered intravenously at a dose of 1.3 μg/g every two days and repeated seven times. The CG was given normal saline for seven times. At the same time, the body weight and the tumor volume were recorded to evaluate for tumor inhibition efficacy as well as systemic toxicity. Two days after the last administration, the mice were sacrificed by CO_2 asphyxiation, and the tumors were excised from each animal. The tumors were rinsed with normal saline, dried, and photographed, and the average tumor weight of each group was determined.

2.16. *Statistics and Data Analysis*

Data expression was shown as ± SD of the mean. Significant differences between SANs and RSANs were analyzed by Tukey Kramer multiple comparison tests, using GraphPad Prism Software, v.6.01 (GraphPad Software, Inc.). Results with $p < 0.05$ were considered significant and very significant with $p < 0.01$.

3. Results and Discussion

3.1. Characterization and Stability Test

The average size of RSANs was found to be 163.2 ± 4.4 nm (Figure 2A), which increased by about 15 nm as compared to SANs (147.9 ± 5.1 nm). The increment was consistent with the thickness of RBCM [35]. The PDI of SANs and RSANs was 0.24 and 0.27, respectively (Figure 2B), which indicated that NPs were well dispersed. In addition, the particle size of the nanoparticles was stable under 200 nm despite a slight increase in 15 days (Figure 2C) under the storage condition of 37 °C. The PDI remained below 0.3 (Figure 2D). These results revealed that NPs were stable. The structure and distribution were also uniform after 15 days. However, while the nanoparticles were stored under condition of 4 °C, the particle size increased significantly within 4 days and the flocculation occurred because of the obvious drug precipitation. It suggested that 37 °C may be a better storage condition. Moreover, after incubating the nanoparticles with serum for three days, no significant precipitation was observed, indicating that the nanoparticles can remain stable under serum conditions as well.

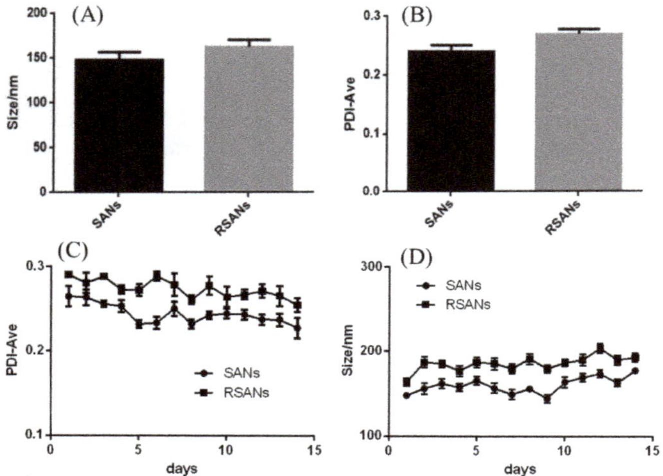

Figure 2. Particle size (**A**) and PDI (**B**) of SANs and RSANs. Changes of particle size (**D**) and PDI (**C**) within 15 days. Data are shown as ± SD (n = 3) of mean (n = 3).

3.2. Morphological Observation

TEM showed that RSANs (Figure 3) were mostly spherical and uniformly distributed. RSANs presented a complete core-shell structure, which demonstrated that the RBCM successfully wrapped SANs. The particle size measured by TEM was about 50–100 nm, and it was smaller than the results obtained by Zetasizer. The difference might be due to the elimination of the water film outside the nanoparticles after the samples were dried.

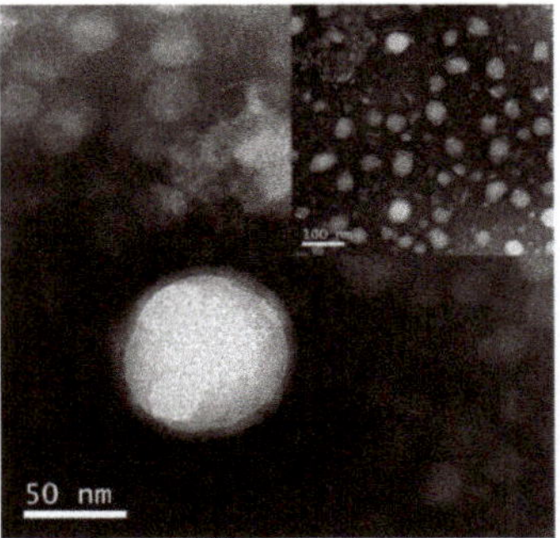

Figure 3. The TEM images of RSANs.

3.3. In Vitro Release Study

Analyzed by ICP-AES, the average EE and DL were 14.31% and 4.98%, respectively (Table S1). As shown in Figure 4, the release of 3 groups was fast for 4 h. For the ATO group, the highest burst release of 98.61% was observed within 2 h, and all ATO was released in 4 h. After 4 h, the release of the SANs was slightly slow down with a cumulative release rate of 91% in 12 h, and the drug was completely released at 36 h. The release of RSANs was further slowed down to 67% at 12 h. The drug was then gradually released until the cumulative release rate of 95% was achieved at 84 h. Resultantly, RSANs showed more significant sustained release than SANs. The result revealed that forming a physical barrier around the nanoparticles by RBCM can significantly control the drug release.

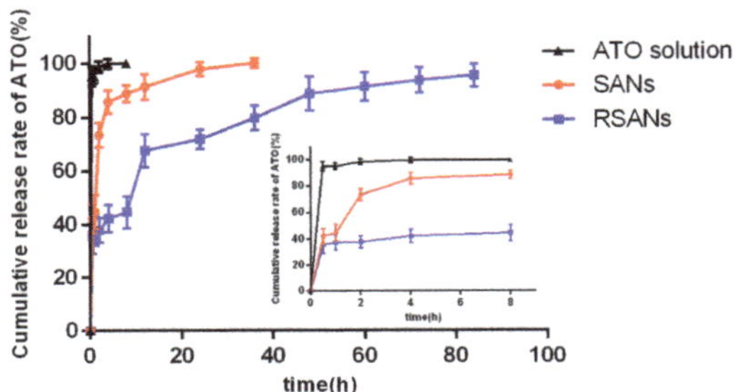

Figure 4. In vitro release curve of ATO, SANs and RSANs at 37 °C for 84 h. Data are shown as ± SD of the mean ($n = 3$).

3.4. Hemolysis Test

As RSANs are administered intravenously, the nanoparticles must not cause the rupture of RBCs. As shown in Figure 5A,B, the positive CG (Tube 1) was completely red and transparent with no cell. This confirmed that the RBCs were ruptured and causes hemolysis. For RSANs group, with the increase in concentration, RBCs sank, and the supernatants become colorless and clear from tube 1 to tube 7 similar to the negative CG (Tube 2). Further observation showed that no erythrocyte was aggregated under the inverted microscope. However, when the concentration increased gradually, slight hemolysis and hemagglutination phenomena were observed (tube 8 to tube 12). Results of hemolysis rate measurement were consistent with visual observation. The percentage of hemolysis was below 0.5% at high concentration, while it was below 0.01% at low concentration. The hemolysis test implied that RSANs were safe and suitable for the intravenous injection at final concentration of ATO lower than 0.0375 mg/mL.

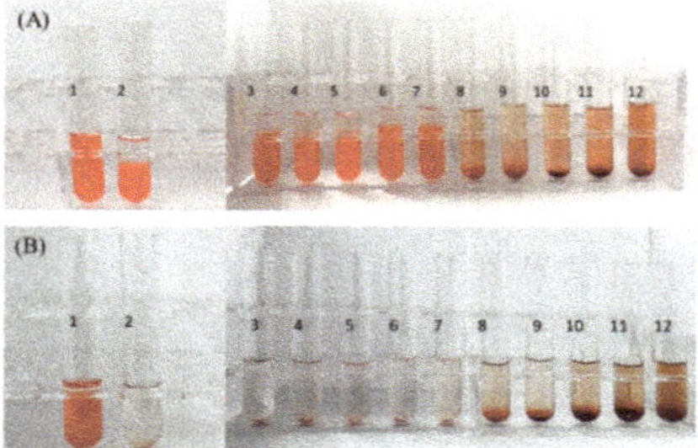

Figure 5. In vitro hemolysis results of RSANs after incubation of 3 h (**A**) and 24 h (**B**). Tube 1 was ultrapure water group, and tube 2 was the physiological saline group.

3.5. Macrophage Uptake Study

After the preparation of RSANs, it was necessary to confirm whether the biological activity of RBCM retains. The character of immune escape is one of the essential biological characteristics. The ability of RSANs for immune evasion was verified using confocal microscopy and flow cytometry by investigating the uptake between SANs and RSANs on RAW 264.7 cells. As shown in Figure 6A, the fluorescence intensity of RSANs groups observed by confocal microscopy was significantly weaker than that of SANs group. Meanwhile, the average fluorescence intensity of SANs and RSANs determined by flow cytometry were 4158.5 and 2045 respectively ($p < 0.01$), which was consistent with confocal results. The study confirmed that encapsulation by RBCM could help the nanoparticles to avoid the uptake of macrophages that can be related to special proteins such as CD47 [36,37]. It can be speculated that RSANs can also be prevented from being ingested by macrophages in vivo, and thus avoid the premature elimination of the drug.

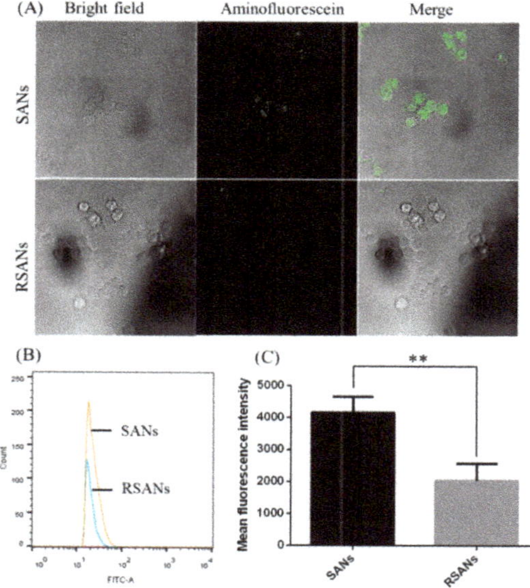

Figure 6. Cellular uptake of SANs and RSANs on RAW 264.7 macrophages. (**A**) Confocal image 630×. The SANs were labeled with 5(6)-Aminofluorescein (green), (**B**) Fluorescence intensity detection by flow cytometry. (**C**) Quantitative analysis of the fluorescence intensity. Data are shown as ± SD of the mean ($n = 3$). ** correspond to $p < 0.01$.

3.6. In Vitro Cellular Uptake

To further confirm the in vitro cellular uptake and structure of RSANS, RBCM, SANs core and cell nucleus of NB4 cells as well as 7721 cells were labeled with Dil, 5(6)-Aminofluorescein and Hoechst 33342, respectively. As shown in Figure 7A,D, the red, green, and blue fluorescence represent the stained RBCM, SANs, and nuclei, respectively. After incubation of NB4 cells and 7721 cells with SANs and RSANs, it was detected that both red and green fluorescence overlapped around the nuclei. This indicated that both SANs and RSANs could be taken up by NB4 cells as well as 7721 cells. In addition, it could be preliminarily speculated the integrity of the basic shell-core structure was also maintained during the process. According to flow cytometry results, the average fluorescence intensity of SANs and RSANs ingested by NB4 cells were 813 and 941 (Figure 7B,C), while those of 7721 cells were 5111 and 5211 (Figure 7E,F) respectively. Both confocal and flow cytometry results showed that the uptake of RSANs was slightly increased without any significant difference. It can be predicted that the change of the negative charge on the surface of the nanoparticles after being coated by RBCM may cause the slight difference, which changed the repulsive force between membranes.

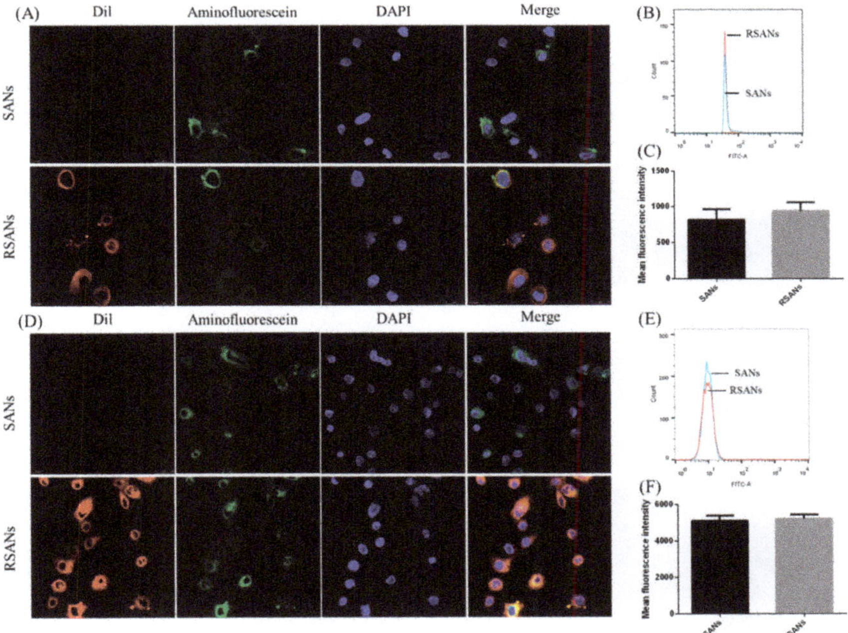

Figure 7. Cellular uptake of SANs and RSANs on NB4 cells and 7721 cells. (**A**,**D**) Confocal images 630×. The nucleus of cells was labeled by Hoechst 33342 (blue), SANs were labeled by 5(6)-Aminofluorescein (green) and RBCM were labeled by Dil (red). (**B**,**E**) Fluorescence intensity detection by flow cytometry. (**C**,**F**) Quantitative analysis of the fluorescence intensity. Data are shown as ± SD of the mean (n = 3).

3.7. ATO Nanoparticles Cytotoxicity

In vitro cytotoxicity assays were performed to evaluate the safety of nanoparticles initially. As shown in Figure 8A, the cell viability of SNs and RSNs were both higher than 95% at SA concentration of 8–60 μg/mL. It indicated that SA and RBCM were not significantly toxic and had excellent biocompatibility. Free ATO revealed potent cytotoxicity at different concentration; however, the toxicity was reduced significantly after being encapsulated in nanoparticles (Figure 8B). It can be predicted that the nano-delivery system newly designed can remarkably improve drug safety during medical treatment.

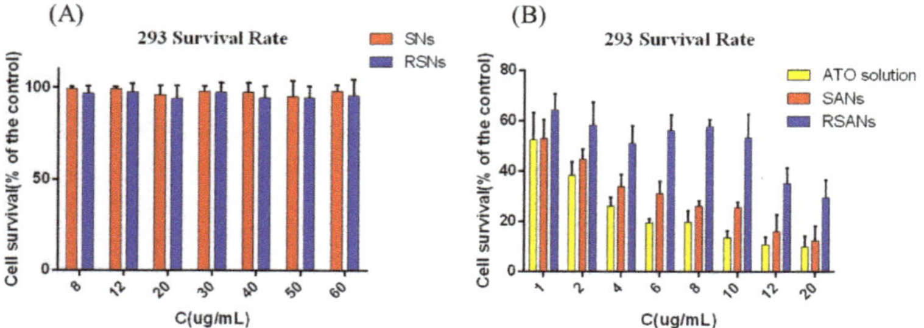

Figure 8. Cell survival of 293 cells after administration with different concentrations of (**A**) SNs and RSNs, (**B**) ATO, SANs and RSANs for 24 h. Data are shown as ± SD of the mean ($n = 3$).

3.8. *In Vitro Efficacy Study*

Figure 9A,B showed that the inhibitory effects of free ATO, SANs, and RSANs on NB4 cells and 7721cells were all dose-dependent. With the increase of ATO concentration, the inhibition was significantly enhanced after 24 h. From Figure 9C,D, potent inhibition was observed at a lower concentration of 1 μg/mL. NB4 cells and 7721 cells were almost completely inhibited at 60 h and 72 h, respectively. At the same time point, the inhibitory intensity of the free ATO group, SANs group, and RSANs group decreased. However, because of the closed system of culture plate and sustained release nanoparticles, drugs in SANs and RSANs could be gradually released over time and eventually reaching the same effect as the free group. According to the results, it can be speculated that RSANs can avoid rapid elimination by the immune system in vivo with the inhibitory effect maintained.

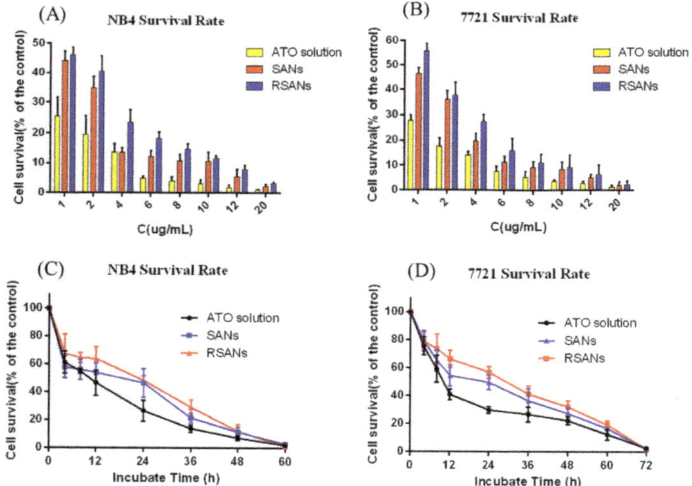

Figure 9. The inhibition effects on NB4 and 7721 cells after administration with different formulations. Cell survival of NB4 (**A**) and 7721 (**B**) cells after administration with different concentrations of ATO, SANs and RSANs for 24 h. Cell survival of NB4 (**C**) and 7721 (**D**) cells after administration with multiple groups at the ATO concentration of 1 μg/mL for 72 h. Data are shown as ± SD of the mean ($n = 3$).

3.9. In Vivo Toxicity and Safety Test

In acute toxicity test, some mice in the high concentration group died, and the surviving mice were inferior to other groups in terms of mental vitality, drinking, and feeding. During continuous administration, the average body weight of the mice in each group showed an increasing trend with no abnormal change, as shown in Figure 10B,C. It indicated that all the agents have no significant systemic toxicity. All hematological analysis results (Figure 10D–F) were within the normal range except for a slight decrease in the number of WBC in the free ATO group. Analysis of the main organ tissue sections after H & E staining showed that the SANs group and RSANs group were similar to the saline group, while the ATO group developed certain lesions (Figure 10A). The specific manifestation including a large number of cardiomyocytes was cytoplasmic loosely stained. Inflammatory infiltration was observed around the local portal area of the liver. The white pulp of spleen part conglutinated to each other with irregular shape and a small number of apoptotic bodies were seen. Moreover, local interstitial congestion could be observed in the kidney. According to the above results, it can be speculated that free ATO group can cause chronic toxicity if administered continuously in this concentration as compared to SANs group and RSANs group. The transient blood concentration of the nano-groups was reduced due to the sustained release, with a lower in vivo toxicity.

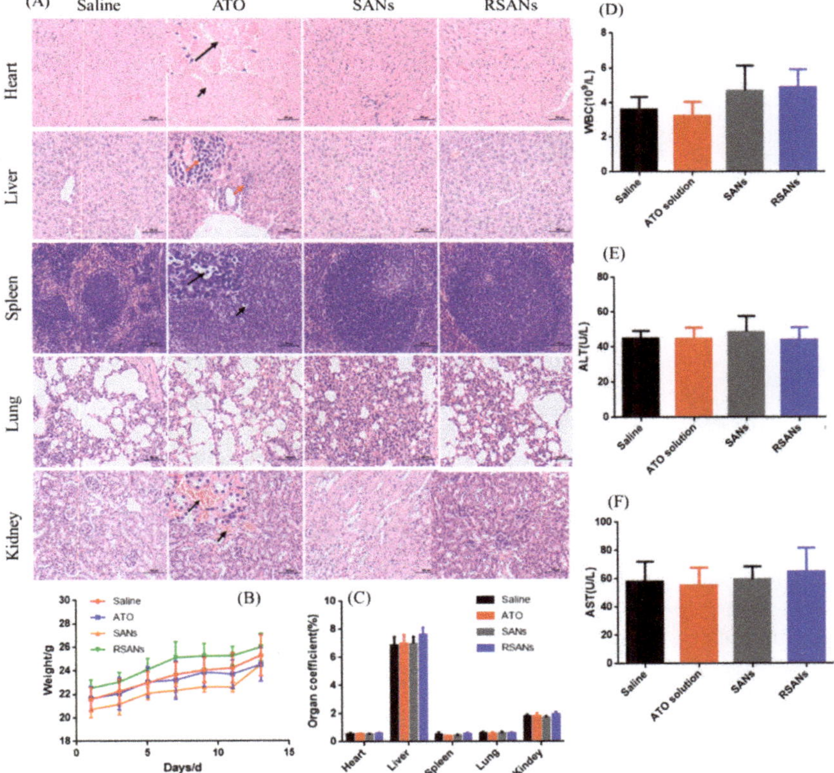

Figure 10. In vivo toxicity and safety evaluation. (**A**) H & E staining of primary organs under 100 μm. (**B**) Weight changes of nude mice during the experiment. (**C**) Organ coefficients, (**D**) Number of white blood cells, (**E**) ALT, and (**F**) AST of each group at the end of the experiment. Data are shown as ± SD of the mean (n = 5).

3.10. In Vivo Anti-Tumor Study

7221 tumor bearing nude mice were used to determine the anti-tumor efficacy of formulations. Figure 11 shows the results of different anti-tumor studies. Figure 11A showed the change in tumor volume during administration. In the saline group, the volume was gradually increased while the change trends of ATO group, SANs group, and RSANs group were first increased and then decreased ($p < 0.01$). The tumor mass and size at the end of the study were monitored as Figure 11C,D. According to the results, it could be speculated that the free ATO was quickly cleared in the body, the amount of drug reaching the tumor site was less than that of the other two groups, and that caused the lowest tumor inhibition efficacy. Over the whole treatment period, SANs group and RSANs group can achieve better tumor inhibition effects with the drug gradually released from the nanoparticles. At the same time, because of the wrapping of RBCM, RSANs were less likely to be cleared by the immune system than SANs, allowing more drugs to reach the tumor site. Figure 11B showed changes of body weight. By the end of the study, body weight of the saline group increased significantly, while the other three groups showed little variation, which was consistent with the results of the in vivo toxicity and safety test. It was further supported that the safety and low toxicity of SANs and RSANs, and the DDS can still improve efficacy of ATO.

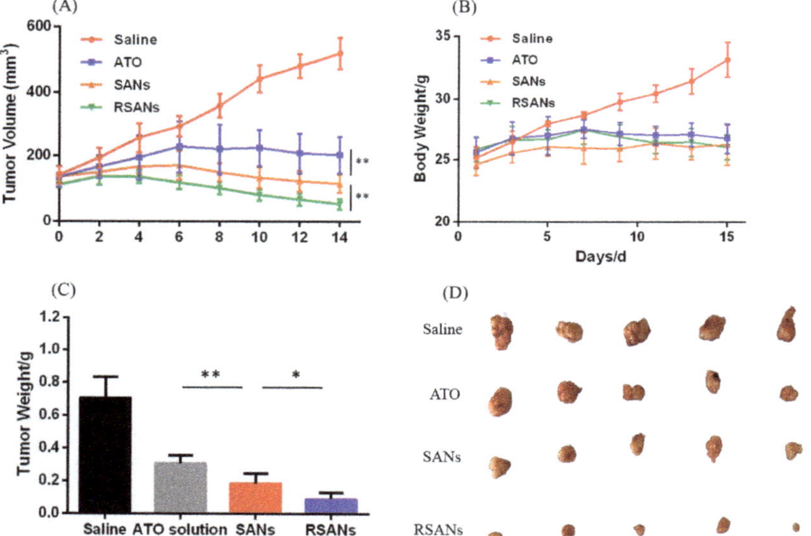

Figure 11. In vivo anti-tumor efficacy of formulation in the 7721 tumor bearing nude mice study. Changes of (**A**) tumor volume and (**B**) body weight during the period of treatment. (**C**) Average tumor weight after treatment. (**D**) Photograph of tumors collected after treatment. Data are shown as ± SD of the mean ($n = 5$). ** correspond to $p < 0.01$, * correspond to $p < 0.05$.

4. Conclusions

The objective of this project was to realize sustained release of ATO and ensure the safety and efficacy of the DDS. SANs were prepared by coating ATO with SA, and then RSANs with a shell-core bilayer structure was obtained by wrapping the RBCM on the surface of the SANs. The nanoparticles were homogeneous with spherical structures and stable for 15 days with an average EE and DL of 14.31% and 4.98%, respectively. Compared to free ATO, the SANs and RSANs showed excellent sustained release in vitro for 3 days. Generally, biological characteristics of RBCM could be used to avoid the recognition of macrophages. Even when ingested by NB4 cells and 7721 cells, RSANs still could maintain its nuclear-shell structure. At the same time, in vitro safety test showed the safety of carrier materials and without observable toxicity to 293 cells or hemolysis and aggregation of red blood cells. Free drug can kill nearly half of cells at a low concentration, while RSANs can significantly reduce the toxicity of ATO. After continuous administration of ATO formulations through IV tail injection, the results revealed that the SANs and RSANs had no significant systemic, blood, and organ toxicity. Moreover, the RSANs could maintain inhibition of NB4 cells and 7721 cells. After in vivo administration, RSANs can avoid rapid clearance and exert their sustained release properties, thereby enough drugs can be transported to the treatment site and achieve the effect. This may allow RSANs greater advantages in the treatment of tumors. Therefore, it could be concluded that the Nano-drug delivery system can decrease the toxicity of ATO with high safety and the potential for treating APL as well as anti-hepatocarcinoma. Meanwhile, RSANs can prolong the duration of ATO in vivo, thus reducing administration times and enhancing patient compliance. In consequence, the DDS can be developed into a safe and sustained release delivery system for ATO.

Supplementary Materials: The following are available online at http://www.mdpi.com/1999-4923/12/1/21/s1, Table S1: Investigation of encapsulation efficiency and drug loading capacity.

Author Contributions: Y.L. (Yumei Lian) participated in the experimental study, including the investigation, methodology, data curation, formal analysis of data, research administration, and drafting the original manuscript. X.W. participated in research administration and revising the manuscript. P.G. participated in research administration. Y.L. (Yichen Li) and F.R. participated in revising the manuscript. J.S. and M.Q., as corresponding authors, acquired the funding, participated in the study conceptualization, data analysis, reviewing and editing the manuscript, and finalized this manuscript. All authors have read and agreed to the published version of the manuscript.

Funding: This research was funded by the National Natural Science Foundation of China (No. 81573617 & 81973487), National Major Science and Technology Projects of China (2018ZX09711003-008-002), and Shanghai Association for Science and Technology (18ZR1419700).

Conflicts of Interest: The authors declare no conflict of interest.

References

1. Evens, A.M.; Tallman, M.S.; Gartenhaus, R.B. The potential of arsenic trioxide in the treatment of malignant disease: Past, present, and future. *Leuk. Res.* **2004**, *28*, 891–900. [CrossRef] [PubMed]
2. Antman, K.H. Introduction: The history of arsenic trioxide in cancer therapy. *Oncologist* **2001**, *6* (Suppl. 2), 1–2. [CrossRef] [PubMed]
3. Zhang, Q.; Vakili, M.R.; Li, X.F.; Lavasanifar, A.; Le, X.C. Polymeric micelles for GSH-triggered delivery of arsenic species to cancer cells. *Biomaterials* **2014**, *35*, 7088–7100. [CrossRef] [PubMed]
4. Stevens, J.J.; Graham, B.; Dugo, E.; Berhaneselassie-Sumner, B.; Ndebele, K.; Tchounwou, P.B. Arsenic Trioxide Induces Apoptosis via Specific Signaling Pathways in HT-29 Colon Cancer Cells. *J. Cancer Sci. Ther.* **2017**, *9*, 298–306. [CrossRef]
5. Akhtar, A.; Wang, S.X.; Ghali, L.; Bell, C.; Wen, X. Recent advances in arsenic trioxide encapsulated nanoparticles as drug delivery agents to solid cancers. *J. Biomed. Res.* **2017**, *31*, 177–188.
6. Tchounwou, P.B.; Wilson, B.A.; Ishaque, A.B.; Schneider, J. Atrazine potentiation of arsenic trioxide-induced cytotoxicity and gene expression in human liver carcinoma cells (HepG2). *Mol. Cell. Biochem.* **2001**, *222*, 49–59. [CrossRef]
7. Cai, B.; Wei, W.; Liu, A.; Jin, Z.; Lin, S. In vivo inhibitory effect of arsenic trioxide on transplanted tumor of pancreatic cancer BXPC-3 cells. *Chin. Tradit. Herb. Drugs* **2010**, *41*, 90–93.
8. Murgo, A.J. Clinical Trials of Arsenic Trioxide in Hematologic and Solid Tumors: Overview of the National Cancer Institute Cooperative Research and Development Studies. *Oncologist* **2001**, *6* (Suppl. 2), 22–28. [CrossRef]
9. Kritharis, A.; Bradley, T.P.; Budman, D.R. The evolving use of arsenic in pharmacotherapy of malignant disease. *Ann. Hematol.* **2013**, *92*, 719–730. [CrossRef]

10. Huang, A.H.; Han, S.P.; Lu, Y.P.; Ma, R.; Zheng, H.S.; Li, F.Z. Preparation and in vitro evaluation of arsenic trioxide glioma targeting drug delivery system loaded by PAMAM dendrimers co-modified with RGDyC and PEG. *Chin. J. Tradit. Chin. Med.* **2018**, *43*, 1618–1625.
11. Fu, X.; Liang, Q.R.; Luo, R.G.; Li, Y.S.; Xiao, X.P.; Yu, L.L.; Shan, W.Z.; Fan, G.Q.; Tang, Q. An arsenic trioxide nanoparticle prodrug (ATONP) potentiates a therapeutic effect on an aggressive hepatocellular carcinoma model via enhancement of intratumoral arsenic accumulation and disturbance of the tumor microenvironment. *J. Mater. Chem. B* **2019**, *7*, 3088–3099. [CrossRef]
12. Swindell, E.P.; Hankins, P.L.; Haimei, C.; Miodragovi, D.U.; O'Halloran, T.V. Anticancer activity of small-molecule and nanoparticulate arsenic(III) complexes. *Inorg. Chem.* **2013**, *52*, 12292–12304. [CrossRef] [PubMed]
13. Huang, S.Y.; Chang, C.S.; Tang, J.L.; Tien, H.F.; Kuo, T.L.; Huang, S.F.; Yao, Y.T.; Chou, W.C.; Chung, C.Y.; Wang, C.H.; et al. Acute and chronic arsenic poisoning associated with treatment of acute promyelocytic leukaemia. *Br. J. Haematol.* **1998**, *103*, 1092–1095. [CrossRef] [PubMed]
14. Hu, X.; Ma, L.; Hu, N. Ailing No. I in Treating 62 Cases of Acute Promyelocytic Leukemia. *Chin. J. Integr. Tradit. West. Med.* **1999**, *19*, 473–476.
15. Sarbari, A.; Sanjeeb, K.S. PLGA nanoparticles containing various anticancer agents and tumour delivery by EPR effect. *Adv. Drug Deliv. Rev.* **2011**, *63*, 170–183.
16. Sheldon, L.A. Inhibition of E2F1 Activity and Cell Cycle Progression by Arsenic via Retinoblastoma Protein. *Cell Cycle* **2017**, *16*, 2058–2072. [CrossRef]
17. Jadhav, V.; Ray, P.; Sachdeva, G.; Bhatt, P. Biocompatible arsenic trioxide nanoparticles induce cell cycle arrest by p21 WAF1/CIP1 expression via epigenetic remodeling in LNCaP and PC3 cell lines. *Life Sci.* **2016**, *148*, 41–52. [CrossRef]
18. Kopel, P.; Wawrzak, D.; Milosavljevic, V.; Moulick, A.; Vaculovicova, M.; Kizek, R. Nanotransporters for Anticancer Drug Delivery. *Nanotechnol. Res. J.* **2015**, *2*, 32–38.
19. Wang, X.; Li, D.; Ghali, L.; Xia, R.; Munoz, L.P.; Garelick, H.; Bell, C.; Wen, X. Therapeutic Potential of Delivering Arsenic Trioxide into HPV-Infected Cervical Cancer Cells Using Liposomal Nanotechnology. *Nanoscale Res. Lett.* **2016**, *11*, 94. [CrossRef]
20. Xiao, X.; Liu, Y.; Guo, M.; Fei, W.; Zheng, H.; Zhang, R.; Zhang, Y.; Wei, Y.; Zheng, G.; Li, F. pH-triggered sustained release of arsenic trioxide by polyacrylic acid capped mesoporous silica nanoparticles for solid tumor treatment in vitro and in vivo. *J. Biomater. Appl.* **2016**, *31*, 23. [CrossRef]
21. Otto, D.P.; Otto, A.; de Villiers, M.M. Differences in physicochemical properties to consider in the design, evaluation and choice between microparticles and nanoparticles for drug delivery. *Expert Opin. Drug Deliv.* **2015**, *12*, 763–777. [CrossRef] [PubMed]
22. Chen, L.; Luo, Z.G.; He, X.W. Application Research of Sodium Alginate in Medical Engineering. *Chin. Med. Equip. J.* **2008**, *29*, 33–35.
23. Zhang, X.H.; Cui, Y.D. Study on the Properties and Structure of Sodium Alginate Superabsorbent. *Polym. Mater. Sci. Eng.* **2006**, *22*, 91–94. [CrossRef]
24. Zhong, J.J.; Wang, D.K.; Zhang, C.X.; Gao, H.; Zhang, X. Pharmaceutical significance of sodium alginate. *Chin. J. New Drugs* **2007**, *16*, 591–594.
25. Li, Z.L.; Wang, J.; Chen, L.M.; Zhang, W.C.; Zhao, C.C. Preparation and properties of ceftriaxone chitosan-sodium alginate (calcium) microspheres. *Chin. Antibiot. J.* **2008**, *33*, 355–358.
26. Daemi, H.; Barikani, M. Synthesis and characterization of calcium alginate nanoparticles, sodium homopolymannuronate salt and its calcium nanoparticles. *Sci. Iran.* **2012**, *19*, 2023–2028. [CrossRef]
27. Kumar, S.; Bhanjana, G.; Sharma, A.; Sidhu, M.C.; Dilbaghi, N. Synthesis, characterization and on field evaluation of pesticide loaded sodium alginate nanoparticles. *Carbohydr. Polym.* **2014**, *101*, 1061–1067. [CrossRef]
28. Sun, Y.; Su, J.; Liu, G.; Chen, J.; Zhang, X.; Zhang, R.; Jiang, M.; Qiu, M. Advances of blood cell-based drug delivery systems. *Eur. J. Pharm. Sci.* **2017**, *96*, 115–128. [CrossRef]
29. Fang, R.H.; Hu, C.M.; Zhang, L. Nanoparticles disguised as red blood cells to evade the immune system. *Expert Opin. Biol. Ther.* **2012**, *12*, 385–389. [CrossRef]
30. Tan, S.; Wu, T.; Zhang, D.; Zhang, Z. Cell or Cell Membrane-Based Drug Delivery Systems. *Theranostics* **2015**, *5*, 863–881. [CrossRef]

31. Hu, C.M.; Zhang, L.; Aryal, S.; Cheung, C.; Fang, R.H.; Zhang, L. Erythrocyte membrane-camouflaged polymeric nanoparticles as a biomimetic delivery platform. *Proc. Natl. Acad. Sci. USA* **2011**, *27*, 10980–10985. [CrossRef] [PubMed]
32. Su, J.H.; Sun, Q.; Meng, Q.; Yin, P.; Zhang, Z.; Zhang, H.Y.; Li, Y. Bioinspired Nanoparticles with NIR-Controlled Drug Release for Synergetic Chemophotothermal Therapy of Metastatic Breast Cancer. *Adv. Funct. Mater.* **2016**, *26*, 7495–7506. [CrossRef]
33. Su, J.; Sun, H.; Meng, Q.; Yin, Q.; Tang, S.; Zhang, P.; Chen, Y.; Zhang, Z.; Yu, H.; Li, Y. Long Circulation Red-Blood-Cell-Mimetic Nanoparticles with Peptide-Enhanced Tumor Penetration for Simultaneously Inhibiting Growth and Lung Metastasis of Breast Cancer. *Adv. Funct. Mater.* **2016**, *26*, 1243–1252. [CrossRef]
34. Su, J.; Sun, H.; Meng, Q.; Zhang, P.; Yin, Q.; Li, Y. Enhanced Blood Suspensibility and Laser-Activated Tumor-specific Drug Release of Theranostic Mesoporous Silica Nanoparticles by Functionalizing with Erythrocyte Membranes. *Theranostics* **2017**, *7*, 523–537. [CrossRef] [PubMed]
35. Hochmuth, R.M.; Mohandas, N.; Blackshear, P.L. Measurement of the elastic modulus for red cell membrane using a fluid mechanical technique. *Biophys. J.* **1973**, *13*, 747–762. [CrossRef]
36. Oldenborg, P.A.; Zheleznyak, A.; Fang, Y.F.; Lagenaur, C.F.; Gresham, H.D.; Lindberg, F.P. Role of CD47 as a marker of self on red blood cells. *Science* **2000**, *288*, 2051–2054. [CrossRef]
37. Weiwei, G.; Hu, C.M.J.; Fang, R.H.; Luk, B.T.; Jing, S.; Liangfang, Z. Surface functionalization of gold nanoparticles with red blood cell membranes. *Adv. Mater.* **2013**, *25*, 3549–3553.

Article

Development and Evaluation of a Reconstitutable Dry Suspension Containing Isoniazid for Flexible Pediatric Dosing

Oluwatoyin A. Adeleke [1,2,*], Rose K. Hayeshi [2,3] and Hajierah Davids [4]

1 Division of Pharmaceutical Sciences, School of Pharmacy, Sefako Makgatho Health Sciences University, Pretoria 0208, South Africa
2 Council for Scientific and Industrial Research, Pretoria 0001, South Africa
3 DST/NWU Preclinical Drug Development Platform, Faculty of Health Sciences, North-West University, Potchefstroom 2531, South Africa; rose.hayeshi@nwu.ac.za
4 Department of Physiology, Nelson Mandela University, Port Elizabeth 6031, South Africa; Hajierah.Davids@mandela.ac.za
* Correspondence: Oluwatoyin.adeleke@fulbrightmail.org or oluwatoyin.adeleke@smu.ac.za; Tel.: +27-12-521-4111; Fax: +27-12-560-0086

Received: 31 December 2019; Accepted: 15 March 2020; Published: 23 March 2020

Abstract: Tuberculosis (TB) is a major cause of childhood death. Despite the startling statistics, it is neglected globally as evidenced by treatment and clinical care schemes, mostly extrapolated from studies in adults. The objective of this study was to formulate and evaluate a reconstitutable dry suspension (RDS) containing isoniazid, a first-line anti-tubercular agent used in the treatment and prevention of TB infection in both children and adults. The RDS formulation was prepared by direct dispersion emulsification of an aqueous-lipid particulate interphase coupled with lyophilization and dry milling. The RDS appeared as a cream-white free-flowing powder with a semi-crystalline and microparticulate nature. Isoniazid release was characterized with an initial burst up to 5 min followed by a cumulative release of 67.88% ± 1.88% (pH 1.2), 60.18% ± 3.33% (pH 6.8), and 49.36% ± 2.83% (pH 7.4) over 2 h. An extended release at pH 7.4 and 100% drug liberation was achieved within 300 min. The generated release profile best fitted the zero order kinetics (R^2 = 0.976). RDS was re-dispersible and remained stable in the dried and reconstituted states over 4 months and 11 days respectively, under common storage conditions.

Keywords: pediatric drug delivery; tuberculosis; reconstitutable dry suspension; isoniazid; polymer-lipid; microparticulate; direct emulsification

1. Introduction

Tuberculosis (TB) remains a major global health problem present in every country in the world, regardless of the availability of standard treatment guidelines [1,2]. It is the leading cause of death from a single infectious agent, ranking above the Human Immunodeficiency Virus (HIV) with about 10 million new active infections and 1.5–2 million fatalities annually [1,3–5]. TB is an airborne, highly contagious disease often spread by coughing and sneezing. It is caused by strains of bacteria known as *Mycobacterium tuberculosis* (Mtb), which primarily infects the lungs (pulmonary TB) and, occasionally, other body parts (extra-pulmonary TB) [1,5–7]. TB has been identified as a key cause of economic devastation, revolving poverty and illness that has entrapped families, societies, and even entire countries, with women, children, and HIV patients being the most vulnerable [2].

Tuberculosis is a major cause of childhood death. The World Health Organization (WHO) recently estimated that 10%–11% of the global population infected with TB are children, with about 233,000 childhood deaths each year [1,3,6]. The TB mortality rate is 70% higher in children under the age of five than it is in adults in high burden areas [1,8]. Research indicates that children serve as reservoirs for active TB infection later in life, evidenced by the fact that globally, about 67 million children under the age of 15 have latent TB [1,9]. Latent TB infection is known as a state of persistent immune responses to stimulation by Mtb antigens with no evidence of clinical manifestations associated with active infection or symptoms of illness. Nevertheless, latent TB can develop into full-blown, active infection later [1].

Thus far, the greatest challenge to the successful treatment of TB in children is the significant shortage of efficient pediatric pharmaceutical formulations [1,10–12]. Despite the alarming statistics reported on the number of active TB cases in children, it is indeed shocking to note that to date, childhood TB has been generally neglected worldwide, evidenced by treatment and clinical care schemes mostly extrapolated from studies in adults [1,13–15]. Over the years, children have been largely excluded from clinical trials resulting in weak evidence-based treatment of pediatric TB infection. With the shortage of suitable child-friendly anti-TB pharmaceutical formulations, it is common global practice to split adult fixed-dose combination (FDC) preparations: (i) into fractions; (ii) crushed to be taken with food, milk, and other liquids; or (iii) extemporaneous compounding to allow for easy use as needed per child. These practices can lead to dose inaccuracies, reduced active drug potencies, impaired dosage stability, irregular bioavailability, and poor compliance [14–16]. Consequently, there is an urgent need for innovative treatment strategies that can contribute towards combating the TB epidemic in pediatric patients.

Isoniazid is the most widely used first-line anti-tubercular agent for the treatment and prevention of TB infection in both children and adults. It is a drug of choice as it is bactericidal, easily administered, inexpensive, and relatively non-toxic in children. Isoniazid is almost completely absorbed from the gastrointestinal tract and penetrates all body fluid cavities, in which drug levels are similar to serum levels [12,15]. Although subject to considerable hepatic metabolism (or first-pass effect) after oral dosing, it reaches concentrations well above the minimum inhibitory levels of Mtb in most tissues and TB lesions when given in standard doses [17]. Isoniazid is considered a class III drug according to the Biopharmaceutics Classification System (BCS) [18], meaning that it is highly hydrophilic (aqueous solubility = 125 mg/mL at 25 °C) [18,19] but not very permeable (log P = −0.64 at 25 °C) [19]. It is weakly basic, crystalline in nature, and does not display polymorphism [20]. The aim of this study was, therefore, to develop and characterize a microparticulate reconstitutable dry suspension containing isoniazid as a model drug. To date, there is not much detailed in literature on the use of anti-tubercular micro-suspensions for pediatric TB treatment or prophylaxis. Conventional dry suspensions are powder mixtures that require the addition of water at the time of dispensing. They are often widely acceptable, intended for oral administration, and usually choice alternatives when drug stability is a major concern. Dry suspensions are easy to use by any age group (particularly children) and, therefore, enhance patient compliance [13,21]. In this study, microparticulate dry suspensions as micro-structuring remains an ideal way of manufacturing highly efficient, rate-modulated pharmaceutical formulations that are beneficial to patients. Generally, microparticulate drug carriers are known to have high stability with excellent drug loading capacities for hydrophobic and hydrophilic drug moieties, enhanced bioavailability, and decreased toxicity [22]. The reconstitutable dry suspension (RDS) was prepared by blending liquid and solid interphases of drug and excipient into a homogenous mix that was lyophilized and pulverized to produce an isoniazid-loaded free-flowing powdery formulation. Formulation evaluation involved zeta potential and polydispersity index analyses, particle sizing, drug loading, dissolution testing, thermal behavior, structural transitions, surface morphology, crystallinity determinations, cytotoxicity, hydro- and environmental- stability testing.

2. Materials and Methods

2.1. Materials

Polyethylene glycol, poly (vinyl alcohol) 87%–89% hydrolyzed, coconut oil, ethylcellulose, hydrochloric acid, anhydrous sodium phosphate dibasic, 3-(4,5-dimethylthiazol-2-yl)-2,5-diphenyltetrazolium bromide (MTT), Trypsin—EDTA, trypan blue, dimethylsulfoxide (DMSO), camptothecin, potassium dihydrogen phosphate monobasic buffer salt, isoniazid, and phosphate buffered saline powder were purchased from Sigma Chemical Company (St. Louis, MO, USA). D-fructose was obtained from Merck Chemicals (Darmstadt, Germany). The human breast cancer (MCF-7) cell line was purchased from American Type Culture Collection (ATCC, Manassas, VA, USA). Dulbecco's Modified Eagle's Medium (DMEM) and fetal bovine serum (FBS) were procured from BD Biosciences (San Jose, CA, USA). All other reagents utilized were of analytical grade and used as received from the supplier.

2.2. Formulation of the Reconstitutable Dry Suspension Employing a Direct Dispersion Emulsification Method

The RDS was prepared using a direct dispersion emulsification technique coupled with lyophilization and dry milling. This involved emulsification of the aqueous phase, which contained 2%$^{w}/_{w}$ polyethylene glycol, 9%$^{w}/_{w}$ poly (vinyl alcohol), 4%$^{w}/_{w}$ D-fructose, 10%$^{w}/_{w}$ isoniazid dispersed in water with the non-aqueous phase made up of 3%$^{w}/_{w}$ ethylcellulose in coconut oil under constant mechanical blending (Silverson Machines, Inc., East Longmeadow, MA, USA) at 6000 rpm over 5–10 min until a monophasic emulsion was produced. With D-fructose as a lyoprotectant and sweetener, the mono-phased emulsion was exposed to liquid nitrogen for 15 min until completely frozen. Thereafter, the frozen sample was lyophilized (Benchtop Pro Freeze Dryer, VirTis, SP Scientific, Stone Ridge, NY, USA) at a temperature of −60 ± 2 °C and pressure of 124 ± 2 mTorr over 96 h to produce a solid lyophilisate that was then dry milled using a laboratory scale grinder (Kinematica GMBH, Eschbach, Germany). The RDS was stored away in airtight, opaque containers until further testing.

2.3. Formulation Yield

The formulation yield was calculated by relating the total formulation weight (actual yield) to the mass of the component excipients used in the preparation of the formulation (theoretical yield). The constituting excipient plus the model drug and produced RDS formulation were weighed as separate entities using a laboratory scale balance (Kern EG 620-3NM, Kern and Sohn, GmbH, Balingen, Germany). The percentage yield was calculated using Equation (1) below:

$$\%\text{Yield} = \frac{\text{Actual yield (g)}}{\text{Theoretical yield (g)}} \times 100 \quad (1)$$

2.4. Physicochemical Characterization of the Reconstitutable Dry Suspension

2.4.1. Particle Size, Polydispersity Index, and Zeta Potential

Measurement of particle size (PS) and polydispersity index (PDI) was based on the principle of dynamic light scattering using the Nano series Zetasizer (Malvern Instruments Ltd, Malvern, UK). For each quantification, RDS samples were re-dispersed in deionized water, appropriately diluted, and sonicated (Table-type Supersonic Cleaner KQ118, Nanjing T-Bota Scietech Instruments and Equipment, Co, Ltd., Nanjing, China) for 15 min at 37 ± 0.5 °C. All measurements were performed as three independent replicates with 10 readings per sample at a measurement angle of 173° and temperature of 25 °C. Zeta potential (ZP) is an indicator of particle surface charge, which determines particle stability in dispersions [23,24]. It was computed based on the Smoluchowski equation [25] using the Nano series Zetasizer with RDS samples dispersed in deionized water and placed in disposable folded capillary zeta cells maintained at 25 °C. For statistical relevance, samples were measured in triplicate with 20 runs per measurement cycle.

2.4.2. Visualization of Surface Morphology

The surface microstructure of the RDS was observed under a JEOL Transmission Electron Microscope (TEM) (JEM-2100 LaB6 200 kV Transmission Electron Microscope, JEOL, MA, USA). About 1 mg of the test sample was dispersed in ethanol and spotted on a carbon coated copper grid. Ethanol was allowed to evaporate under atmospheric conditions prior to sample loading into the TEM viewing stage (JEOL JEM-2100 LaB6 200 kV Massachusetts, USA). The Gatan Digital Micrograph Software was used to facilitate sample viewing.

2.4.3. Thermal Analysis

The thermal behavior of the drug loaded RDS and pure isoniazid (reference) was characterized using a differential scanning calorimetry (DSC) machine equipped with a 50-position automatic sampler (Q2000 Differential Scanning Calorimeter TA Instruments, New Castle, DE, USA). Separate aluminum crucibles containing about 9 mg of each sample were analyzed under an inert nitrogen flow of 25 mL/minute and heating rate of 10 °C.min^{-1} while an empty crucible subjected to the same testing condition served as the reference. All measurements were performed in triplicate per sample under three heating cycles at a temperature range of 0–220 °C.

2.4.4. Determination of Crystallinity

The crystalline nature of the RDS formulation and pure isoniazid were evaluated under ambient conditions (25 ± 2 °C) using an X-ray diffractometer equipped with the X'Pert PRO data collector software (PANalytical Inc. MA, USA). Each sample (about 3 mg) analysis was performed at a voltage and current of 45 kV and 40 mA, respectively, and a 2θ range of 5 to 90° using a continuous scan mode with a scan step size of 0.03.

2.4.5. Structural Elucidation

The RDS formulation and pure isoniazid chemical backbone structural transitions were determined using the Fourier transform infrared (FTIR) spectrophotometric approach. The FTIR spectra of each test sample was generated on a Perkin Elmer Spectrum 100 Series FTIR Spectrophotometer coupled with Spectrum V 6.2.0 software (Beaconsfield, UK). Initially, blank background measurements were taken, and then, about 5 mg of each test sample was placed on a clean holder situated on the test stage of the machine for structural analysis that was recorded at a frequency range of 650–4000 cm^{-1}, scan time = 32 scans and resolution of 4 cm^{-1}.

2.5. Drug Loading Efficiency

To determine the amount of isoniazid loaded in the RDS formulation, 10 mg sample was placed in 100 mL phosphate buffered saline (pH 7.4) with continuous stirring over 4 h (Five-Position Hot Plate/Stirrer, Model 51450 series, Cole-Parmer, IL, USA), to ensure complete dissolution and release of all entrapped drug molecules. The resulting solution was then appropriately diluted with distilled water and passed through a 0.45 μm nylon membrane Corning® syringe filter (Corning Incorporated, NY, USA). The actual drug content was analyzed using Ultraviolet (UV) Spectrophotometry (PerkinElmer Lambda 35, UV/Vis Spectrometer, Perkin Elmer, Singapore) at a maximum wavelength of absorption for isoniazid, λ_{max} = 262 nm. The percentage of isoniazid load in the RDS formulation was mathematically computed with reference to the originally added amount (Equation (2)). Tests were conducted as three replicate samples.

$$\%\text{Drug loading efficiency} = \frac{\text{Actual drug amount (g)}}{\text{Theoretical drug amount (g)}} \times 100 \quad (2)$$

2.6. In Vitro Dissolution Studies

In vitro dissolution studies were performed on an RDS formulation sample size containing an equivalent of 100 mg isoniazid. The study was carried out using the dissolution tester (Ewreka GmbH, DT 820 series, Heusenstamm, Germany). The RDS formulation was contained in a gelatin capsule (CapsCanada, Ontario, Canada), which was placed in a basket holder attached to the dissolution tester stirring shaft (United States Pharmacopeia Apparatus 1). The basket was then immersed into separate dissolution jars containing 500 mL of either pH 7.4 or pH 6.8 or pH 1.2 pre-heated buffered solution maintained at 37 ± 0.5 °C with continuous rotation at 100 rpm. Samples (5 mL) were removed at the end of the 2-h time-point for quantification. Furthermore, at pH 7.4, drug loaded RDS formulation underwent an extended release testing up to a point when 100% isoniazid liberation was observed. Briefly, samples (5 mL) were removed at pre-determined time intervals and replaced with 5 mL freshly prepared preheated, buffered solution (pH 7.4, 37 ± 0.5 °C). All experiments were performed in triplicate under sink conditions and isolated samples were appropriately diluted, filtered using a 0.45 μm nylon membrane Corning® syringe filter (Corning Incorporated, NY, USA) and analyzed spectrophotometrically (PerkinElmer Lambda 35, UV/Vis Spectrometer, Perkin Elmer, Singapore) at λ_{max} of 262, 263, and 267 nm for pH 7.4, 6.8, and 1.2 buffered solutions, respectively, to determine the amount of isoniazid contained in each sample per time against buffer blanks. The actual quantity of isoniazid released per unit time was calculated using the linear polynomial calibration equations as follows: (a) pH 7.4: $y = 39.38x$; $R^2 = 0.98$; (b) pH 6.8: $y = 32.96x$; $R^2 = 0.99$, and (c) pH 1.2: $y = 45.77x$; $R^2 = 0.98$ where x and y, respectively, represent concentration (mg/mL) and absorbance. Percentage cumulative drug release [26] was computed relative to the total amount of isoniazid present in each medium at the end of the 2-h period. To get an indication of the release kinetics of the RDS formulation, the pH 7.4 release profile was selected for further mathematical model fitting over an extended period when 100% drug release was achieved. Resulting data were further analyzed using the KientDS version 3.0 open source software, which aided the selection of the model of best fit, which was based on mathematical computations, such as the zero, first, and second order kinetics, Higuchi, Korsmeyer-Peppas, Michaelis-Menten and Weibull approaches [19,24,27].

2.7. *Stability Evaluations*

2.7.1. Short-Term Stability Testing under Different Environmental Conditions

The physicochemical stability of the RDS formulation (500.0 mg ± 2 mg per test) was monitored under different storage conditions over four months. A set of samples were placed in an enclosed glass holder that was transferred into the stability tester (Labcon PSIE RH 40 Chamber Standard Incubator, Laboratory Marketing Services, Maraisburg, South Africa) fixed at 30 °C ± 2 °C and a relative humidity of 65% ± 3% adapted from the WHO stability testing scheme for pharmaceutical products containing well established drug substances [28]. Another sample group was stored in airtight, opaque glass vials under standard room conditions (temperature: 25 °C ± 5 °C and humidity: 55% ± 5%). The last sample set was stored in airtight, opaque glass vials and placed in the refrigerator (Sanya Labcool Pharmaceutical Refrigerator, MPR-720R, Sanyo Electrical Biomedical Co. Ltd, CA, USA) at 4 °C ± 1 °C. Formulation stability was monitored at 0, 1, and 4 month intervals for changes in particle size, polydispersity index, zeta potential, and drug content. Samples were also physically examined for any changes in physical appearance or color changes. All tests were conducted in triplicate.

2.7.2. In Vitro Aqueous Stability Assessment

The stability of the RDS formulation in aqueous solution (mimicking the re-constitution process) was evaluated. RDS samples (500 ± 2 mg) were placed in airtight, opaque containers and dispersed in 50 mL sterile deionized water. Hydrated samples were prepared in triplicate per test conditions. Test conditions included placement under ambient conditions (25 ± 5 °C), as well as in the refrigerator (4 ± 2 °C) for 11 days. Hydrated samples (2 mL) were collected from each test vessel for isoniazid content quantification at 0, 1, 5, and 11 days using UVspectrophotometry, as described earlier. Samples were manually agitated daily and before collection to ensure uniform re-dispersion. All tests were carried out in triplicate.

2.8. Preliminary Formulation Toxicity Assessment

The human breast adenocarcinoma (MCF-7) cell line was used for preliminary evaluation of the RDS formulation biocompatibility. The sample was dissolved in double-distilled water to a final stock concentration of 2 mg/mL, filter-sterilized through a 0.22 μm Cameo acetate membrane filter (Millipore Co., Bedford, Massachusetts), and stored at 4 °C until used. The negative control was 1% DMSO while 100 μM camptothecin, a known chemotherapeutic agent, was used as the positive control. The MCF-7 cell line was routinely maintained in Dulbecco's Modified Eagle's Medium (DMEM) supplemented with 10% heat-inactivated fetal bovine serum (FBS), at 5% carbon dioxide and 37 °C ± 0.1 °C. Cells were seeded into 96-well plates at a final concentration of 15,000 cells/well. Test samples were added at final well concentrations of 100, 50, 25 μg/ml, and incubated for 24 h. Thereafter, the spent medium was replaced with 200 μL of 0.05 mg/mL 3-(4,5-dimethylthiazol-2-yl)-2,5-diphenyltetrazolium bromide) (MTT) solution and incubated at 37 °C ± 0.1 °C for 3 h. The MTT was then aspirated from the cells and replaced with 200 μL DMSO to dissolve the formazan crystals. The absorbance was read at a maximum wavelength of 540 nm against DMSO as a blank using a microtiter plate reader (BioTek Powerwave XS, Winooski, VT, USA). The absorbance readings obtained were used to compute the number of viable cells present in the media. All results are presented as the mean reading ± standard deviation, and the statistical significance of all experimental data was evaluated using the GraphPad Prism 7 software (GraphPad, San Diego, CA, USA) using a two-way Analysis of Variance (ANOVA). The number of viable cells in both test and control samples were determined using Equation (3), and all tests were conducted in triplicate.

$$\%\text{Cell viability} = \frac{\text{Number of viable test cells}}{\text{Number of viable control cells}} \times 100 \quad (3)$$

3. Results and Discussions

3.1. Formulation Synthesis and Yield

The isoniazid-loaded RDS formulation was developed using a combination of the direct dispersion emulsification technique, lyophilization, and dry milling. A cream-white, free flowing isoniazid loaded RDS powder with an average yield of 87.43% ± 0.13% was obtained.

3.2. Size, Polydispersity Index, and Zeta Potential Determination and Morphology

The RDS powder formulation showed an average particle size of 1.63 ± 0.20 μm, indicating its intrinsic micro-structure. The PDI can be described as a ratio that provides information about homogeneity of particle size distribution as it relates to a particular system serving as a useful reflection of the quality of the particulate system/dispersion ranging from 0.0–1.0 with values ≤0.1 relating to the highest quality of particulate dispersion, ≤0.3 as optimum, ≤0.5 as generally acceptable [24]. A generally acceptable PDI value of 0.37 ± 0.04 was recorded for the RDS powder, indicating that the particles were mostly homogenous and well dispersed within the formulation. The measured ZP was −41.10 ± 5.57 mV, showing a stable system, where a ZP value of ±30 mV is considered a stable and satisfactory formulation [24,29]. Representative graphs based on an independent measurement of the zeta potential and particle size distribution are shown in Figure 1a,b. Researchers have shown that TEM imaging is an effective and powerful tool for characterizing the morphology of nano- and micro-structured biomaterials and drug carriers as it uses more powerful beams to produce higher resolution images with more details and information [30,31]. The TEM micrographs showed minimally aggregated, dispersed RDS particles that appeared as darker areas relative to the background, with rounded outer morphologies (Figure 1c,d).

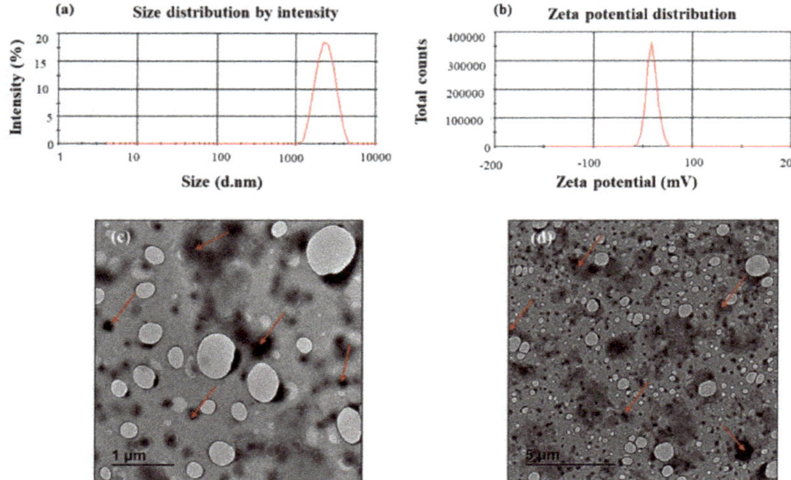

Figure 1. Representative graphs displaying: (**a**) particle size distribution, (**b**) zeta potential distribution, as well as TEM micrographs showing different surface topographies and characteristics of the reconstitutable dry suspension (RDS) particles at different scales: (**c**) 1 µm and (**d**) 5 µm, respectively.

3.3. Thermal Behavior

Salient changes in the thermal behavior of isoniazid relative to that of the RDS formulation were studied using conventional DSC methods. The recorded melting peak of pure, unformulated isoniazid was 171.7 °C, which compared well to literature [32]. The melting peak of isoniazid (Figure 2a) was characterized by its sharp and defined geometry confirming its purity and crystalline nature. The RDS thermogram was typified by the presence of multiple distinct sharp or broad peaks showing its physicochemically stable and multi-component state. The RDS formulation showed an intermittent appearance of sharp and blunt peaks representing its semi-crystalline nature (Figure 2b,c). The isoniazid melting peak identified on the RDS formulation thermogram (171.1 °C) was similar to that of the pure drug (171.7 °C), showing its stability within the excipient matrix. However, the formulated isoniazid peak presented as a broad band, which can be related to physical transitioning from crystalline into amorphous structures attributable to solvation and encapsulation of drug molecules into the processed polymeric carrier. A significant change in heat flow from −36.4 mW (unformulated isoniazid) to −3.7 mW (isoniazid in RDS) (Figure 2b,c) was seen, further supporting the likely drug encapsulation in the amorphous state, coupled with stable molecular dispersions within the polymeric chains [24]. Overall, the thermograms (Figure 2a–c) demonstrate the absence of any destructive/irreversible physicochemical interactions between isoniazid and the excipients during the RDS preparation process.

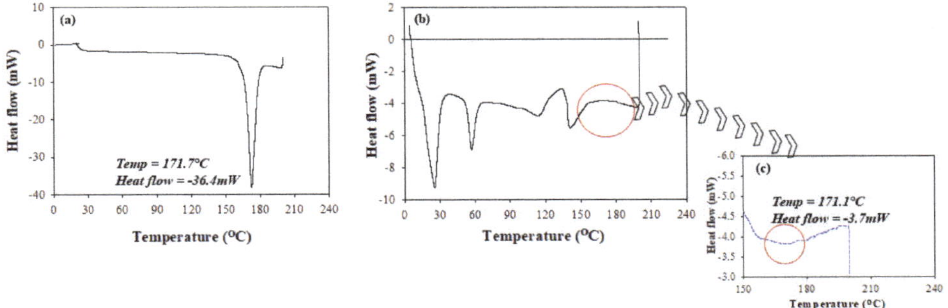

Figure 2. Differential scanning calorimetry (DSC) thermograms of (**a**) unformulated isoniazid, (**b**) RDS formulation, and (**c**) an expanded segment of the RDS thermograms showing the transitions that occurred with formulated isoniazid.

3.4. X-ray Diffractometry

The changes in matrix crystallinity between pure isoniazid and the RDS formulation was further confirmed using X-ray diffraction analysis (XRD) with recorded diffractograms presented in Figure 3a,b. Diffractograms recorded for pure isoniazid showed high intensity, well-defined sharp peaks between 9.3θ and 32.3θ, with intensities as high as 39,885.7 validating the crystalline nature of isoniazid (Figure 3a). This trend differs from that observed for the RDS formulation, which presents broader, not so well-defined peaks between 13.9θ and 28.1θ at a maximum intensity of 11,142.9, which is much lower relative to the isoniazid diffractogram. Furthermore, low intensity (less than 4288.1) plateau-like, broad-banded sections were also identified between 5.3θ—14.6θ and 28.5θ—90.1θ, further confirming that the RDS formulation consists of more than one component, characterized with intermittent, mostly amorphous and minimal crystalline domains, i.e., a semi-crystalline nature (Figure 3b). These findings agree with the outcomes of the DSC analysis.

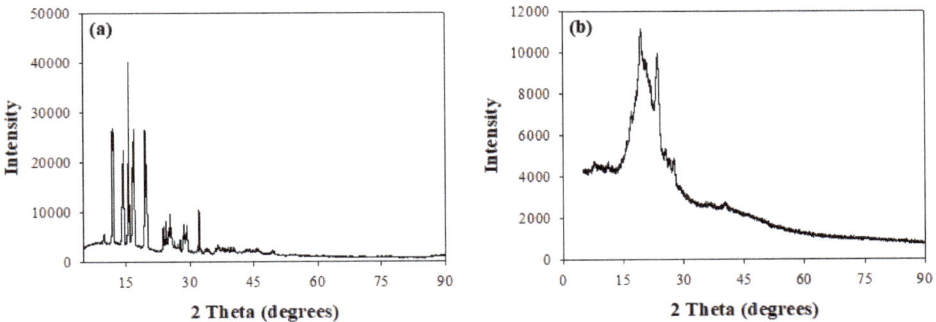

Figure 3. X-ray diffractograms of (**a**) unformulated isoniazid and (**b**) isoniazid loaded RDS formulation.

3.5. Structural Analysis

FTIR spectrum depicting characteristic vibrational frequencies of isoniazid in its unformulated state and encapsulated within the RDS formulation were compared (Figure 4). The vibrational band assignments for formulated and unformulated isoniazid were executed based on their positions, nature, magnitudes, and intensities relative to each characteristic functional group. The key bands plus wavenumbers identified and compared for both the RDS formulation and isoniazid in its pure state are presented in Table 1. The vibrational frequencies of key functional groups as it relates to the chemical backbone structure of isoniazid compares well for both pure drug and RDS formulation, with characteristic peaks presenting with varying intensities. The isoniazid loaded RDS formulation appeared overly dominant in most regions with some of its peaks suppressing those of pure isoniazid. This suggests physical solid–liquid/solid–solid dispersion and significant drug encapsulation. The O–H stretch recorded at 3676 cm^{-1} for the RDS formulation further supports the physically dispersed state in the water/oil medium, followed by snap-freezing and lyophilization.

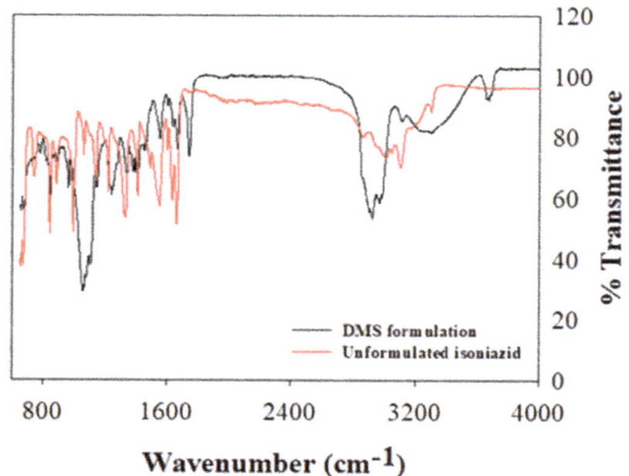

Figure 4. The Fourier transform infrared (FTIR) spectrum of the unformulated isoniazid and RDS formulation, showing different characteristic vibrational bands.

Table 1. FTIR vibrational frequencies representing formulated and unformulated isoniazid key functional groups.

Functional Groups and Remarks	Vibrational Frequencies (cm^{-1})	
	Unformulated Isoniazid	Isoniazid Loaded RDS Formulation
C–C–C–C and C–N–C–C torsion	653	665
NH_2 rock	677	677
C–C–C out of plane bending	746	757
C–C–H out of plane bending	850 and 1022	850 and 1023
C–N–C and C–C–C in plane bending	1063	1062
Aromatic C–N stretching	1344	1330
C–C–H in plane bending	1210	1212
O=C-N in plane bending	1330	1333
C–N–H in plane bending	1411	1408
C–C stretching	1477	1471
C=O stretching	1558	1552
NH_2 scissoring	1632	1635
Aromatic C–H stretching	3052	3054
N–H stretching	3307	3308
O–H stretching	-	3676

3.6. Drug Content

The percentage content of isoniazid encapsulated within the RDS formulation matrix was computed relative to the theoretical drug content, totaling 94.12% ± 2.10%. This indicates that there was minimal drug loss during the RDS manufacturing and processing phases.

3.7. Drug Release Behavior

To understand the in vitro drug release behavior and kinetics, dissolution studies were carried out on the RDS formulation containing an equivalence of 100 mg isoniazid in pH 7.4, 6.8, and 1.2 buffered solution over two hours under biorelevant conditions. Percentage cumulative drug release (%CDR) was calculated as the total amount of isoniazid liberated from the RDS formulation matrix, with an increase or decrease in %CDR representing a respective rise or decline in the release rates. %CDR varied for each dissolution media (pH 1.2 = 67.88% ± 1.88%, pH 6.8 = 60.18% ± 3.33%, and pH 7.4 = 49.36% ± 2.83%). Isoniazid release decreased as media pH increased. In other words, a reduction in the pH of the aqueous micro-environment seemed to impact the processes of formulation wetting, pore formation and closure, water penetration, phase transitioning, drug-excipient dissolution and degradation, changes in drug-excipient physical interaction and solubility process, and eventual diffusion of drug molecules coupled with gradual matrix geometry transitions (erosion or swelling) [33]. The RDS formulation demonstrated the potential to stabilize and release the isoniazid molecules in different dissolution media with significant differences in the percentage of drug released under these different conditions.

In addition, isoniazid release at pH 7.4 was selected for further mathematical model fitting to understand the RDS formulation release kinetics over a period when 100% drug liberation was achieved (Figure 5). In vitro isoniazid release from the RDS formulation at pH 7.4 was characterized with an initial burst at 5 min (7.51 ± 0.72 %), followed by a relatively consistent increase in drug release over time, until complete release (100%) was recorded at approximately 300 min.

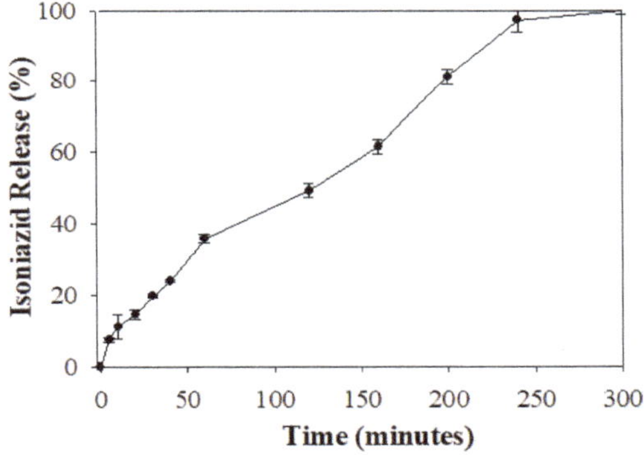

Figure 5. Graphical profile showing 100% isoniazid release from the RDS formulation over time.

The generated profile (Figure 5) was further analyzed using mathematical kinetic models employing the KinetDS, version 3.0 open source freeware. Release profile analysis and model of best fit selection was based on a combination of robust validation quantities, namely the correlation coefficient (R^2) closest to one and lowest Akaike Information Criterion (AIC) numerical value (Table 2) [31]. On this basis, the zero order kinetic model provided the best fit parameters (R^2 = 0.976 and AIC = 74.080) for the isoniazid release data depicted in Figure 5. This indicates that isoniazid release from the RDS formulation is consistent over time, irrespective of the initial drug concentration. In principle, zero order drug release mechanism is beneficial for achieving continuous drug plasma and biological fluid levels, reducing dosing frequency, and improving patient compliance, aiding desired pharmacotherapeutic efficacy [34].

Table 2. Representative mathematical models and their respective fit parameters.

Mathematical Models	R^2	AIC
Zero order	0.976	74.080
First order	0.159	176.881
Second order	0.091	129.633
Korsmeyer-Peppas	0.997	110.291
Weibull	0.990	92.616
Michaelis-Menten	1.000	120.737
Higuchi	-0.341	127.138

3.8. Short-Term Formulation Stability Assessment

3.8.1. Stability Evaluation under Varying Storage Conditions

Evaluation of formulation stability was performed in triplicate per sample under multiple storage conditions: Stability tester—30 °C ± 2 °C and 65% ± 3%; Room—25 °C ± 5 °C and 55% ± 5%; and Refrigerator—4 °C ± 1 °C; over four months employing particle size, polydispersity index, zeta potential, and drug content as indicators of stability. Results showed minimal numerical differences in indicators measured, indicating that the isoniazid loaded RDS formulation can be described as stable under the prescribed environmental storage conditions (Table 3). A slight alteration in physical appearance relating to a color change from white to cream-white was observed at four months, under accelerated storage in the stability chamber (30 °C ± 2 °C; 65% ± 3%) and room conditions (25 °C ± 5 °C; 55% ± 5%). Despite the fact that stability indicators remain closely related, the slight color change is undesirable, considering patient acceptance, making these environments unsuitable for the storage of the RDS formulation. Therefore, the most ideal storage setting for the RDS formulation is, thus, at 4 ± 1 °C as stability indicators are comparable with no color change observed (Table 3).

Table 3. Stability indicators measured under the different storage conditions.

RDS Formulation Test Conditions	Time (Months)	Stability Indicators				
		PDI [a]	ZP [b] (mV)	PS [c] (μm)	DC [d] (%)	Color
Stability tester	0	0.37	−41.10	1.63	94.12	No change
	1	0.40	−42.59	1.58	94.07	No change
	4	0.39	−43.60	1.82	93.76	Slight change
Room	0	0.37	−41.10	1.63	94.12	No change
	1	0.44	−42.71	1.69	93.95	No change
	4	0.46	−40.51	1.97	91.05	Slight change
Refrigerator	0	0.37	−41.10	1.63	94.12	No change
	1	0.39	−41.65	1.61	93.95	No change
	4	0.38	−40.91	1.70	94.02	No change

[a] Particle Size (standard deviation ≤ 0.13 μm in all cases), [b] Polydispersity Index (standard deviation ≤ 0.03 in all cases), [c] Zeta Potential (standard deviation ≤ 1.03 mV in all cases), [d] Drug Content (standard deviation ≤ 0.99% in all cases).

3.8.2. Formulation Stability in an Aqueous Environment

The stability of the drug loaded RDS formulation in an aqueous medium was investigated under select storage conditions (room and refrigeration) over 11 days, mimicking storage duration for commonly reconstituted antibiotic solutions/suspensions. Percentage drug content was chosen as the hydrostability indicator, and overall, there were insignificant changes in its numerical values under room or refrigerated storage conditions, with no visible color changes (Table 4). Summarily, the RDS formulation is stable in the aqueous medium over 11 days, either refrigerated or stored under ambient conditions, suggesting that the RDS formulation is a potentially useful preparation for re-constitution purposes, especially in pediatrics.

Table 4. Hydrostability indicators at different time-points, under ambient and refrigerated conditions.

Test Conditions	Time-Points (Days)	Drug Content (%)	Discoloration
Room/Ambient	0	94.12	None
	1	93.18	None
	5	93.61	None
	11	92.89	None
Refrigerator	0	94.12	None
	1	93.33	None
	5	93.79	None
	11	92.56	None

3.9. Cell Viability Assessment

Preliminary assessment of the effects of the RDS formulation on viability was performed on MCF-7 cell lines. Tests were conducted over 24 h using different concentrations of the RDS ranging from 0 µg/mL (negative control) to 100 µg/mL. This cell line is routinely used as a prototype for assessing the biocompatibility of drugs, delivery systems, and biologicals that are non-specific for treating carcinomas [35–37]. Graphical representations of the impact of different test formulation concentrations on cell viability relative to camptothecin (positive control) and the negative control are presented in Figure 6. The RDS lowest concentration (25 µg/mL) caused significant MCF-7 cell proliferation ($p < 0.0500$: $p = 0.0001$). The increased viability may signify that the RDS formulation supported cell growth at a low concentration (25 µg/mL) and can be considered an indication of biocompatibility. This may be attributed to the drug or excipient concentration or a combination of both. In contrast, the 50 and 100 µg/mL reduced cell viability, although not significantly ($p > 0.0500$: 50 µg/mL—$p = 0.0056$; 100 µg/mL—$p = 0.0001$). Nevertheless, higher concentrations of the RDS (50 and 100 µg/mL) did not reduce cell viability as much as the camptothecin that decreased it to 48.93% ± 3.99%. In general, the RDS was well-tolerated by the MCF-7 cells at all test concentrations. Cellular responses following exposure to the formulation can be described as biphasic and dose-dependent, a phenomenon associated with hormesis (a two-phased adaptive response of cells and organisms to increasing or decreasing amounts of external stress, e.g., drug, chemical substances, and disease state) [38,39]. It is not unusual for cells to exhibit hormetic effects as they are biological systems known to be dynamic and constantly evolving. These initial findings form a baseline for future work on understanding the biocompatibility of the RDS formulation and its components (active drug and excipients) in vitro and in vivo.

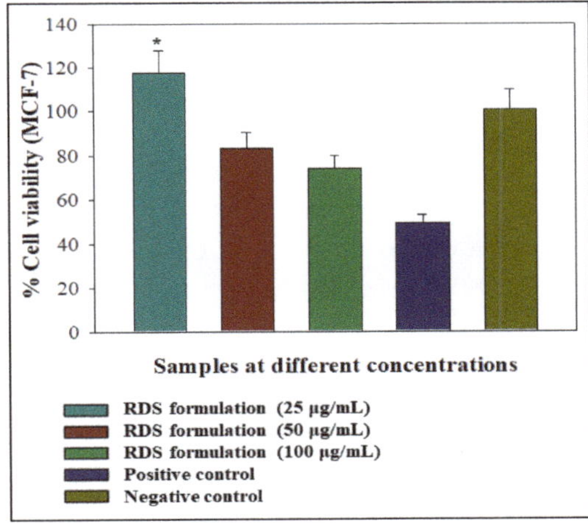

Figure 6. Graphical illustration of the percentage cell viability 24-h post-exposure to the RDS formulation in the human breast cancer (MCF-7) cell lines as a prototype. Statistical difference between two data sets was considered significant when $p < 0.05$.

4. Conclusions and Future Work

In this study, isoniazid loaded reconstitutable dry suspension was prepared using the direct dispersion emulsification technique, coupled with lyophilization and dry milling. The direct dispersion technique produced a relatively high yield (87.43% ± 0.13%) of RDS particles with good drug loading capabilities (94.12% ± 2.10%). The RDS formulation showed no significant evidence of toxicity supported by outcomes from viability studies in the MCF-7 cells. The formulation was physicochemically stable, mostly amorphous, with marginal intermittent crystalline domains, and had no irreversible alterations in its backbone chemical structure. It demonstrated the ability to regulate isoniazid release in a controlled, zero order manner, and was environmentally stable under common storage conditions, either as a dry powder or in the hydrated form. The findings from this work may contribute towards improving flexible pediatric dosing for tuberculosis drug treatment, considering the current global shortage of such preparations, especially for the first-line anti-tubercular drugs.

To establish a course for further investigations, we identified the need to extend biocompatibility testing for the RDS formulation and its components to normal cells and tissue isolates from animal models (e.g., mice, rabbit, or pigs) employing cytotoxicity assays and histopathological techniques, respectively. Further preclinical evaluation of pharmacokinetics, efficacy, and eventual optimization of isoniazid dosing and absorption from the RDS formulation, in animal models similar to humans, such as pigs, are important next steps. Considering the hydrophilic/hydrophobic and particulate nature of the RDS, we anticipate that it can find extensive use as an effective carrier for other anti-tubercular drugs (e.g., pyrazinamide, rifampicin) suitable for pediatrics irrespective of their solubilities and molecular weights. For implementation purposes, a range of these bioactives would need to be tested in vitro/in vivo and optimized to ensure desirable drug loading, controlled release, and absorption for the intended pharmacotherapeutic application. Overall, the RDS formulation reported herein has the potential for improving tuberculosis treatment within the pediatric population.

Author Contributions: Conceptualization, O.A.A.; methodology, O.A.A., R.K.H., H.D.; formal analysis, O.A.A., R.K.H.; investigation, O.A.A., R.K.H., H.D.; writing-original draft preparation, O.A.A.; writing, review and editing, O.A.A., R.K.H., H.D.; project administration, O.A.A., R.K.H.; resources, O.A., R.K.H., H.D.; funding acquisition, O.A.A., R.K.H. All authors have read and agreed to the published version of the manuscript.

Funding: This research was funded by the South African National Research Foundation/Department of Science and Technology (Grant Number-85110 and 113143) and Sefako Makgatho Health Sciences University DHET Research Development Grant (Grant Number: D112-RDG).

Acknowledgments: We thank the National Centre for Nano-structured Materials, Council for Scientific and Industrial Research, South Africa, for sample characterization. We thank the CSIR National Centre for Nano-structured Materials for sample characterization. Opinions, findings and conclusions or recommendations expressed in this publication are that of the authors. The funders accept no liabilities whatsoever in this regard.

Conflicts of Interest: The authors declare no conflict of interest.

References

1. World Health Organization. WHO Global TB Report. 2018. Available online: http://www.who.int/tb/publications/global_report/en/ (accessed on 10 December 2019).
2. TB Alliance. The Pandemic. 2019. Available online: https://www.tballiance.org/why-new-tb-drugs/global-pandemic (accessed on 2 December 2019).
3. Suárez-González, J.; Santoveña-Estévez, A.; Soriano, M.; Fariña, J.B. Design and optimization of a child-friendly dispersible tablet containing Isoniazid, Pyrazinamide and Rifampicin for treating Tuberculosis in pediatrics. *Drug Dev. Ind. Pharm.* **2020**, *46*, 309–317. [CrossRef] [PubMed]
4. Venturini, E.; Turkova, A.; Chiappini, E.; Gali, L.; de Martino, M.; Thorne, C. Tuberculosis and HIV co-infection in children. *BMC Infect. Dis.* **2014**, *14*, 1–10. [CrossRef] [PubMed]
5. Sinha, P.; Shenoi, S.V.; Friedland, G.H. Opportunities for community health workers to contribute to global efforts to end tuberculosis. *Glob. Public Health* **2020**, *15*, 474–484. [CrossRef] [PubMed]
6. Chan, J.G.Y.; Chan, H.K.; Prestidge, C.A.; Denman, J.A.; Young, P.M.; Traini, D. A novel dry powder inhalable formulation incorporating three first-line anti-tubercular antibiotics. *Eur. J. Pharm. Biopharm.* **2013**, *83*, 285–292. [CrossRef]
7. Hussain, A.; Shakeel, F.; Singh, S.K.; Alsarra, I.A.; Faruk, A.; Alanazi, F.K.; Christoper, G.P. Solidified SNEDDS for the oral delivery of rifampicin: Evaluation, proof of concept, in vivo kinetics, and in silico GastroPlusTM simulation. *Int. J. Pharm.* **2019**, *566*, 203–217. [CrossRef]
8. Hamzaoui, A.; Yalaoui, A.; Cherif, F.A.; Slim, L.; Barraies, A. Childhood tuberculosis: A concern for the modern world. *Eur. Respir. Update Tuberc.* **2014**, *23*, 278–291. [CrossRef]
9. Jenkins, H.E. Global burden of childhood tuberculosis. *Pneumonia* **2016**, *8*, 24. [CrossRef]

10. Peña, M.J.M.; García, B.S.; Baquero-Artigao, F.; Pérez, D.M.; Pérez, R.P.; Echevarría, A.M.; Amador, J.T.R.; Durán, D.G.P.; Julian, A.N.; Tuberculosis, W.G.; et al. Tuberculosis treatment for children: An update. *An. Pediatr.* **2018**, *88*, 52.e1–52.e12.
11. Pérez, R.P.; García, B.S.; Fernández-Llamazares, C.M.; Artigao, F.B.; Julia, A.N.; Péna, M.J.M. The challenge of administering anti-tuberculosis treatment in infants and pre-school children. *An. Pediatr.* **2016**, *8*, 4–12.
12. Samad, A.; Sultana, Y.; Khar, R.K.; Aqil, M.; Chuttani, K.; Mishra, A.K. Reconstituted powder for formulation of antitubercular drugs formulated as microspheres for paediatric use. *Drug Discov. Ther.* **2008**, *2*, 108–114.
13. Shanbhag, P.P.; Bhalerao, S.S. Development and evaluation of oral reconstitutable systems of cephalexin. *Int. J. Pharm. Tech. Res.* **2010**, *2*, 502–506.
14. Pouplin, T.; Phuong, P.N.; Van Toi, P.; Pouplin, J.N.; Farrar, J. Isoniazid, pyrazinamide and rifampicin content variation in split fixed-dose combination tablets. *PLoS ONE* **2014**, *9*, e102047. [CrossRef] [PubMed]
15. Lopez, F.L.; Ernest, T.B.; Tuleu, C.; Gul, M.O. Formulation approaches to paediatric oral drug delivery: Benefits and limitations of current platforms. *Expert Opin. Drug Deliv.* **2015**, *12*, 1727–1740. [CrossRef] [PubMed]
16. Long, S.S.; Prober, C.G.; Fischer, M. *Principles and Practice of Pediatric Infectious Diseases E-Book*, 5th ed.; Elsevier Health Sciences: Amsterdam, The Netherlands, 2017.
17. Grange, J.M.; Schaaf, H.S.; Zumla, A. *Tuberculosis: A Comprehensive Clinical Reference*; Saunders/Elsevier: Philadelphia, PA, USA, 2009.
18. Moretton, M.A.; Cagel, M.; Bernabeu, E.; Gonzalez, L.; Chiappetta, D.A. Nanopolymersomes as potential carriers for rifampicin pulmonary delivery. *Colloids Surf. B Biointerfaces* **2015**, *136*, 1017–1025. [CrossRef]
19. Adeleke, O.A.; Tsai, P.C.; Karry, K.M.; Monama, N.O.; Michniak-Kohn, B.B. Isoniazid-loaded orodispersible strips: Methodical design, optimization and in vitro-in silico characterization. *Int. J. Pharm.* **2018**, *547*, 347–359. [CrossRef]
20. Brennan, P.J.; Young, D.B. Handbook of Anti-Tuberculosis Agents. *Tuberculosis* **2008**, *88*, 112–116.
21. Fülöp, V.; Jakab, G.; Bozó, T.; Tóth, B.; Endrésik, D.; Balogh, E.; Kellermayer, M.; Antal, I. Study on the dissolution improvement of albendazole using reconstitutable dry nanosuspension formulation. *Eur. J. Pharm. Sci.* **2018**, *123*, 70–78. [CrossRef]
22. Anselmo, A.C.; Mitragotri, S. An overview of clinical and commercial impact of drug delivery systems. *J. Control. Release* **2014**, *190*, 15–28. [CrossRef]
23. Yin Win, K.; Feng, S. Effects of particle size and surface coating on cellular uptake of polymeric nanoparticles for oral delivery of anticancer drugs. *Biomaterials* **2005**, *26*, 2713–2722. [CrossRef]
24. Öztürk, A.A.; Yenilmez, E.; Özarda, M.G. Clarithromycin-Loaded Poly (Lactic-co-glycolic acid) (PLGA) Nanoparticles for Oral Administration: Effect of Polymer Molecular Weight and Surface Modification with Chitosan on Formulation, Nanoparticle Characterization and Antibacterial Effects. *Polymers* **2019**, *11*, 1632. [CrossRef]
25. Kodama, T.; Tomita, N.; Horie, S.; Sax, N.; Iwasaki, H.; Suzuki, R.; Maruyama, K.; Mori, S.; Manabu, F. Morphological study of acoustic liposomes using transmission electron microscopy. *J. Electron Microsc.* **2009**, *59*, 187–196. [CrossRef] [PubMed]
26. Jermy, B.R.; Alomari, M.; Ravinayagam, V.; Almofty, S.A.; Akhtar, S.; Borgio, J.F.; AbdulAzeez, S. SPIONs/3D SiSBA-16 based Multifunctional Nanoformulation for target specific cisplatin release in colon and cervical cancer cell lines. *Sci. Rep.* **2019**, *9*, 1–12. [CrossRef] [PubMed]
27. Yang, X.; Trinh, H.M.; Agrahari, V.; Sheng, Y.; Pal, D.; Mitra, A.K. Nanoparticle-based topical ophthalmic gel formulation for sustained release of hydrocortisone butyrate. *AAPS Pharm. Sci. Tech.* **2016**, *17*, 294–306. [CrossRef]
28. World Health Organization (WHO). Guidelines for Stability Testing of Pharmaceutical Products Containing Well-Established Drug Substances in Conventional Dosage Forms. WHO Technical Report Series 863, Annex 5. 1996. Available online: https://apps.who.int/medicinedocs/pdf/s5516e/s5516e.pdf (accessed on 24 February 2020).
29. Dodiya, S.; Chavhan, S.; Korde, A.; Sawant, K.K. Solid lipid nanoparticles and nanosuspension of adefovir dipivoxil for bioavailability improvement: Formulation, characterization, pharmacokinetic and biodistribution studies. *Drug Dev. Ind. Pharm.* **2013**, *39*, 733–743. [CrossRef] [PubMed]
30. Manaia, E.B.; Abuçafy, M.P.; Chiari-Andréo, B.G.; Silva, B.L.; Junior, J.A.O.; Chiavacci, L.A. Physicochemical characterization of drug nanocarriers. *Int. J. Nanomed.* **2017**, *12*, 4991. [CrossRef]

31. Maciel, V.B.; Yoshida, C.M.; Pereira, S.M.; Goycoolea, F.M.; Franco, T.T. Electrostatic self-assembled chitosan-pectin nano-and microparticles for insulin delivery. *Molecules* **2017**, *22*, E1707. [CrossRef]
32. Afinjuomo, F.; Barclay, T.G.; Parikh, A.; Chung, R.; Song, Y.; Nagalingam, G.; Triccas, J.; Wang, L.; Liu, L.; Hayball, J.D.; et al. Synthesis and Characterization of pH-Sensitive Inulin Conjugate of Isoniazid for Monocyte-Targeted Delivery. *Pharmaceutics* **2019**, *11*, 555. [CrossRef]
33. Siepmann, J.; Siepmann, F. Mathematical modeling of drug delivery. *Int. J. Pharm.* **2008**, *364*, 328–343. [CrossRef]
34. Paolino, D.; Tudose, A.; Celia, C.; Di Marzio, L.; Cilurzo, F.; Mircioiu, C. Mathematical Models as Tools to Predict the Release Kinetic of Fluorescein from Lyotropic Colloidal Liquid Crystals. *Materials* **2019**, *12*, 693. [CrossRef]
35. Bácskay, I.; Nemes, D.; Fenyvesi, F.; Váradi, J.; Vasvári, G.; Fehér, P.; Vecsernyés, M.; Ujhelyi, Z. Role of Cytotoxicity Experiments in Pharmaceutical Development. In *Cytotoxicity*; Çelik, T.A., Ed.; InTech: London, UK, 2018; ISBN 978-1-78923-430-5.
36. Senthilraja, P.; Kathiresan, K. In vitro cytotoxicity MTT assay in Vero, HepG2 and MCF-7 cell lines study of Marine Yeast. *J. Appl. Pharm. Sci.* **2015**, *5*, 80–84. [CrossRef]
37. Garcia-Fernandez, J.; Turiel, D.; Bettmer, J.; Jakubowski, N.; Panne, U.; Rivas García, L.; Llopis, J.; Sánchez González, C.; Montes-Bayón, M. In vitro and in situ experiments to evaluate the biodistribution and cellular toxicity of ultrasmall iron oxide nanoparticles potentially used as oral iron supplements. *Nanotoxicology* **2020**, *20*, 1–16. [CrossRef] [PubMed]
38. Calabrese, E.J. Hormesis: Path and progression to significance. *Int. J. Mol. Sci.* **2018**, *19*, 2871. [CrossRef] [PubMed]
39. Kloner, R.A. Remote Ischemic Conditioning as a Form of Hormesis. In *The Science of Hormesis in Health and Longevity*; Academic Press: Amsterdam, The Netherlands, 2019.

MDPI
St. Alban-Anlage 66
4052 Basel
Switzerland
Tel. +41 61 683 77 34
Fax +41 61 302 89 18
www.mdpi.com

Pharmaceutics Editorial Office
E-mail: pharmaceutics@mdpi.com
www.mdpi.com/journal/pharmaceutics

www.ingramcontent.com/pod-product-compliance
Lightning Source LLC
LaVergne TN
LVHW071034170726
843515LV00006B/1283

* 9 7 8 3 0 3 9 3 6 3 9 2 6 *